Ignition Systems for Gasoline Engines

Michael Günther · Marc Sens
Editors

Ignition Systems for Gasoline Engines

3rd International Conference, November 3–4, 2016, Berlin, Germany

Springer

Editors
Michael Günther
IAV GmbH
Chemnitz
Germany

Marc Sens
IAV GmbH
Berlin
Germany

ISBN 978-3-319-45503-7 ISBN 978-3-319-45504-4 (eBook)
DOI 10.1007/978-3-319-45504-4

Library of Congress Control Number: 2016955791

Printed on acid-free paper

This Springer imprint is published by Springer Nature
The registered company is Springer International Publishing AG
The registered company address is: Gewerbestrasse 11, 6330 Cham, Switzerland

Contents

Requirements for Ignition Systems

Challenges to the Ignition System of Future Gasoline Engines – An Application Oriented Systems Comparison

Martin Schenk(✉), Franz Xaver Schauer, Christina Sauer, Gerhard Weber, Joachim Hahn, and Christian Schwarz

BMW AG, 80788 Munich, Germany
{Martin.Schenk, Franz-Xaver.Schauer, Christina.Sauer, Gerhard.WD.Weber, Joachim.Hahn, Christian.Schwarz}@bmw.de

Abstract. Recent advancements of the gasoline engines combustion processes, e.g. a rightsizing or an alternative combustion process approach, make significantly higher demands on ignition systems than previous engine generations, especially with respect to charge dilution and gas exchange strategies. Among others, it is especially the optimization of the ignition system, which allows an extension of the current thermodynamic and combustion limits without drawbacks in engine performance. Besides the demands of the stationary operation one has to especially bear in mind the requirements originating from the transient operation of the engine.

In this presentation the potentials and limitations of ignition innovations are examined. With reference to a current series ignition system, both an optimization of the conventional spark ignition as well as alternative ignition concepts are analyzed. Both stationary and transient aspects will be covered. Furthermore realization relevant aspects and needs for the automotive implementation will be analyzed.

Keywords: Spark ignition · Corona ignition · Non-thermal discharge · Combustion

1 Introduction

The challenges as well as the chances of ignition have accompanied the last 150 years of the development of the gasoline engine. Hereby the activities focused for a long time on the development of low-cost, robust and as maintenance-free as possible ignition components to assure a reliable engine operation under all climatic conditions.

With rising awareness towards fuel consumption and exhaust emissions and for this reason increasing legislative requirements, the functional demands on the ignition increased considerably since the 1970's. The legislative restrictions with respect to the emissions of pollutants – gaseous as well as particulate emissions – are continuously tightened ever since. This is accompanied by the necessity of a continuous increase in fuel efficiency to match the steadily tightened CO_2 fleet emission limits.

M. Günther and M. Sens (eds.), *Ignition Systems for Gasoline Engines*,
DOI 10.1007/978-3-319-45504-4_1

The resulting advancements and developments of the gasoline engine combustion processes, e.g. as a rightsizing or as an alternative combustion process approach, make significantly higher demands on modern and alternative ignition systems than previous engine generations, especially with respect to charge dilution and gas exchange strategies. Therefore, in addition to the measures to ensure a stable, fast and complete main combustion phase, such as an increased charge motion, it is especially the optimization of the ignition systems, which allows an extension of the current thermodynamic and combustion technology operation limits without drawbacks in engine smoothness, acoustic acceptability, as well as dynamic performance. These optimizations also help to ensure the compliance with respect to the enforced emission limit regulations in the complete range of the engine operation map. Besides the demands of the stationary operation one has to especially bear in mind the requirements originating from the transient operation of the engine as well as from the global market usability and from the increasing fuel quality diversification correlated with it.

Within the frame of this presentation the potentials and limitations of various ignition innovations are examined. With reference to current series ignition systems, both optimizations of the spark ignition as well as alternative ignition concepts, i.e. corona ignition and, for selected examples, passive pre-chamber ignition are analyzed. Besides the evaluation of the stationary potentials for selected combustion strategies with respect to their key operation parameters, one main focus of our presentation will be the transient and realization relevant aspects. The resulting needs as well as the potentials for the automotive implementation will be analyzed.

2 Challenges to Ignitions Systems Within the Engine Map

With increasing restrictions with respect to fuel consumption and emission-legislation the challenges to modern combustion strategies rise steadily. Hereby the thermodynamic and combustion necessities go significantly beyond the operation conditions of current engines.

Generally there is the demand for even higher power densities. However, this demand still requires the same engine smoothness quality for lower part load and idle operation, as current engines with less power densities do. This clearly enlarges the gap between the ignition capability in full load and the reliability of inflammation in lower part load. On the one hand we have to handle even higher pressures at the time of ignition, which can be covered by conventional ignition systems primarily by a reduction of the gap between ground and the center electrode. On the other hand higher power densities are naturally accompanied by lower overall compression ratios, which makes the part load and idle ignitability even more challenging. Thus the demands to an optimal ignition system rise significantly with part load.

Beyond the demand for even higher power densities there is a strong demand for a steady and sustainable reduction in fuel consumption synchronous with continuously more severe emission limitations. Considering the ideal thermodynamic cycle process for SI engines, there are two main pathways to rise the ideal efficiency of the engine: Firstly to rise the compression ratio and secondly to rise the dilution of the mixture either by air or by EGR, to achieve an overall better polytropic coefficient.

To achieve an overall high dilution ratio, both stratified and homogeneous approaches are known and are quite distinct in their demands with respect to the inflammation process.

For the stratified operation, the spray formation provides in principle very favorable ignition conditions at the time and the location of the ignition event, however in general dominated by a high temporal and spatial fluctuation of these ignition conditions. Thus, a suitable ignition system ideally covers a considerable ignition space and a considerable ignition time (multiple bursts or long duration), to average these fluctuation properly and guarantee a safe inflammation process in each combustion cycle [1, 2].

For homogeneous diluted mixtures, the demands appear in principal different. Due to the generally early injection timing, mixture gradients are by far on a larger scale in time and space. However the mixture itself at the time and location of the ignition event imposes by far harder thermodynamic inflammation requirements compared to the "merely" statistical coverage of the stratified operation. We again have the demand of an as large as possible inflammation area and the demand of sufficient long ignition duration, to assure a reliable initiation of the flame kernel formation, without the risk of the flame kernel being extinguished again. In the case of homogeneous dilution the size of the inflammation area reduces the burn delay and therefore improves the inflammation statistic and the ignition time necessary. For homogeneous inflammation conditions we in generally see a strong correlation between the average burn delay (within this paper defined as 0–5 % MFB duration) and the 5 % MFB timing statistics of the initial flame kernel.

To achieve a reasonable fast total burn rate and stable overall combustion, in general a high level of charge motion and turbulence is favorable. However, to create this charge motion level, efficiency losses in the gas exchange process have to be accepted. Furthermore, the heat transfer with the combustion chamber walls rises by the increase in charge motion. Higher compression ratios do ease the combustion and the inflammation process, however, do as well increase the heat losses due to wall interaction again.

For the ignition process, the ability to deal with and the necessity for a strong charge motion are important aspects regarding the overall efficiency. Ideally spoken, the ignition system should be capable to deal with both, very little and very high charge motion levels.

Besides the efficiency gained by better thermodynamic properties, diluted combustion offers considerable potentials for a reduction in fuel consumption in part load operation due to significant reduction of gas exchange losses. Furthermore, strategic valve train measures such as early inlet closing offer significantly increased advantages with respect to gas exchange losses. However, both measures impose severe challenges on the conventional ignition process, on the one hand by pure dilution effects, on the other hand by the effect of significantly lowering the charge motion level as well as the temperature level at the time of ignition. Especially the lower part load below 0.4 MPa IMEP poses considerable challenges to the concepts mentioned above. These challenges increase in general even more with rising revolution rate and with lower loads, for example for effective zero load or idle.

The other efficiency measure, the increase in the compression ratio, is generally correlated with the necessity to extend both, the knocking as well as the component protection limitation of the SI engine. Furthermore one has to consider the compliance with the worldwide available fuel quality standards. In the context of engine smoothness at full load, inflammation speed is one essential aspect, as well as short flame distances and a fast flame propagation. Further aspects are a moderate acoustic and vibration excitations have to fulfill the so called NVH requirements. Dilution and Miller concepts partly enforce these challenges, as they in principal make the inflammation process more difficult and thus less stable. On the other hand charge motion measures do increase pressure gradients and heat losses due to wall interaction. Thus ideally spoken, the perfect ignition concept would be capable of addressing at least a part of these challenges.

Another key challenge, which gains especially importance for diluted combustion concepts, is the capability to match the demands of transient operation, either as a drop or a rise in load. With respect to emission legislation requirements strongly rising challenges have to be met by the ignition systems especially during the warm up phase. This is particularly important with respect to particle emissions limitations, which gain notably importance due to engine start stop strategies. Aspects of particle free combustion, emission optimized warm up, resistance with respect to fouling of the ignition components as well as a high tolerance with respect to load change by means of a late center of combustion are of eminent importance for the dynamic operation of the engine.

An even further aspect of innovative ignition systems is the energy consumption of the ignition system itself. This aspects gains especially importance, since with the introduction of the new driving cycles such as WLTP, the state of charge of the vehicle energy sources is monitored before and after the specific test cycle. The overall system weight has, last but not least, also to be taken into account.

Figure 1 summarize the main challenges to modern ignition systems with respect to the combustion strategies (left side) and to the car integration (right side).

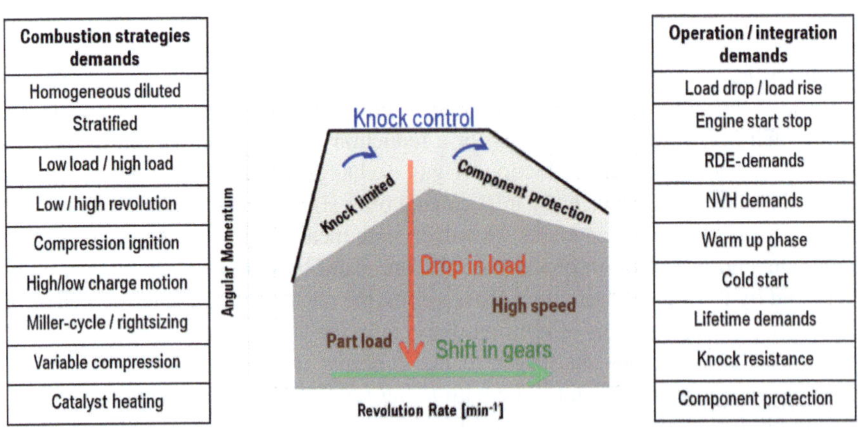

Fig. 1. Challenges and demands of modern SI combustion systems to alternative ignition systems.

3 Test Setup and Procedure

3.1 Engines for Thermodynamic Testing

The experiments described below were carried out both on a single-cylinder [3], 4-stroke research engine and research inline multi-cylinder engines of three and four cylinders, with principal basic geometries based on the new BMW inline engine generation, respectively. Specific changes were, among others, especially the piston geometry, which was optimized to both, corona as well as charge motion. Within this publication we focus on the initiation and the main phase of the combustion process. The experiments and analysis were designed accordingly. The engines used were equipped with variable camshaft phasing (VANOS) on inlet and exhaust side and variable valve lift (Valvetronic) on the inlet side. The compression ratio was chosen between values of $\varepsilon = 9.5$ up to $\varepsilon = 12$ and will be stated accordingly with the results. Standard premium fuel with RON = 98 was used for the testing. The engine parameters are listed in Table 1.

Table 1. Engine data

Engine	1-Cyl.	3-Cyl. (2x)	4-Cyl. (2x)
Displacement	$\sim 500\ cm^2$	$\sim 1500\ cm^3$	$\sim 2000\ cm^3$
Bore/Stroke	82 mm/95 mm		
	12:1	9.5:1/12:1	10.5:1/11:1
No. of valves/valve-lift/cam-phasing	Four variable on inlet variable cam-phasing on inlet and outlet		
Cooled EGR	Yes	None/Yes	Yes/None
Injection system	Direct multi-hole injection central position		

3.2 Ignition Systems

Transistor Coil Ignition. A state-of-the-art transistor coil ignition system (TCI) was used as reference [3]. If not mentioned otherwise the spark plug had a single-electrode air gap with 0.75 mm distance. The energy of the plug top ignition coil on the secondary side amounted to 100 mJ.

Advanced Spark Ignition Systems. The advanced spark ignition systems (ASI) consisted in principle of a coil-based spark ignition system, which offers the option to vary the secondary current during the spark discharge both in current strength and current duration. Figure 2 displays a scheme of the principle discharge characteristics of a conventional high energy spark and an ASI. The approaches towards the technical realization for such kind of ignition concepts are manifold [4–6]. They have in common however the variability of the physical parameter duration and the secondary current of the elongated glow discharge.

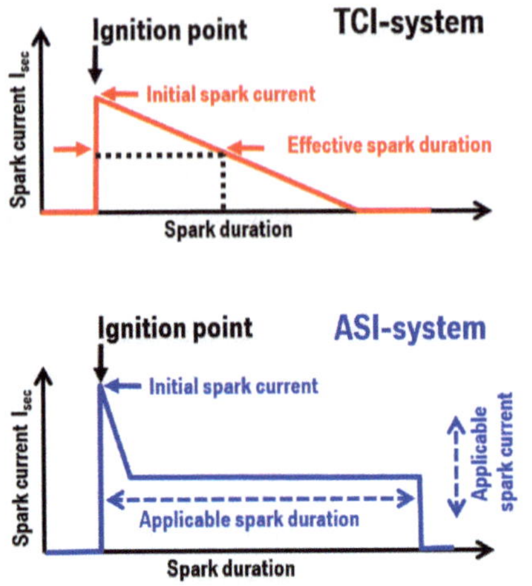

Fig. 2. Schematic diagram of a single coil spark discharge (TCI; top) and a spark discharge obtained by an advanced spark ignition system (ASI; bottom)

Radio Frequency Corona Ignition. The corona ignition systems used here generate high AC voltages in the frequency range of appr. 1 MHz to 5 MHz at the tip of the igniters. The frontend electrodes of the igniters were formed to a regular star with four or five equally shared tips. The combustion chamber acts as a counter electrode. Compared to a standard plug, the tips of the corona igniter have to be sharp for building a so called point-plane arrangement as illustrated at the top of Fig. 3. This electrode configuration generates a highly inhomogeneous electric field decreasing strongly from the tip towards ground. Plasma channels are formed in the area, where the ionization field strength is exceeded. Since the areas with the field strength below the ionization limit remain almost unchanged, no complete plasma channel is build up as for spark ignition (Fig. 3, bottom) and the energy input mainly accelerates electrons and does not heat up the gas considerably [7]. More details about the setup of a corona ignition system can be found in [2, 8, 9].

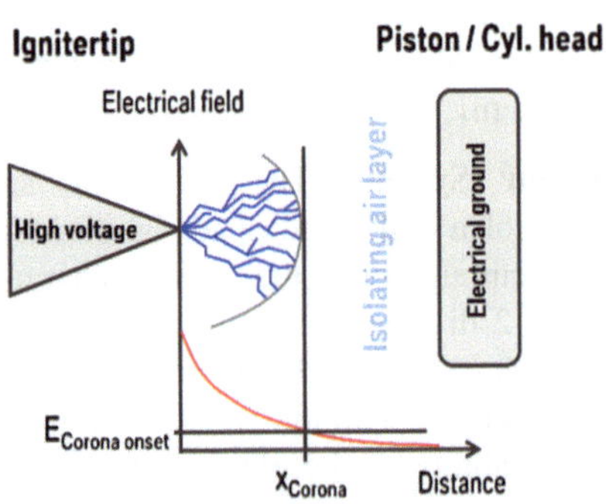

Fig. 3. Schematic diagram of a corona discharge [7]

Pre-chamber Ignition System. For the experiments presented we used a passive pre-chamber ignition sparkplug with orifices oriented almost tangential to the chamber walls of the pre-chamber. It was operated in passive gas exchange mode with a 100 mJ plug-top coil, as it was used for the TCI as well.

4 Results

Within this section we compare the performance and capabilities of an ASI system and a corona ignition system with respect to a series TCI system. We start with the comparison of the part load behavior (Sect. 4.1), where the largest advantages for both systems are sited. However, the two alternative ignition system display quite distinct differences with respect to the series reference and towards each other. These aspects will also be pointed out in the first part. In a second step (Sect. 4.2) we compare the corresponding performance characteristics in the medium load region, which offers advantages especially for the corona system. However, regarding emissions in lean operated mode, also the ASI system still do offer NOx emission advantages. The third subsection (Sect. 4.3) puts emphasis on the advantages of the corona system in the high load region of the engine map, especially in the context with high charge motion and restrictions due to knock and component protection needs. Especially for diluted and strongly Millerized combustion, the corona ignition system offers specific advantages especially within the knock limited region of the engine map. In interdependency with the engine map and the need for worldwide fuel compatibility these advantages do offer additional chances, such as enabling a general rise in the maximum possible compression ratio or options for acoustical improvements. Additionally a short prospect to a comparison with a passive pre-chamber spark plug is given within this context.

The next section (Sect. 4.4) highlights briefly some important transient aspects, such as torque reserves, optimized catalyst heating and an improved engine warmup.

The last section of this chapter (Sect. 4.5) is devoted to important aspects, which have to be addressed, when integrating such kind of ignitions systems into a series car environment.

In the subsequent result sections, the interconnecting lines among the measurement points of a data series serve merely the clarity of the illustration and do not necessarily represent a specific underlying correlation.

4.1 Potentials in Part Load

The demands in the lower part load region of the engine map regarding the ignition system are manifold and strongly dependent on the combustion strategies. One of the main sources of efficiency loss in the part load region of a conventional SI engine is the so called gas exchange loss or throttle loss.

For qualitative load control, as realized in the stratified operation [10], the engine can operate almost with open throttle, thus reducing the gas exchange losses to a minimum. The demands of stratified part load with respect to the ignition system have been widely studied and result in the need of a wide statistical coverage of the "ignition window", the time and space area, when an ignitable mixture is available at the point of ignition. Therefore we often see the usage of so called multi-spark solutions, which provide a reasonable long, but intersected coverage of the temporal "ignition window". The statistical coverage of this stratified ignition window by means of a spacious inflammation system, such as corona, offers significant advantages, which has been widely proven in several publications [2, 9]. It appears to be among one of the key performance areas of spacious ignition systems, which allows to extend the stratified operation of the engine to low loads with high EGR-rates [9] and very low NOx emission values, but also enables the conventional catalyst heating points [9]. The freedom to decouple in part the location of the ignition from the ignitor hardware location, as can be done in part with the corona principle, offers optimal boundary conditions to significantly avoid or reduce particle raw emissions, which might occur due to igniter wetting.

However, by far the vast majority of current SI engines operate on the basis of a quantitative load control, which allows a stoichiometric operation and therefore the use of a three way catalyst (TWC). To reduce the gas exchange losses for this operation mode of the engine, both strategies employing external EGR as well as modern variable

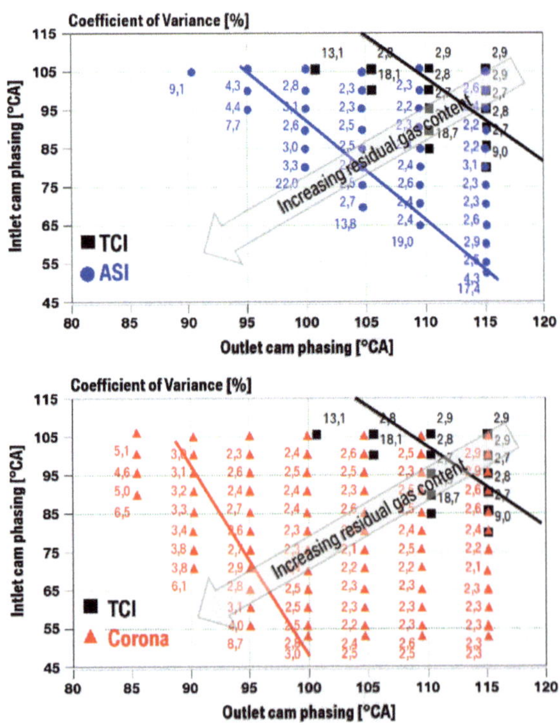

Fig. 4. Variation of cam-phasing for TCI (black) and ASI (blue) on the top and TCI (black) and corona (red) on the bottom. Depicted are the values for the CoV. The residual gas fraction increases significantly towards the lower left end of each plot. The solid lines symbolize the approximated limit for a maximum CoV of 3 % (1000 rpm, IMEP = 1 bar, CR = 12).

valvetrains can be used. Especially effective is the use of fully variable valvetrains such as the BMW Valvetronic, which allow the strategy of simultaneous late outlet valve closing combined with an early inlet valve closing (on an fully variable inlet valve lift) [11], which leads effectively to a strongly Millerized condition in lower part load and significantly reduces the gas exchange work loop of the p-V-diagram. This strategy however imposes significant challenges towards the ignition process, basically due to two reasons: Firstly the high internal residual gas fractions leads to a lowered laminar burning velocity, secondly the early inlet closing leads to a reduced charge motion as well as reduced temperature at the time of ignition. Both effects reduce the initial burning velocity significantly. To achieve a higher flame kernel stability for this operating condition, both the elongated spark duration (ASI) as well as the corona ignition do offer significant advantages. Figure 4 displays a TVDI-cam-phase variation for 1000 rpm and an IMEP of 1 bar. The outlet valve lift is 9.8 mm, the inlet vale lift is variable and in the range below 1 mm, thus regulating the amount of gas flow through the inlet valves. The upper side compares the conventional TCI ignition with an ASI, the lower side displays the comparison with a corona system.

The mechanism, however, by which the higher residual gas tolerance is achieved, is fundamentally different for corona and ASI. The corona ignition profits both from significant size, a freely adjustable duration and an inflammation mechanism rather independent of temperature and charge motion [3, 12]. Thus the burn delay is significantly reduced and so is the statistical standard deviation of the 5 % MFB point. This in average results in a better CoV and an improved fuel consumption even for the same amount of residual gas [3] as the TCI System. The ASI gains its advantage by providing a longer spark duration at sufficient current level. This helps to ignite even strongly diluted mixtures without the occurrence of heavily retarded cycles, thus enabling us to operate stable even for long burn delays. However, in contrast to the corona ignition, it does not significantly reduce the individual burn delay, besides statistical effects in the average burn delay. This principle mechanism is confirmed by the fact, that a multi-spark operation, with moderate intermediate interruptions of the spark, did not offer the same advantageous result, as a sufficiently long spark of the same total on time duration. Figure 5 illustrates these correlations by depicting the burn delay statistics as a function of the 5–50 % MFB burn duration for all three ignition systems. In Fig. 5 (top) plots the standard deviation for all TCI and all ASI points of the cam-phase variation of Fig. 4. The points match within the specific distribution range, since not every 5–50 % MFB burn duration is uniquely correlated to fixed burn delay (and thus std. dev.). However the distribution range of TCI and ASI becomes instable for longer main combustion durations, thus bending upwards for appr. 12°CA and 14° CA, respectively. With Corona we see almost no scattering and the stable region lasts until appr. >16°CA 5–50 % MFB combustion duration.

This results in the fact, that corona ignition can extend stable combustion to significantly higher residual gas fractions than ASI, which in average provides half the benefits corona offers in this conditions. Figure 6 illustrates the achievable residual gas fractions this for an operation point of 1250 rpm and an IMEP of 2 bar. The results where gained by means of a 1-dimensional gas exchange analysis (CR = 12).

The part load challenge to the ignition as well as its fuel efficiency benefits are especially high for the above mentioned TVDI operation. However for TVDI operation

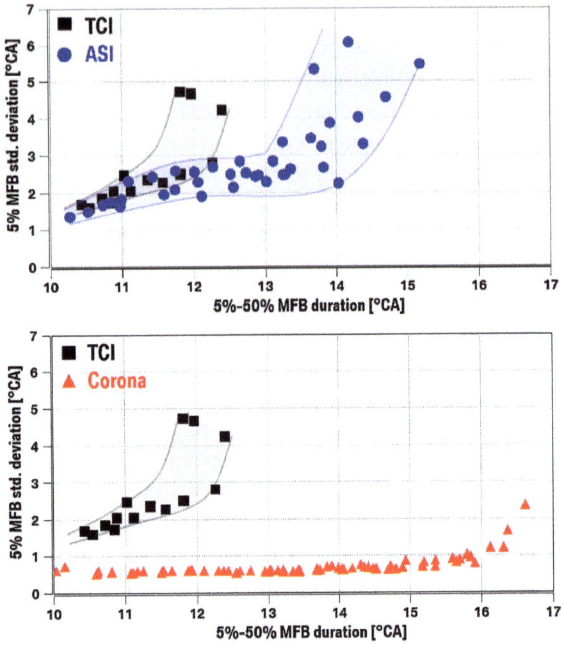

Fig. 5. Variation of cam-phasing for TCI (black) and ASI (blue) on the top and TCI (black) and a corona (red) on the bottom. Depicted are the values for the standard deviation of the 5 % MFB point. (1000 rpm, IMEP = 1 bar, CR = 12).

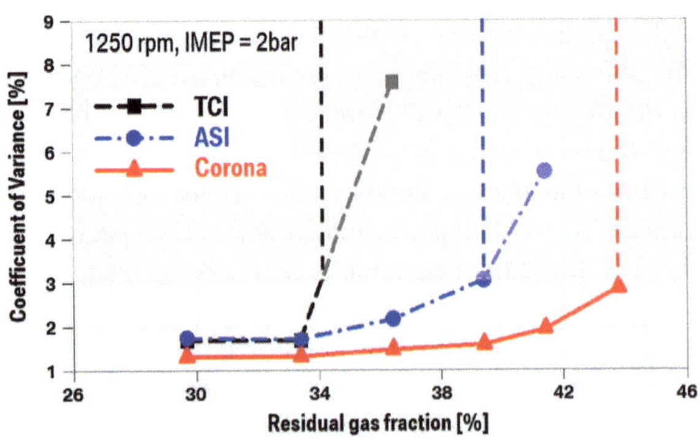

Fig. 6. Engine smoothness as a function of the internal residual gas rate for selected TVDI cam-phasings for TCI (black), ASI (blue) and corona (red). Depicted are the values for the CoV. Setting a CoV limit of 3 %, corona and ASI show both a considerable increase in the residual gas tolerance. Corona does display the maximum in the range of 44 % residual gas fraction (1250 rpm, IMEP = 2 bar, CR = 12).

the fuel efficiency is also significantly improved in comparison with the conventional TGDI operation [11].

Considering the challenges towards the ignition systems within the engine operation map, Wolf et al. [3] reported the ignition conditions for spark ignition to be considerable more challenging for lower load and for reduced charge motion. That naturally results in the highest spark duration demands for these operating conditions. As already stated in [13], the early inflammation process is to a considerable extent dependent on time based phenomena, where as the main combustion is, due to its turbulent character, predominately dependent on crank-angle based phenomena, since charge motion rises predominately with the average piston speed [14]. Thus the average burn delay rises on the crank angle scale with engine speed, resulting in continuously earlier and thus less favorable ignition timings [13]. Looking upon the demands towards the ASI duration, the minimal duration demand decreases significantly with speed, since the same spark duration covers for rising engine speeds an increasingly larger crank angle interval. This can be seen in Fig. 7, depicting the results gained for varying the spark duration until the misfire limit for a variation of the revolution rate at an effective zero load of the engine. A proper estimate is, that not more than a 30°CA equivalence is necessary as spark duration. Since the spark duration of the fully loaded TCI system is rather independent of the revolution rate, the relative advantages of an ASI regarding the spark duration have to decrease with rising revolution rate. Due to the overall shortened burn delay of the corona system, the relative advantages with respect to the TCI system rise with increasing revolution rate [1].

For engine operation employing external EGR, the EGR tolerance at zero effective load is a very important measure in dynamic operation. The reason is, that usually due to the volume of the inlet air/EGR path of a low pressure EGR-system, the EGR rate changes only comparatively slow. Thus, for a sudden drop in load, the engine will remain to be operated for a short transient period at a comparatively high EGR-level, but at significant lower load levels than before. However, since a misfire-free operation has to be assured under all circumstances, the EGR tolerance at low loads partly corresponds to the EGR levels allowed at higher loads. Here both, the corona as well as

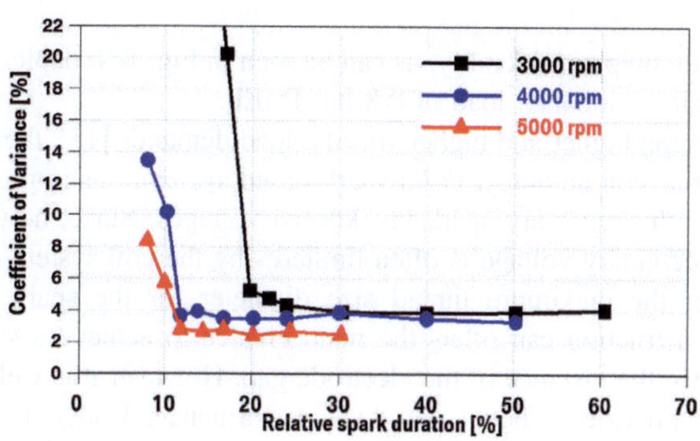

Fig. 7. Variation of engine speed for effectively zero load. The cam-timing was set for low residual gas content. The duration of the ASI System was varied until the misfire limit. The viewgraph displays the CoV as a function of the relative spark duration.

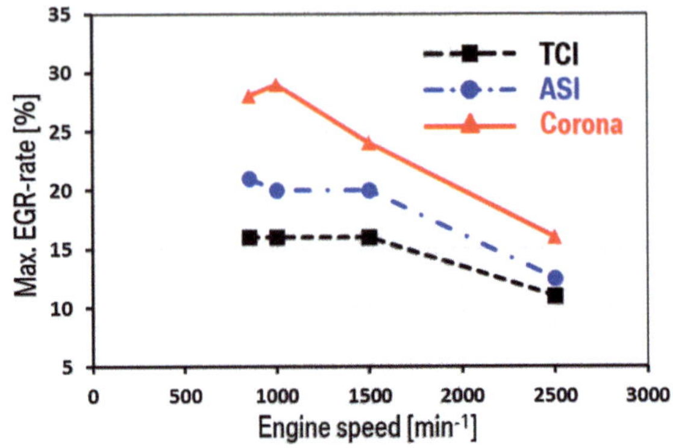

Fig. 8. Comparison of the maximal tolerable EGR-Rate as a function of revolution rate (IMEP = 0.8 bar).

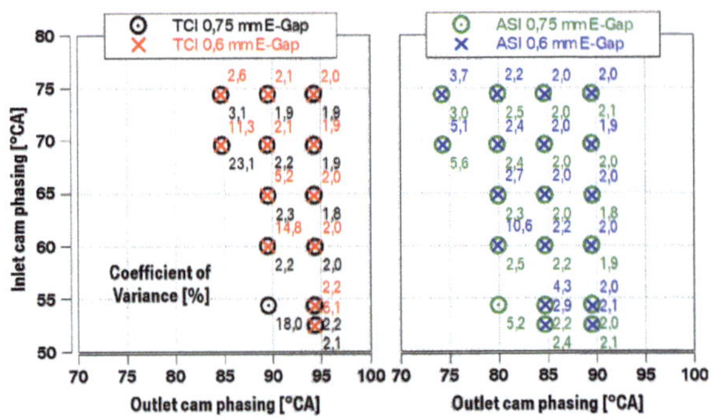

Fig. 9. Comparison of anelectrode gap variation (0.75 vs. 0.6 mm; 1000 rpm; IMEP = 2 bar). For both gaps a cam phase variation was performed, both for TCI and ASI, respectively. The resulting CoV's are depicted. The ASI outperforms the TCI Systems and for each electrode gap and can even overcompensate the disadvantage due to a reduced electrode gap.

the ASI ignition do help considerably, as can be seen in Fig. 8. It depicts the maximum tolerable EGR-Rate at a const. load of 0.8 bar IMEP.

With proclaiming higher and higher specific load demands [15], the densities at the time of ignition rise continuously. For a spark based system, this imposes even higher demands towards the necessary spark breakdown voltages. Since, however, the maximum bearable secondary voltage is often limited - by the coil system or by the spark plug ceramics or the maximum thread size diameter for the spark plug, which a cylinder head construction can offer- the main chance to achieve a stable spark formation is to reduce the distance of the electrode gap. However it is well known, that a larger spark gap provides a better part load performance. Under this constraint we investigated in the performance degradation involved in reducing the electrode gap of our TCI based from 0.75 mm to 0.6 mm, i.e. by 20 %, respectively. Figure 9 displays the results for a cam phase variation for 1000 rpm and an IMEP = 2 bar. It displays the important capability and advantage of the ASI System to even over-compensate a decrease in part load performance one might have to accept due to the need of a reduction in spark gap distance. This is an important advantage of the ASI System in the prospect of continuously rising ignition pressure demands.

4.2 Potentials in Upper Part Load

In the medium and upper part load region of the engine map, the combustion and ignition conditions become less critically, thus the differences among the various alternative ignition systems become less pronounced. One reason for the better inflammation condition might be seen in the higher level of turbulence. This effect has been already reported by Wolf et al. [3] for the comparison of TCI and corona systems. On the other hand, the higher the speed of the engine, the more pronounced are the relative differences between the corona and the spark ignition systems, even with the absolute overall dilution tolerance decreasing with rising engine speed [3].

How do advanced ignitions systems as the ASI fit in such a comparison? As already stated in the previous section, the ASI System does not in principle shorten the burn delay of the spark based inflammation. It simply avoids to a certain extent delayed combustion cycles and misfire. As the inflammation improves towards higher loads and deteriorates towards higher speeds, the in principle unchanged burn delay leads us to the typical results depicted in Fig. 10. It displays the achievable NOx levels, in a homogeneous lean combustion environment. The optimization was done by firstly optimizing for the best fuel consumption, and secondly choosing always the leaner and less NOx producing value of the lean out variation, presumed the fuel consumption does not deteriorate significantly. The principle tendencies of Fig. 10 are clearly pronounced. The corona system in generally provides the best lean burn results regarding the NOx emissions. It is followed by the ASI system. The poorest performance is displayed by our TCI reference system. However, progressing towards higher loads, the ASI System and TCI approach each other much faster with respect to the ignition performance, than it is the case for the corona system. The relative differences between corona and TCI become more pronounced for higher speeds, however the differences between ASI and TCI are reduced towards higher speeds for constant load. This is due to the fact that the less spark duration becomes necessary with increasing speed and therefore higher turbulence levels.

This behavior is confirmed when looking upon the results of an external EGR variation at 2500 rpm and an IMEP = 6 bar (Fig. 11). Here the TCI and the ASI perform already almost identically, the corona system however is still capable of providing an extended dilution limit and thus a fuel consumption benefit. However, taking into account the dynamic restrictions, i.e. the ability to perform a sudden load drop without the risk of misfires, the EGR limits are less than in the stationary thermodynamic optimum and often restricted by the low load tolerance (see prev. section). Thus indirectly we gain fuel efficiency benefits in the medium load operation also for the ASI system.

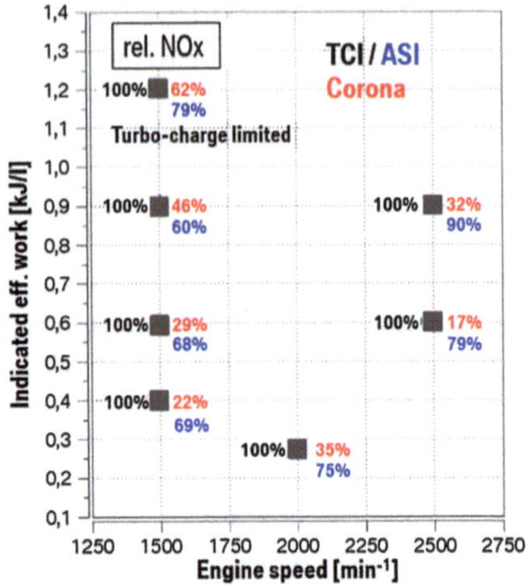

Fig. 10. NOx levels achieved for an air dilution variation at various points in the engine map. For each single comparison the cam-phasings were kept identical, only varying the air dilution level, the optimization was done with the focus on the best fuel consumption in combination with the lowest achievable NOx level. Below an IMEP of 3 bar we tolerated a CoV of 3 % in maximum, in the medium load region we reduced this limit to 2 %. The point at IMEP = 12 bar was primarily restricted by the amount of air available rather than by the lean limit.

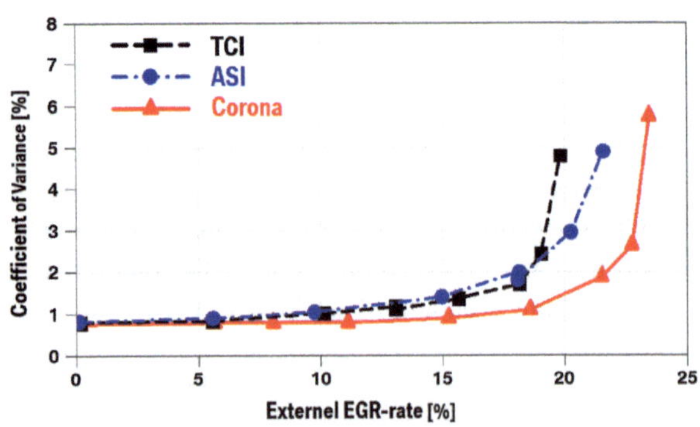

Fig. 11. Results of an external EGR-variation at 2500 rpm and an IMEP = 6 bar. Whereas TCI and ASI perform almost identical, the corona system still offers some advantage, as by a higher dilution limit and in an overall better CoV under diluted conditions, which also corresponds to a fuel consumption benefit.

4.3 Potentials at High Load

In the knock limited higher and highest load area of the engine map, the differences between the corona ignition system and the TCI reference are again more pronounced. The reason for this is the strong limitation of the combustion by knock on one hand and by the component protection on the other hand. Due its significantly better

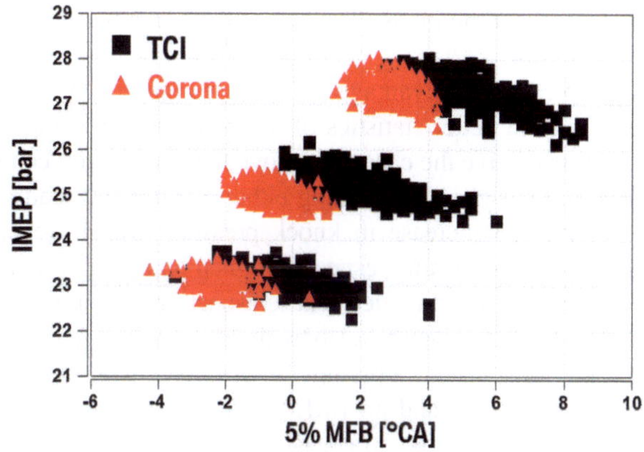

Fig. 12. Scatter diagram comparing spark and corona ignition at high load. The spark plug is colored black, corona ignition red. (4000 min^{-1}; IMEP = 23 bar, 25 bar and 27 bar respectively; CR = 9.5) [3]

inflammation statistics, corona does offer significant potential in the high load region, even at highest charge motion levels. This is clearly displayed in Fig. 12, which displays a load variation at 4000 rpm and for a strongly tumble oriented charge motion (CR = 9.5).

One should emphasize, that the corona application window for a given frequency is, among other parameters, primarily a function of particle density, distance between ground and electrode prong and the peak sharpness. Our results show, that these demands can be met properly even up to very high loads. It can be expected, that proper technological solutions are available to preserve the advantage of corona ignition even for very high load application. To our experience [3, 13] high charge motion did not prove to be an obstacle. It even seems in part to help to prevent the risk of power arc. Figure 13 displays as an example the accessible application window for a full load operation point at 2500 rpm and an IMEP = 27 bar. The upper voltage limit

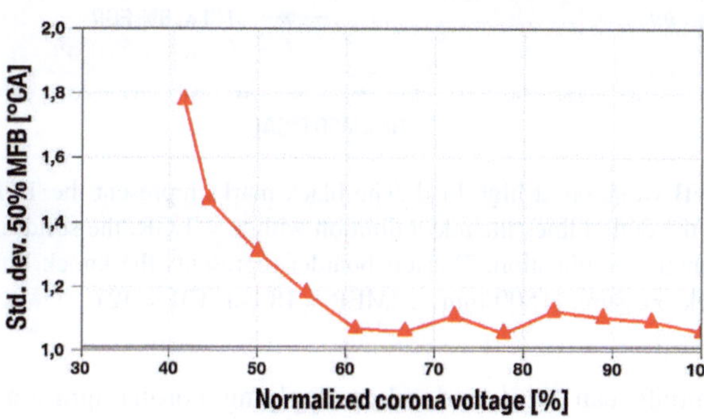

Fig. 13. 50 % MFB Std. dev. as a function of the applied voltage. (2500 min^{-1}; IMEP = 27 bar; CR = 9.5)

was due to internal system limitations. The lower limit represented the corona onset voltage.

Up to now, when presenting advantages of spacious ignition systems at high loads, publications focused on the better statistics of the 5 % and 50 % mass fraction burnt point. These improvements give the chance to move the combustion center closer to the knock limit as for a TCI system, thus gaining either a better full load efficiency or an improved torque without an increase in knock pressure [3]. However the full load advantages are even more farreaching, especially when applying corona to diluted and strongly Millerized combustion strategies. These aspect gain even more importance in combination with the need to comply also with low quality fuels. The underlying arguments become obvious, when applying for example EGR at an LET operation point, as depicted for 1500 rpm and an IMEP = 18 bar. Figure 14 displays the corresponding CoV's for a 50 % MFB sweep. Black lines represent the TCI results, the red the corona results, the dotted lines include a dilution with 8 % EGR and the solid lines represent an undiluted stoichiometric combustion. The left boarder represents the knock limit, defined by a specific level of knock pressure.

It can be seen clearly that the corona with EGR is even capable of reaching better CoV-levels than the TCI without dilution. The differences become even more pronounced for later 50 % MFB positions. If one considers the engine smoothness at LET as a possible limit for the choice of the compression ratio (bearing in mind hot country and low fuel quality environments), the corona can serve as an enabler for an overall higher compression ratio, especially in combination with dilution measures. This results in direct benefit in cycle relevant fuel consumption and customer acceptance regarding NVH behavior of the engine.

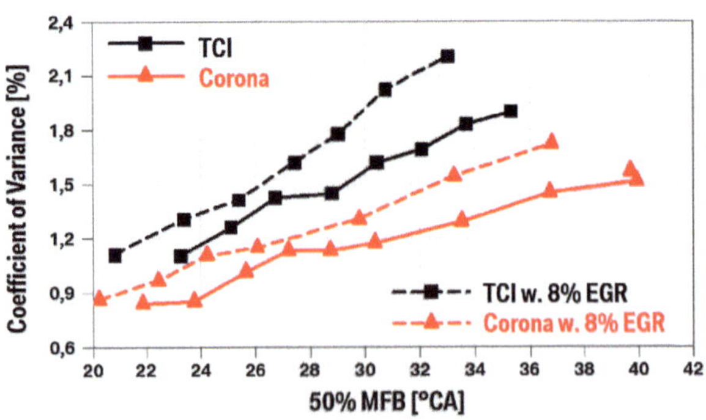

Fig. 14. 50 % MFB variation at high load. The black marks represent the TCI results, the red the corona results, the dotted lines include a dilution with 8 % EGR, the solid lines represent an undiluted stoichiometric combustion. The left boarder represents the knock limit, defined by a fixed level of knock pressure. (1500 min^{-1}; IMEP = 18 bar; CR = 12)

Similar potentials can be found when applying corona ignition to Millerized combustion. As an example we compare TCI, corona and a passive pre-chamber spark plug for a valve lift variation with keeping the inlet open timing constant. Figure 15

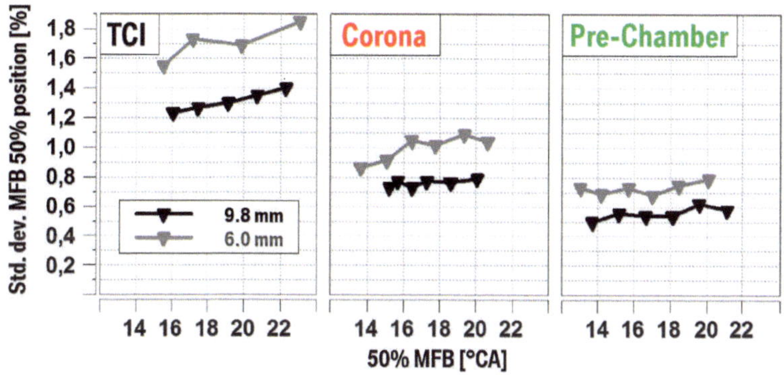

Fig. 15. 50 % MFB variation at high load for a variation in valve lift, with the inlet opening timing being kept constant. The black and gray lines represent 9.8 mm and 6 mm valve lift, respectively. The TCI, corona, and pre-chamber results are depicted from left to right, respectively. (2000 min^{-1}; IMEP = 18 bar; CR = 12)

displays the statistics of the 50 % MFB points achieved for an ignition timing sweep until the knock limit (left side). The results were gained by a single cylinder research engine (2000 rpm, IMEP = 18 bar; CR = 12). The corona displays significant advantages up to a factor of two, outnumbered only moderately by the pre-chamber ignition system.

However, the better statistics of the pre-chamber plug are achieved by an increase in main combustion speed of about 25 % compared to the TCI system. This results in an increase in pressure gradients, as can be seen in the scatter plot diagrams in Fig. 16, which in some cases might be undesired due to NVH reasons. The scatter-plots represent the corresponding cycle of the 50 % MFB sweeps. To illustrate the distribution

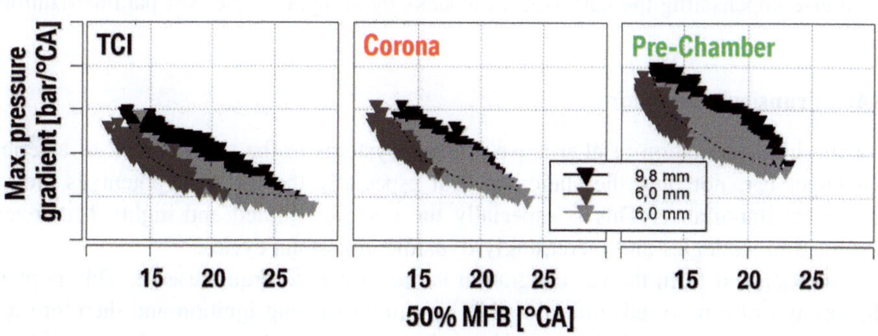

Fig. 16. 50 % MFB variation at high load for a variation in valve lift, with the inlet opening timing being kept constant. The black and gray dots represent the individual cycles of 9.8 mm and 6 mm valve lift, respectively. It has to be pointed out, that all cycle of one 50 % MFB-variation are plotted together in the diagram to pronounce the differences between the parameter variations. The TCI results are depicted on the left, the pre-chamber results on the right side, whereas the corona results can be seen in the middle diagram. (2000 min^{-1}; IMEP = 18 bar; CR = 12)

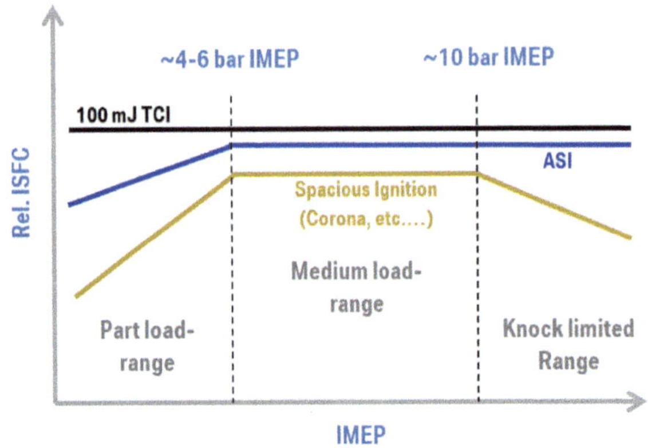

Fig. 17. Simplified scheme of the relative fuel consumption benefits for ASI and corona with respect to a 100 mJ TCI for moderately low revolution rates. In the part load range, we clearly see advantages for both alternative ignition systems. For increasing load towards the medium range primarily the corona system keeps part of its relative benefits, especially with respect to emissions. For high load the ASI system behaves similar to a TCI System. The corona, however, display strong high and full load advantages, especially for a highly Millerized or diluted approach.

for a single ignition timing, the earliest distribution of each 6.0 mm valve-lift sweep is depicted in dark grey. The ASI System is not taken into account in the comparison of this section, since it is expected to behave similar to the TCI System.

To summarize the stationary effects for varying engine loads, we consider Fig. 17 as a useful illustration of the principal relative differences. However, one has to keep in mind that the ASI system allows to achieve a spark breakdown even at operating pressure conditions by reducing the electrode gap of the spark plug and counter-compensating the part load drawbacks by an appropriate ASI parametrization.

4.4 Transient Behavior

For the implementation of alternative ignition systems in the environment of a series passenger car, not only the stationary, but especially the transient potentials are of immanent importance. This is especially the case for diluted and highly Millerized combustion strategies and increasingly dynamic emissions cycles.

One demand from the car integration is the so called torque reserve. This implies the amount of torque reduction possible by simply shifting ignition and therefore the 50 % MFB point towards later positions than the thermodynamic optimum, without making any changes of the relative air filling of the engine. Since the so called air path might be comparatively slow, the torque reserve is among the fastest measures to reduce and to build up momentum. It is limited either by the acceptable engine smoothness limit (CoV) or, at higher loads, by component protection temperature limit of the exhaust gases. As already seen in the case of the 50 % MFB sweeps in Figs. 14 and 15 the corona system offers here considerable advantages compared to spark

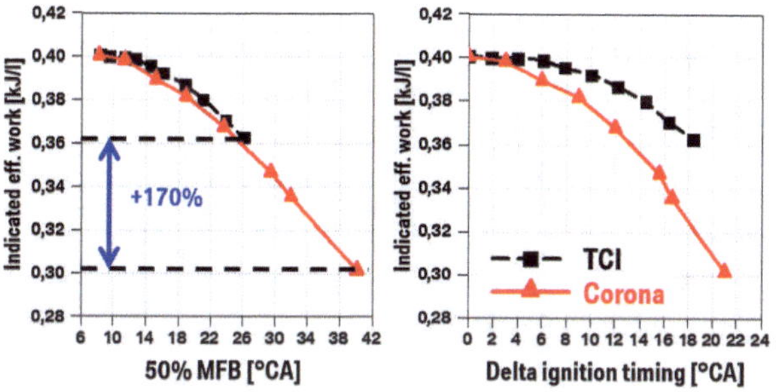

Fig. 18. Torque reserve for a stoichiometric operation points without dilution. The red marks represent the corona results, the black the TCI results. The right viewgraph depicts the IMEP as a function of the 50 % MFB position, the left plots the corresponding CoV. The middle graph depicts the IMEP as a function of the delta ignition timing with respect to 50 % MFB = 8°CA.

ignition systems. The reason is, that by reducing the burn delay for a given 50 % MFB point, the corona system results in a strongly improved 50 % MFB position statistics. Since for late 50 % MFB positions the standard deviation of the 50 % MFB position is directly correlated to the CoV, corona does provide a significant improvement in engine smoothness and therefore in torque reserve, except the case, where the temperature of the exhaust gases is the primary limit for the late shift of the 50 % MFB position rather than the engine running smoothness. These difference is becoming even more pronounced for diluted combustion, as can be seen in the previous chapter. Among others this provides considerable comfort in the handling of high EGR-rates in a diluted combustion concept, since it allows to faster decrease the external EGR rate in case of a sudden load drop. Figure 18 displays the gain in torque reserve for an operating point of 2500 rpm and an IMEP = 4 bar. The relative gain in torque reserve is additional 170 %. The graph in the middle indicates, that the torque structure for a corona application is different from a conventional spark and thus has to be adapted (see also next section).

The next advantage concerns particle emissions. Most conventional catalyst heating points with TCI systems employ a so called ignition injection to improve the catalyst heating stability towards late 50 % MFB positions, to allow a fast warm up of the exhaust after-treatment components. However, this strategy might lead to a moderate soot particle generation in the raw emissions of the engine, since the spray is directly targeted towards the spark gap and the initial fame kernel. Skipping these ignition injection avoids this particle generation, however, leads to unacceptable high CoV-levels. Corona ignition is capable to reduce this CoV to an even better level than with ignition injection, without having to employ an ignition injection at all. Thus corona ignition enables an effectively particle free catalyst heating operation, with even better engine smoothness and further improved potentials for a faster warm up. Figure 19 depicts the described phenomena for a 3-cylinder Engine with CR = 12.

Also in the general engine warm up, the corona ignition systems outnumbers the TCI system with respect to a significantly better idle smoothness, especially for cold environment temperatures (based on corona JIS tests for repetitive −10 °C starts).

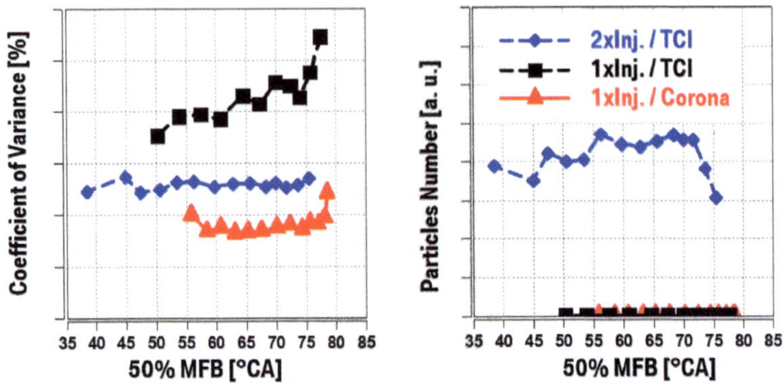

Fig. 19. Catalyst heating point with (blue) and without (black) an ignition injection, using a conventional TCI ignition system and without an ignition injection but with a corona ignition system (red). Clearly the better engine smoothness for the particle free catalyst heating employing the corona system can be seen.

4.5 Functional Integration Aspects

Integrating an advanced ignition system of the kind of ASI or corona does imply a variety of measures to be taken.

Speaking about the parameter space, for an ASI system the spark duration as well as spark current are varied, whereas a corona system needs the corona voltage and corona duration to be specified. Additionally, of course for all systems the ignition timing has to be specified as well. Besides the ignition timing, this leads us to two additional parameters compared with our reference TCI system (exactly speaking to one, if the charging time of the TCI is considered as a parameter as well). Since the ASI does in principle not change significantly the burn delay with respect to the TCI system, the additional application efforts can be restricted to energy parametrization

Table 2. Ignition parameters and their interdependencies.

System	TCI	ASI	Corona
Timing	System specific ignition timing		
Ignition parameters	– Energy	– Spark duration – Spark current	– Corona-Voltage – Corona-Duration
Interdependencies	Burn delay: – Independent of ignition parameters – Dependent on timing and operation point		Burn delay: – Strongly dependent on ignition parameters – Dependent on timing and operation point
Derating	Derating (fouling etc.) increases risk of misfire		Derating changes burn delay and has to be compensated

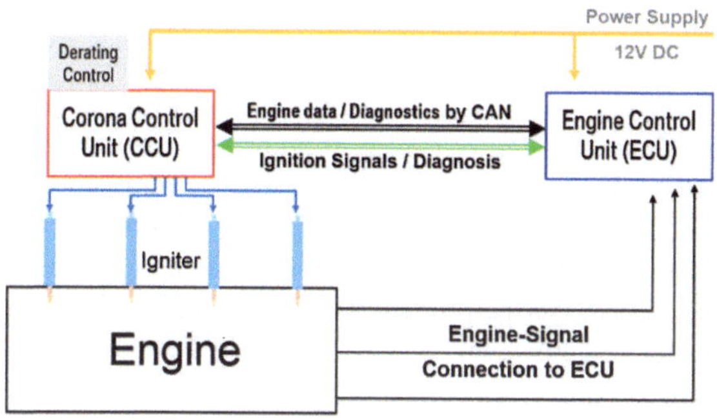

Fig. 20. Possible architectural scheme of a data communication for an implementation of a corona ignition system.

(spark current as well as spark duration). This leaves the maps of ignition timing and the torque structure models in principle unchanged. However, dynamic aspects, especially for diluted combustion, do make a more detailed energy parametrization and eventually a synchronous data communication in advance to each ignition event necessary. Table 2 summarizes these parameters.

In the case of corona system the key input parameters corona primary voltage (determining the corona size), corona duration and ignition timing have to be specified (Table 2). However, as we have seen by the results of previous publications [3, 9, 13], the burn delay strongly depends on the corona size and, to a somewhat weaker extent, on the corona duration as well. In contrast to spark ignition systems the burn delay in a specific operation point is not only a function of the engine operating point and ignition timing, but of the ignition parametrization as well. This offers the chance to parameterize burn delay as well, instead of only the ignition timing, but on the other hand it creates the need to actively control the burn delay (as a function of the ignition timing, duration and the corona size). Since overall Corona induced burn delay is changed significantly with respect to TCI systems, not only the ignition timing, but also the torque model of the engine map has to be adapted, when switching from a spark to a corona based ignition system.

In both cases, ASI and corona, we assume the need for a fast communication of the ignition parameters. One option would be to use the ignition trigger lines for that purpose, thus there would be no need for a significant change in cabling on the ECU side. For a corona system, due to a need of a separate corona control unit (CCU), an additional separate CAN communication would be helpful due to several reasons.

However, compared to the conventional ignition path, the electrical architecture of the corona system does cause necessary changes in the electrical infrastructure. Thus the CCU has to be implemented in the packaging of the engine or the chassis and the HV-coaxial cables have to be fed from the mounting location of the controller to the corona igniters. A scheme of a possible implementation is given in Fig. 20.

Since the size of the corona is essential for the burn delay and thus to the combustion, the corona size, or, better speaking, the resulting burn delay, has to be controlled over the lifetime of the engine. The control system has to be capable to detect fouling as well as eventual prong wear and to take the necessary countermeasures.

5 Conclusion

We presented an overview over the increasing challenges to the ignitions system for future SI engines. Generally higher charge motion and load dynamics as well as an increasingly diluted operation, combined with Miller strategies offer considerable chances in engine performance, however, imposing significant challenges towards future ignition systems. The improvement in conventional spark performance, such as the ASI system presented, can help to meet already a part of the challenges.

The major challenges, however, can be met only by a spacious ignition system, which provides both the improved combustion statistics and the stability needed for an advanced thermodynamic performance of future gasoline engines. From the currently known spacious ignition systems, the corona-ignition does display a highest level performance in the complete engine map, providing an ignition source with only very moderate energy consumption over the complete engine map. This provides ad-vantages in lower fuel consumption as well as lower engine out raw emissions, but also in driving comfort and dynamic behavior.

During the recent year in development the corona system proved itself to be a reliable and exceptional powerful ignition system, opening new pathways to combustion initiation for almost any SI engine combustion strategies. Taking into account our experience gained by successful corona-implementations in prototype demonstrators, by durability tests, both on engines as well as in laboratory setups, we are very positive that the tasks necessary for a series implementation of a corona system can be met in the near future as an enabler for combustion strategies of highest fuel efficiency. Successful results from cold start testing (−10 °C JIS repetitive cold-start) confirm this expectation.

Key demand of course remains the miniaturization of the CCU electronics. The possible future availability of a power supply at a voltage level higher than 12 V and also synergy options by implementing part of the CCU in the ECU could help on this pathway.

List of Abbreviations

5 % MFB	Crank angle position of 5 % mass fraction burnt
50 % MFB	Crank angle position of 50 % mass fraction burnt
ASI	Advanced spark ignition
CCU	Corona control unit
CR	Compression ratio
ECU	Engine control unit
IMEP	Indicated mean effective pressure
LET	Low end torque
NVH	Noise vibration harshness
TCI	Transistor coil ignition
TGDI	Turbo gasoline direct injection, i.e. in our case full valve lift and variable cam phasing
TVDI	Turbo Valvetronic direct injection
Valvetronic	BMW fully variable valvelift system

Acknowledgements. The authors would like to thank our partners for providing the prototypes of the corona ignition systems as well as of the advanced spark ignition systems. Further we would like to express our gratitude to the colleagues who supported the measurements, especially the electrical measuring techniques. Especially we would like to thank Mr. Muhammad Faris Bin Omar, who supported many of the measurements depicted.

References

1. Fessler, M., Wetzel, M., Schenk, M., Rottengruber, H., Huhn, W.: Ermittlung von thermodynamischen Potenzialen alternativer Zündsysteme an einem strahlgeführten Brennverfahren. In: 8th International Symposium on Combustion Diagnostics, Baden-Baden, Germany (2008)
2. Schenk, M., Fessler, M., Rottengruber, H., Fischer, H.: Vergleich der thermodynamischen Potenziale alternativer Zündsysteme für ottomotorische Brennverfahren. In: 10th International Symposium on Combustion Diagnostics, Baden-Baden, Germany (2012)
3. Wolf, T., Schenk, M., Schröter, M., Zellinger, F., Klaus, B., Pfeiffer, D., Fischer, H.: Rf-corona-ignition vs. spark ignition: a comparison for varying thermodynamic conditions and combustion strategies of modern turbocharged gasoline engines. In: 2nd Symposium on Advanced Ignition Systems for Gasoline Engines, Berlin, Germany (2014)
4. Alger, T., Gingrich, J., Roberts, C., Mangold, B. Sellnau, M.: A high-energy continuous discharge ignition system for dilute engine applications. SAE Paper # 2013-01-1628 (2013)
5. Weyand, P., Lorenz, F., Schilling, S., Geringer, B., Hofmann, P., Graf, J., Hofherr, T.: Adaptive continuous spark ignition as enabler for high dilution EGR operation. In: Günter, M., Tröger, R. (eds.) Ignition Systems for Gasoline Engines, Tagungsband, IAV Berlin (2014)
6. Ruan, G., Kienzle, D., Roth, D., Becker, M.: Doppelspulenzündung für Ottomotorenkonzepte mit AGR, MTZ 03/2013, pp. 212–216 (2013)
7. ETH Zürich: Elektrische Energiesysteme, Lecture on high-voltagetechnology FS09: Durchschlag in Gasen
8. Freen, P.: Radio frequency electrostatic ignition system feasibility demonstration. Energy Innovations Small Grant Program, Final report (2005)
9. Schenk, M., Fessler, M., Wolf, T., Klaus, B., Fischer, H.: Comparison of the thermodynamic potential of alternative ignition systems for SI engines. In: 1st Symposium on Advanced Ignition Systems for Gasoline Engines, Berlin, Germany (2012)
10. Schwarz, C., Missy, S., Steyer, H., Durst, B., Schünemann, E., Kern, W., Witt, A.: Die neuen Vier- und Sechszylinder-Ottomotoren von BMW mit Schichtbrennverfahren, MTZ 05/2007, pp. 332–341 (2007)
11. Liebl, J., Landerl, C., Hofmann, R., Melcher, T., Steyer, H.: Der neue BMW Vierzylinder-Ottomotor mit Valvetronic, MTZ06/2001, pp. 450–463 (2001)
12. Starikovskaia, S.M.: Plasma assisted ignition and combustion. J. Phys. D Appl. Phys. **39**, R265–R299 (2006)
13. Schenk, M., Wolf, T., Schröter, M., Zellinger, F., Klaus, B., Pfeiffer, D., Fischer, H.: Corona- vs. spark ignition: a fundamental comparison for varying thermodynamic conditions of modern turbocharged gasoline engines. In: SIA Powertrain Conference, Versailles (2015)
14. Wachtmeister, G.: Lecture Notes on Engine Combustion, Technische Universität München, Lehrstuhl für Verbrennungskraftmaschinen
15. Fraidl, G., Kapus, P., Neubauer, M.: Gasoline Engine 2020: 200 kW/l and 200 g/kWh?. In: SIA Powertrain Conference, Versailles (2015)

Extension of Operating Window for Modern Combustion Systems by High Performance Ignition

Martin Brandt, Alexander Hettinger, Andreas Schneider, Hartwig Senftleben(✉), and Tim Skowronek

Robert Bosch GmbH, Blaichach, Germany
Hartwig.Senftleben@de.bosch.com

Abstract. Upcoming legislation motivates further development of modern combustion systems and engine designs to achieve a reduction of fuel consumption and emissions. Increased power density, charge motion and dilution drives requirements of the ignition system. Bosch prepared a high performance ignition device focused on high energy output and long spark duration while providing high flexibility to select relevant performance parameters. The ignition concept furthermore is designed as a plug and play solution in order to fit to established hardware interfaces and offers high energy efficiency.

Based on this technology several engine tests were completed at global OEMs and inside Bosch. The following paper summarizes the main test results which cover extension of EGR rates, lean limit and reduction of emissions as well. Based on this results possibilities how the benefits can be used in real applications are derived and discussed.

Investigations were performed on different engines representing diverse combustion concepts especially with regard to charge motion level. The concluded tests also confirm that increased ignition performance only in combination with parameters of the combustion concept itself can lead to the desired benefits on engine level.

Besides the review of mentioned engine tests a new advanced ignition system called CEI (Controlled Electronic Ignition) is introduced. The main properties of the used high ignition device are described including a comparison to the visible market trend for ignition requirements.

1 Introduction

Current development trends for combustion systems in spark ignition engine focus mainly on exhaust gas recirculation and/or on air dilution (e.g. [24]) as possible approach for increasing the engine efficiency, Fig. 1. For homogeneous lean combustion concepts enleanment needs to be strongly expanded in order to reduce raw nitrogen oxide emissions to make this approach a valuable solution also in terms of costs for exhaust gas after treatment.

However, increasing inert gas fraction also means more severe conditions for the ignition and combustion system itself. The negative influence of the rise in air dilution is reflected in the form of cyclic variations at engine operation. Therefore, in practice,

M. Günther and M. Sens (eds.), *Ignition Systems for Gasoline Engines*,
DOI 10.1007/978-3-319-45504-4_2

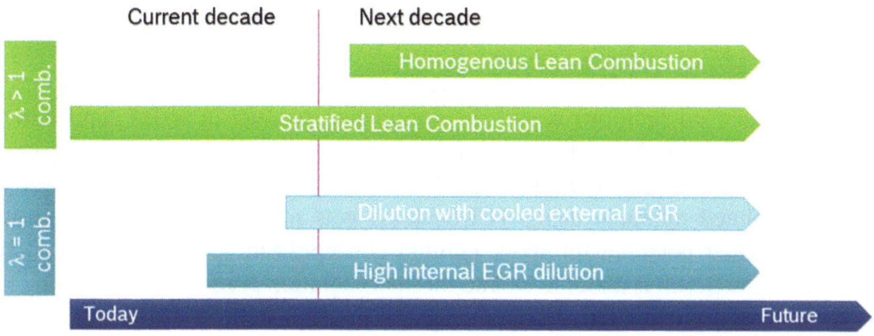

Fig. 1. Roadmap combustion concepts

further mixture dilution is not applicable when exceeding a smooth running limit of about 3–5 % (COV, Coefficient of Variance). The disturbance in running smoothness is mainly caused by cyclic variations in the in-cylinder flow field, mixture homogeneity and flame kernel formation [1, 2]. This leads to sporadic cycles with delayed combustion or even misfires. In order to stabilize the initial stage of combustion in diluted mixtures a wide number of ignition concepts and systems were proposed over time, recently leading to frequent discussions in expert groups. Figure 2 summarizes the variety of different proposals using a main classification based on how the flame kernel volume is activated during the ignition process. For more details about single ignition proposals see [3, 4] or latest IAV Conference Proceedings [5, 6].

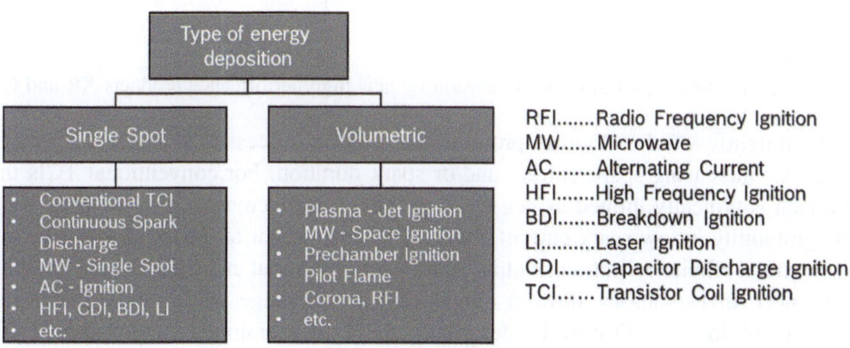

Fig. 2. Classification of ignition systems

"Single spot ignition" strictly means the deposition of ignition energy within a fixed location inside the combustion chamber, whereby the energy distribution is not actively controlled by the ignition system itself. Nevertheless, a spark volume is defined by the spark channel and even enlarged by charge motion, e.g. by deflecting the ignition spark, which essentially leads to an activation of a flame volume. This is one of the key factors, which yields the potential to ignite diluted mixtures using spark ignition systems.

“Volumetric” ignition concepts, such as corona or plasma-jet ignition actively provide a flame kernel volume even in low charge motion conditions at ignition timing.

Despite of the tremendous variety of ignition concepts and their competition, the TCI (Transistor Coil Ignition) is still the dominating product in series application of passenger cars. Its simplicity, low cost and proven robustness, especially compared to innovative, often complex systems made TCI to an unrivaled product. However, the ignition of diluted mixtures using TCI-Systems gets more severe and the question arises how further potential for spark ignition concepts can be accessed. To address these challenges Bosch is about to extend the spark ignition portfolio by proposing new high performance products as shown in Fig. 3.

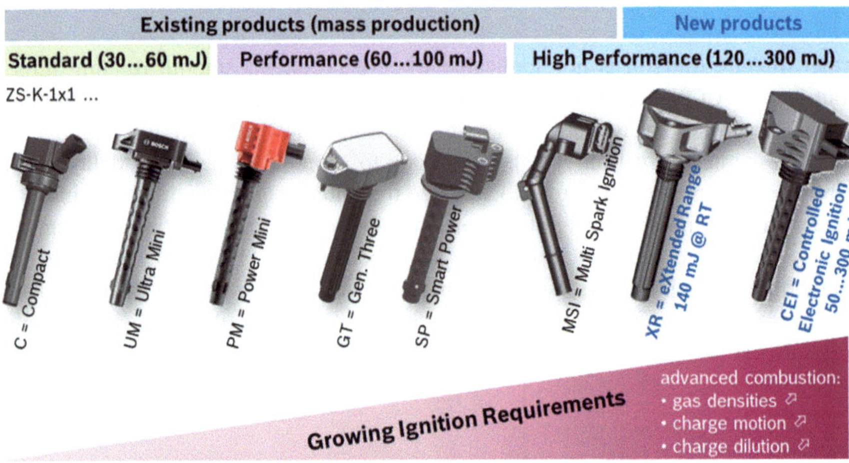

Fig. 3. Spark ignition portfolio incl. extension by new high performance products XR and CEI

A straightforward method to promote the ignition process is to boost the ignition energy by increasing spark current and/or spark duration. For conventional TCIs this approach is typically limited as negative side effects do increase e.g. spark plug wear, low ignitability in low spark current phases, size and weight of the ignition device and the electrical efficiency (ratio of output spark energy to input energy delivered from the board net). Investigations showed efficiencies in the range of 20–30 % under real engine conditions [7]. Due to the described conflict it is assumed a conventional TCI can be designed for a spark energy of some 140 mJ but not significantly higher. This is covered by a new proposed XR coil (see Fig. 3).

To solve some of the drawbacks as listed above connected with conventional TCIs and also to avoid the negative impact of decreasing spark current over time on flame kernel formation a new high performance ignition approach called CEI (Controlled Electronic Ignition) is described in this paper. This includes in a detailed way: identification of requirements for advanced combustion concepts (main aspect: highly diluted mixtures), discussion of target requirements for high performance ignition, review of present state of the art ignition systems, description of CEI with key properties, summary of test results and benefits for different engine/combustion types. At the end of the article short summary and final remarks are presented.

2 Requirements on Ignition System for Modern Combustion Concepts

2.1 Requirements for Efficient Combustion

Depending on the combustion concept and engine operation point different requirements on ignition system may arise. For stratified combustion one of the key factors is to cover the temporal and geometrical tolerances with regards to local mixture composition at spark plug location [8, 18]. A particular importance is assigned to the injection process, in order to create a preferably stoichiometric mixture in the vicinity of the spark plug. However, temporal fluctuations in the mixture formation process need to be covered, which can be achieved by long spark durations. Similar requirements on the ignition system apply at catalyst heating operation, so called homogeneous split injection (HSP). As mentioned before, long spark durations are limited when using conventional TCI. Even though in-house tests showed advantages of CEI for stratified operation the focus was put on homogeneous combustion concepts to derive requirements, since no market indication is obvious implying stratified operation as a mainstream trend.

At full load operation main requirements arise from high gas densities at ignition timing, leading to high ignition voltage demands of up to 40 kV. After spark breakdown the conditions to promote the flame kernel formation are more favorable than at low load conditions, since high gas densities and high flow velocities are present, leading to efficient energy transfer into the surrounding mixture. One straightforward method to decrease the ignition voltage demand would be to decrease the electrode gap distance. However, this would result in negative consequences at low load conditions due to quenching losses, especially with higher dilution of the mixture. Here, ignition

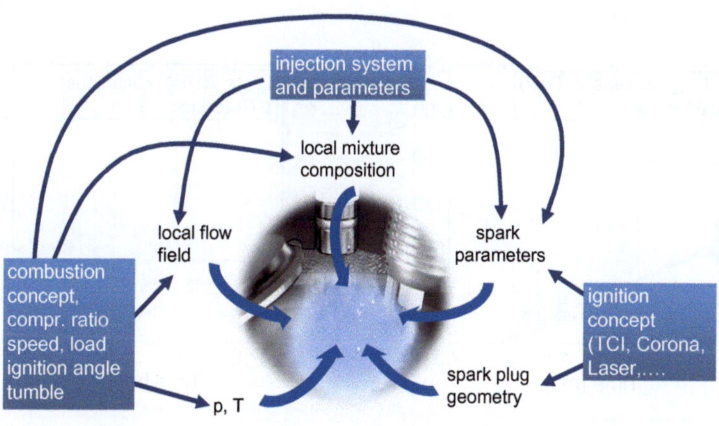

Fig. 4. Influencing factors on flame kernel formation

voltages are rather low but the requirements on ignition system to promote the flame kernel formation are high. Besides the influence of ignition system, the initial stage of combustion is influenced by many other factors, see Fig. 4.

To exploit further potentials of spark ignition systems at diluted engine operation, the interaction of all shown parameters needs to be considered. Despite global homogeneous operation, local mixture inhomogeneity may arise at spark plug location. These can be covered by long spark durations [16]. The local flow field is of particular importance, since it affects the spark deflection. High local flow velocities lead to long plasma arcs, which help to distribute the thermal spark power in the mixture. This leads to high energy transfer efficiencies [9, 17].

Basic investigations in an optical combustion vessel showed disadvantages with regard to maximum spark deflection when using a TCI System with its typical decreasing shape spark current. Figure 5 shows the evolution of spark length as function of time at local flow velocity of 10 m/s for 50 consecutive cycles (gray lines) and two different spark current profiles (left & right). The spark length was determined by optical analysis of captured high speed images. Based on single cycles and on mean spark length it can be observed that for current levels below approximately 40…50 mA the spark length steadily decreases. The reason for this is the steady increase of the number of so called "restrikes" with the decreasing spark current. A restrike occurs if a certain value of spark length is exceeded, leading to formation of a short circuit arc between the plasma channels. In this case the deflection process is interrupted and the channel length is abruptly shortened [10]. Single restrike events can be identified in the waveforms (gray lines) showing steep spark length gradients. With regard to ignition at engine operation it can be expected that below a certain current level the ignition plasma remains in the immediate vicinity of the spark plug electrodes preventing good access to fresh mixture. Furthermore, higher heat losses from the flame kernel to the relatively cold electrodes are promoted [10]. The right picture of Fig. 5. shows the evolution of spark length if the spark current is kept constant over the entire discharge duration. This shape was achieved by using a laboratory ignition system. The number of restrikes decreases significantly, leading to stable spark deflection. Therefore,

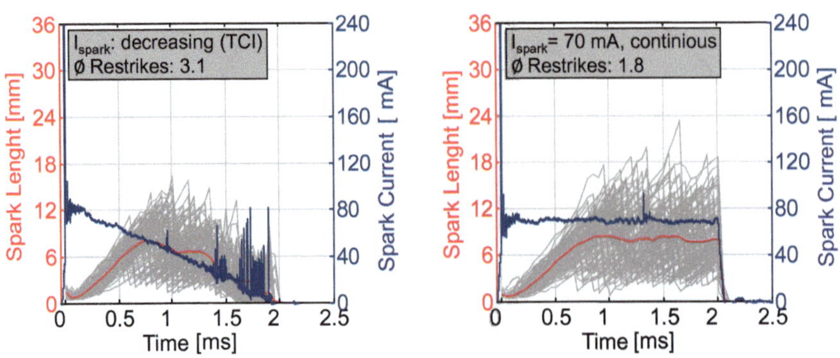

Fig. 5. Evolution of spark length for a decreasing (left) and continuous (right) spark current for 50 cycles (gray lines) at v = 10 m/s. Red line indicates mean spark length.

an ignition system with continuous spark current level would be favorable. The maximum spark deflection also depends on local value of flow velocity.

High flow velocities lead to long plasma arcs. However, the deflection process cannot be sustained endlessly, since a restrike becomes more probable as the plasma arc becomes longer. The reason for that is the increasing spark voltage (burn voltage) with spark length, leading to high electric field strengths between the plasma channels [9]. Engine measurements at different operating points revealed, that restrike voltages higher than 6 kV are less likely. However, at highest flow velocities a so called "spark blowout" is possible. In such cases the spark plasma can detach from the electrodes, therefore requiring new spark breakdown between the electrodes. The ignition system therefore needs to establish high voltage build up and provide a new spark breakdown within a short time after spark blowout.

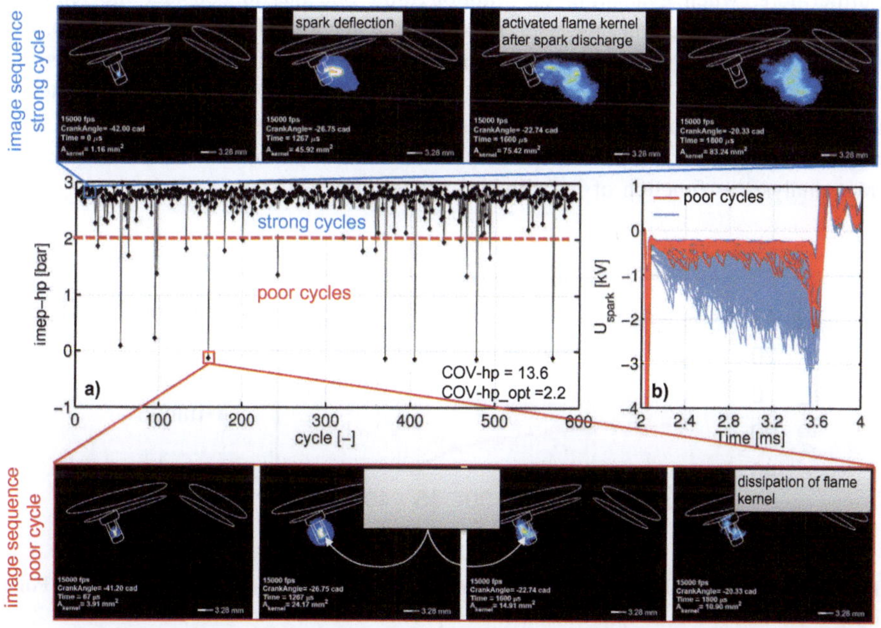

Fig. 6. (a) imep-hp as f(cycle #) at IgnAng = 42° before TDC (Top Dead Center), dashed line separates poor cycles from strong cycles, (b) U_{spark} as f(t). Image sequence at the bottom: formation of a misfire. Image sequence at the top: flame kernel formation in case of strong cycle

The meaning of flow velocity at spark plug location on spark deflection and flame kernel formation in condition of lean combustion had been discussed in recent paper by Schneider et al. [10]. Based on optical measurements on a single cylinder research engine it was shown, that misfires and delayed combustion cycles refer to low velocity of air-fuel-mixture in the vicinity of spark plug. Figure 6(a) shows the indicated mean effective pressure imep-hp (high pressure loop) at lean limit ($\lambda = 1.6$) for 600 cycles, spark timing being at 42° cad before TDC (Top Dead Center). A continuous spark current profile was used for the measurement. A limit (imep-hp = 2 bar) is defined to indicate the delayed combustion cycles and misfires (dashed line). The cycles below this limit are classified as "poor" cycles. The COV-hp (COV of high pressure loop) includes all 600 cycles, while COV-hp_opt includes only those cycles being above the defined limit ("strong" cycles), see figure legend. Figure 6(b) shows the corresponding spark voltage waveforms. Poor cycles (red lines) clearly indicate low spark voltages and so can be referred to cycles with low flow velocities. The image sequence at the bottom shows how a misfire develops. The reduced flow leads to static arcs which in turn lead to high quenching losses. It has to be pointed out that in such conditions, spark ignition systems cannot release their potential, even if ignition energy is increased, either by high spark current levels or long spark durations. According to these findings requirements for the ignition system and also for the Operating Point: imep = 2 bar, n = 2000 rpm, λ = 1.6 [10]. Combustion system can be derived [10]. Furthermore, when evaluating the dilution limit one has to consider the temporal phases of the combustion process, consisting of flame kernel formation (mfb0-5), first half of combustion (mfb5-50, duration in [cad] from 5 % till 50 % of total mass fuel burned) and the last half of combustion (mfb50-95). The COV, often used as criterion for evaluating ignition systems or ignition parameters contains the variation of all three phases. Figure 7 shows the described phases of combustion obtained from pressure trace analysis as function of in cylinder residual gas fraction.

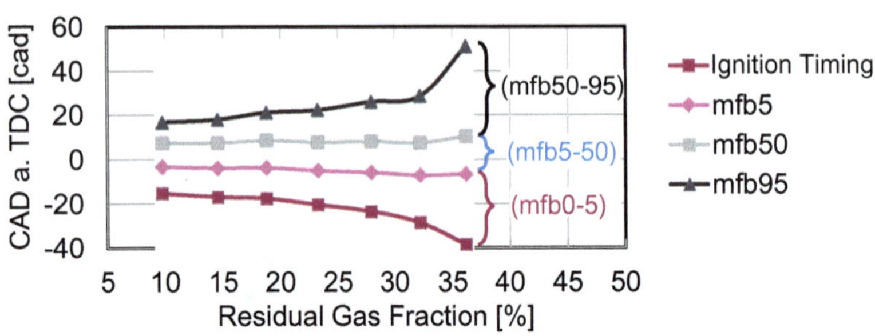

Fig. 7. Different combustion phases as function of residual gas fraction. Operating Point: imep = 3.5 bar, n = 2000 rpm, ext. EGR sweep, High Tumble

The optimum of isfc (indicated specific fuel consumption) is obtained when the combustion phasing (mfb50) is approximately at 6–8 cad a. TDC. The dilution of mixture leads to lower laminar burning velocities and therefore longer overall burn durations. Figure 7 shows, that most time of the combustion process is consumed for flame kernel formation (mfb0-5) and the second half of combustion (mfb50-95). The shortest time is consumed for mfb5-50, since this combustion phase takes place around TDC, where, especially for high tumble concepts, highest TKEs (Turbulent Kinetic Energies) are present. The biggest trade-off in combustion of diluted mixtures arises between the phase mfb50-95 and mfb0-5. The best ignition performance is achieved in case of late ignition timing (typically near TDC), where high gas temperature and high flow velocity are present. However, retarding the ignition timing towards TDC will shift the main combustion phase (mfb5-95) and especially the mfb50-95 as well, leading to delayed combustion cycles. Consequently, the COV increases. The ignition system is only capable to shorten the first phase of combustion (mfb0-5).

However, for spark ignition systems this means that high flow velocities at ignition timing must be present in order to achieve maximum energy transfer efficiency. Since ignition timing cannot be shifted towards TDC for the reasons stated above, the main alternative is to shorten the burn duration (mfb5-95), by e. g. higher tumble level or higher compression ratio [11, 23], which in consequence will allow later ignition timing. Therefore, when evaluating the COV, attention must be laid on the whole combustion process, especially when comparing different ignition systems. Strong fluctuations in the main combustion phase may either mask the potential of ignition systems or can lead to unfavorable ignition conditions in case of long burn durations. In the latter case substantial improvements of spark ignition systems may not be possible. Nakata et al. [11] show that increasing spark current can lead to significant improvements with regard to flame kernel growth if velocities at spark gap are increased by engine set-up. If the above mentioned requirements for combustion system are met, high performance spark ignition systems can provide stable flame kernel formation even in diluted conditions. To sum up, the following requirements for spark ignition systems can be defined:

- Spark current level above 50 mA over the entire discharge duration to stabilize spark deflection by avoiding frequently occurring restrikes in strong flow fields.
- Spark duration of 3 ms at low engine speed (1000–2000 rpm) to provide high spark energy level and to cover possible mixture inhomogeneity at spark plug location.
- Ignition voltages of 40 kV to provide secure ignition at high loads.
- The ignition system must sustain spark voltages (burn voltages) of at least 6 kV, at conditions where high spark elongation is present.
- Fast re-ignition capability in case of spark blowout, typically occurring in strong flow fields.

2.2 Derived Component Requirements

Taking a conventional ignition transformer (so called ignition coil) as a reference, the following requirements for a new ignition component are derived:

1. Target performance values acc. to findings in Sect. 2.1:
   (a) Spark duration: 3 ms
   (b) Minimum spark current: 50 mA
   (c) Secondary voltage (depending on load): 40 kV
2. Definition of spark energy and secondary voltage offer independent from each other: Demand for both parameters has different reasons. In case of conventional ignition coils, both parameters depend on each other since both are delivered from the same inductive energy store. That means for a given value of stored energy the available spark energy is influenced by the energy consumption needed to establish

the breakdown voltage at the spark plug electrodes which further depends on cylinder pressure, spark gap and spark plug fouling.

3. Spark current trace with optimized energy content: A triangle shaped spark current typical for conventional ignition coils represents only 50 % compared to a rectangular trace. Ways to increase total spark energy in conventional coils usually are limited with respect to increased spark plug electrode wear or increasing the low current area not contributing to an efficient combustion process.
4. Definition of spark duration and spark current level independent from each other: acc. to above requirements, a long spark duration with a spark current above a certain threshold enables ignition of diluted mixtures. In the same way high spark current levels/peak values can be limited only for challenging operation conditions of the engine and beyond that be avoided in order to minimize spark plug electrode wear.
5. Spark stabilization: enable spark to withstand turbulences or in case of blowout enable quick re-ignition.

   Besides the requirements for secondary performance driven by the combustion concept there are further aspects to be considered in view of mass production engines:
6. High energy efficiency: important in order to really contribute to reduce fuel consumption and also limiting heat dissipation for high energy ignition. The latter becomes more severe in combination with engine concepts connected with high ambient temperatures for the ignition device.
7. Fit to established hardware and software interfaces: of course this is not an absolute must, but seen as a success factor for acceptance and project implementation. Reasons: ignition components are mainly handled as "commodity"; high requirements for adaptation of a complete new system will lower the chances for market entry.
8. Small package size/high energy density: growing packing density or downsizing trends realized in modern engine concepts will most probably not allow that package size of the ignition device can proportionally grow to the increase of performance output.
9. Robustness and durability: mandatory requirements anyway but efforts to safeguard robustness and durability are expected to be tremendous in case of completely new technological approaches due to technological challenges or risks on component and system level.
10. Cost: Definitely one of the all-time key parameters in the highly competitive market of automotive ignition. In terms of high performance ignition, additional costs for this higher performance usually need to be balanced with according benefits for the combustion system and have to consider implementation efforts in the system environment (see also requirement no. 7).

2.3 Requirements Development vs. Ignition Solutions

Conventional ignition coils clearly dominate the automotive market segment of passenger cars. A rough differentiation may separate clusters acc. to spark energy

- 30…60 mJ used in PFI applications for several decades
- 65…90 mJ used in DI applications for some 20 years
- approx. 100 mJ appearing in latest DI applications with increased requirements.

Besides these standard solutions Bosch started early to work on solutions with higher performance output:

(a) Proposal for a high power conventional ignition coil with 140 mJ spark energy: Over many years (since 2007) only built as a laboratory sample – known as “PowerPlus” – and widely spread in the community as enabler for testing of new combustion concepts due to the outstanding high spark energy. Results were published acc. to a study [4] where PowerPlus was rated as a reference ignition system representing conventional technology. Meanwhile Bosch started a study to prepare industrialization of an ignition coil with 140 mJ nominal output at room temperature, now called XR (=eXtended Range) coil. Focus here is also on providing a reasonable high output of min. 120 mJ at high ambient temperatures. In contrast to CEI, the XR coil can be optionally available with or without integrated power stage/electronics. Motivation is a high number of applications having the power stage placed in the ECU, since high ambient temperature for the ignition device make integration of electronics difficult or even impossible.

(b) First industrialization of MSI (multi spark ignition) for lean stratified DI applications (realized in 2009): Based on a standard ignition coil regularly providing ≈70 mJ in single spark mode, a special closed loop control for multiple discharge and recharge enables a significant increase of spark duration and spark energy output while keeping the small package size unchanged. MSI functionality is achieved with an electronic control inside the ignition coil. Significant fuel consumption benefit could be realized for a serial engine supported by MSI technology [12].

Further progress in development of new high performance ignition concepts in general can be separated in two segments:

- Concepts mainly based on conventional technological approaches, extending the basic idea of MSI but avoiding the spark interruption during the recharge phase.
- Concepts with a significant different technological approach like high frequency, microwave, Laser or Corona ignition (see Fig. 2).

Recent publications [6] report progress in both segments with regard to high performance output and according combustion benefits. In view of the derived component requirements listed in Sect. 2.2 this article is mainly focused on the first segment of conventional concepts. Although own evaluations have been performed including samples from the second segment of concepts, the received results do not indicate that these approaches will be relevant for introduction in mass production for automotive passenger car applications within the next 5 to 10 years.

With respect to the first segment the proposed solutions are connected more to a potential industrialization. In order to allow a continuous, uninterrupted spark, long spark duration and as a result scalable, high spark energy output, main proposals include

- DCI (Dual Coil Ignition) [13],
- DCO (Dual Coil Offset) [14] and
- CMC (Coupled Multi-Charge Ignition) [15].

All three are based on repeated recharge and discharge of an inductive store. Therefore (similar to MSI) there is no limitation to spark duration and spark energy due to the working principle itself. Limitation due to self-heating vs. maximum ratings of used components is of course a relevant aspect.

Brief description of concepts:

- DCI [13]: Combining two ignition coils being charged and discharged alternating and by superposing of both high voltage outputs an uninterrupted spark is achieved.
- DCO [14]: Basically similar to DCI but additionally aiming on reduction of overlapping of secondary current traces coming from two ignition coils.
- CMC [15]: Basically similar to above described concept with two ignition coils but in addition a step-down converter and a more advanced control procedure is used to achieve further reduction of high secondary current peak duration and reduced voltage load for high voltage diodes.

Main pros and cons:

Pro (DCI, DCO, CMC):

- technical feasibility
- high and scalable secondary performance.

Con:

- high spark current peaks: DCO: >0.4 A measured with sample, CMC: 0.2…0.4 A,
- high primary current: DCO: 20 A measured with sample, CMC 35…70 A,
- necessity of high voltage diodes (40…50 kV) for DCI and DCO to decouple ignition coils secondary circuits
- big package size and weight due to double coil design (expected to be significantly higher compared to CEI (Sect. 3)).

Though a lot of good results have been collected in the ignition community, the remaining cons connected to the described solutions above motivate to look for further solutions. A new Bosch approach is described in the following chapter.

3 CEI Working Principle and Sample Status

3.1 CEI Working Principle

Based on the set of requirements above and evaluation of alternative approaches known from publications and own studies a new proposal for high performance ignition is presented in the following. The essential idea for the design is the separation of the two main functions of an ignition device, which is to provide …

- high breakdown voltage and
- spark energy

The name CEI = Controlled Electronic Ignition is chosen according to the working principle. The basic approach is more an evolutionary one, based on technical design components which are in general available and already proven in other applications. This includes:

(a) ignition coil optimized for high voltage generation,
(b) step-up converter to generate the spark energy,
(c) electronic unit to control operation of (a) and (b) triggered by ECU input.

All three functional elements are combined in one package and represent the CEI device (Fig. 4).

The selected functional design shows the benefit of high performance output, which in terms of spark energy and spark duration is not limited due to the working principal. Of course, in practical applications a limitation has to be considered based on heat dissipation, component maximum rating and ambient temperature. Main features of CEI based on the shown design and related to a standard conventional coil are (Fig. 8):

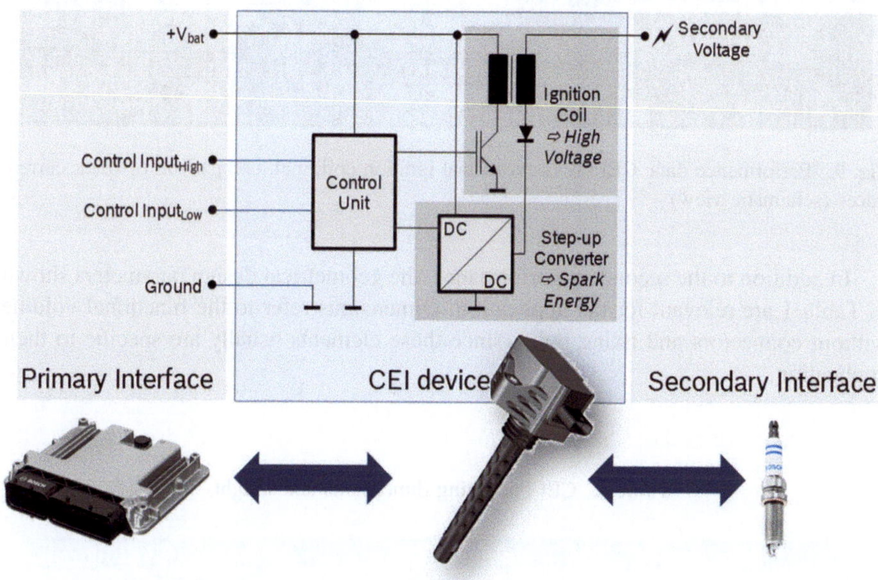

Fig. 8. Block diagram of CEI (Controlled Electronic Ignition) and interface structure suitable for operation with established ECUs and spark plugs

- Widely extended range for applicable spark energy and spark duration.
- Increased ignition voltage and electrical efficiency.

- Flexible spark performance data and ignition voltage applicable on demand and independent from each other.

A more detailed comparison between the performance of a standard conventional ignition coil and CEI is summarized in Fig. 9. Especially the change of the spark current shape by means of the functional principle of CEI leads to several benefits like high performance and the flexibility of selecting the single performance values independent from each other. The electrical performance data are defined acc. to internal evaluations of requirements connected with combustion concepts as discussed above (Sect. 2.1).

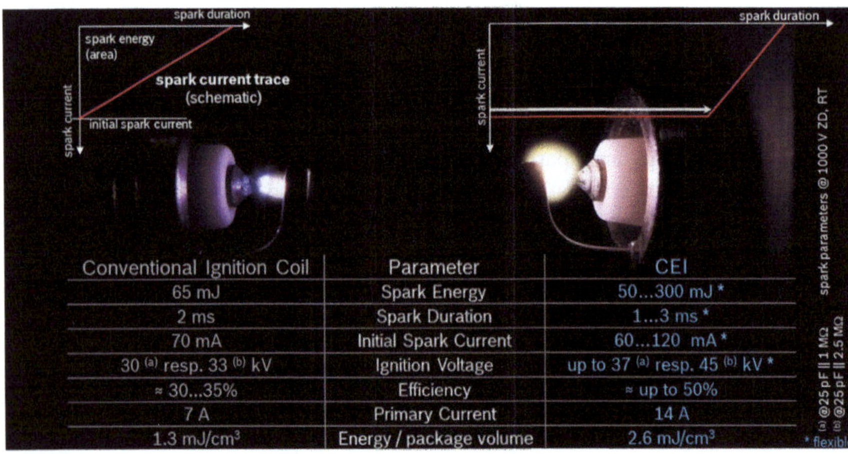

Conventional Ignition Coil	Parameter	CEI
65 mJ	Spark Energy	50...300 mJ *
2 ms	Spark Duration	1...3 ms *
70 mA	Initial Spark Current	60...120 mA *
30 (a) resp. 33 (b) kV	Ignition Voltage	up to 37 (a) resp. 45 (b) kV *
≈ 30...35%	Efficiency	≈ up to 50%
7 A	Primary Current	14 A
1.3 mJ/cm³	Energy / package volume	2.6 mJ/cm³

Fig. 9. Performance data: CEI vs. conventional ignition coil, incl. comparison of spark current traces (schematic view)

In addition to the secondary performance, the geometrical design parameters shown in Table 1 are relevant for the application. Dimensions refer to the functional volume without connectors and fixing point, since these elements usually are specific to their application.

Table 1. CEI Packaging dimensions and weight

#	Parameter	Value for CEI	Package (A3 design)
1	Dimensions (W x H x L)	45 x 45 x 60 mm³	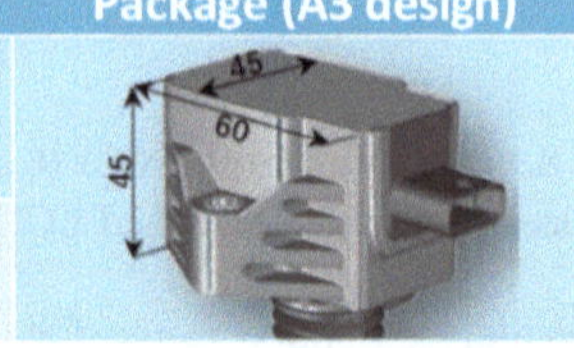
2	Weight	≈300 g	

3.2 CEI Sample Status and Performance Measurement Results

Bosch activities with CEI are running as a study based on an idea created in a brainstorming session in 2012. Besides theoretical considerations, simulations and measurements were started in the early beginning to verify the idea of CEI. On this bases the development and buildup of samples were realized. Figure 10 shows the sample history.

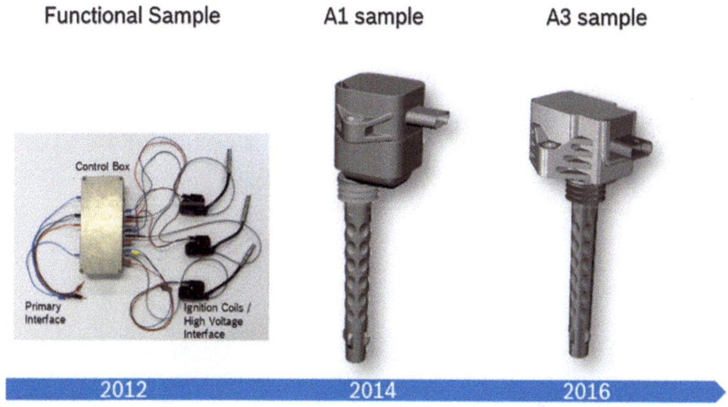

Fig. 10. CEI sample development and present status

Measurements of secondary performance data are presented in Figs. 11, 12 and 13. Figure 11 starts with measurement results for the spark current showing the flexibility to achieve very long spark duration (variation achieved by changing the control signal) and very high spark energy values. In addition the spark current level can also be

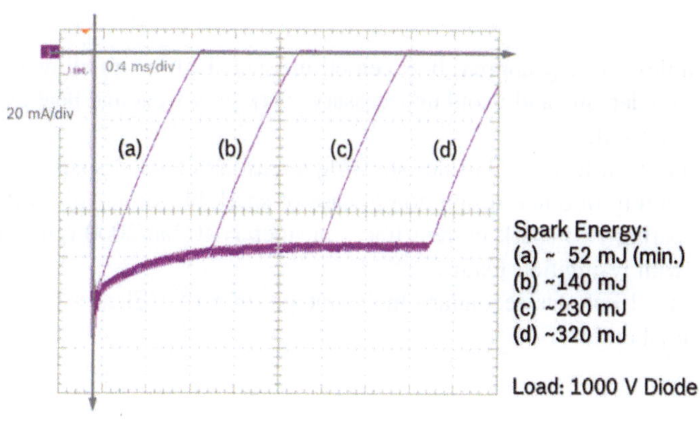

Fig. 11. Measured CEI spark current traces and resulting spark energy levels obtained by variation of input control signal: with CEI spark current is kept at high level for long spark duration

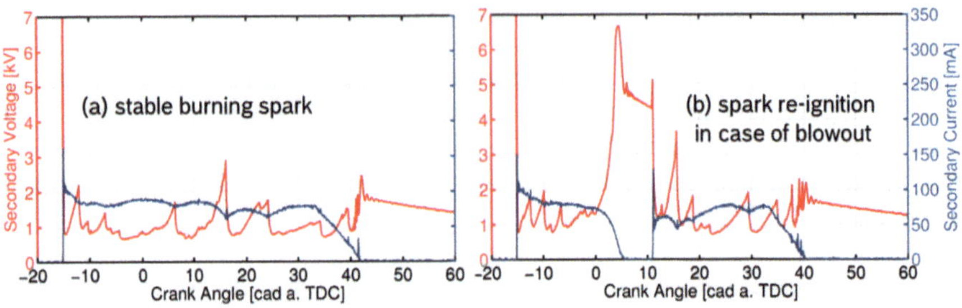

Fig. 12. Measured CEI spark current and voltage traces at imep = 10 bar, n = 2000 rpm showing behavior at (a) moderate and (b) strong turbulence with (a) stable burning spark and (b) spark re-ignition in case of blowout due to special feature of CEI

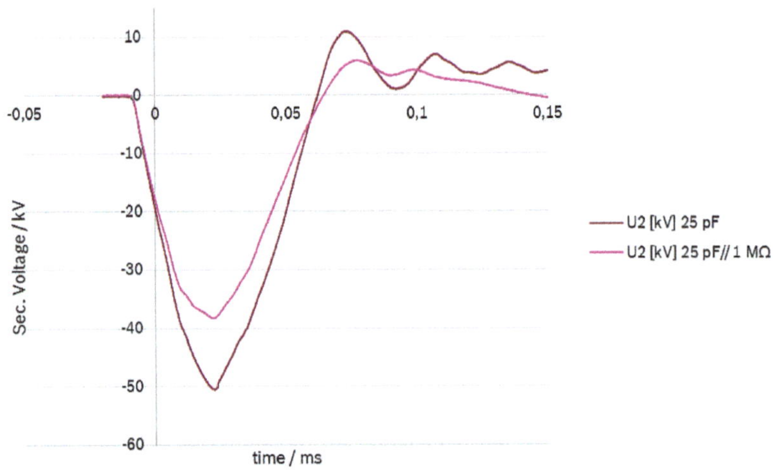

Fig. 13. Measured CEI secondary voltage traces at laboratory load conditions as indicated

adjusted in a flexible way approx. between 60 mA and 120 mA to allow spark energy performance on demand and avoid unnecessary spark plug wear and heat dissipation in the CEI device itself.

As an additional feature - besides showing a reasonable good spark stability - CEI provides the ability of quick re-ignition in case of spark blowout. This is illustrated in Fig. 12 comparing two spark current traces at imep = 10 bar/2000 rpm, one without (a) and one with re-ignition feature.

Finally Fig. 13 shows secondary high voltage offer of CEI acc. to different load conditions in laboratory.

4 Engine Results

Performance tests with CEI have been evaluated on many engines over the last years, to study the impact of continuously fed high ignition energy under different cylinder flow respectively charge motion conditions. A summary of these results will be presented in this section to show the bandwidth of the achieved increase of engine performance.

Tests have been carried out mainly for EGR and homogenous lean combustion process but also for stratified lean and special catalyst heating modes. Nevertheless the focus in this paper will be on EGR and homogenous lean combustion since EGR dilution is already widely used and homogenous lean combustion tends to be a promising development step for gasoline engines.

Table 2 lists the different engines used for evaluation of the CEI potential. All engines are modern turbocharged (T/C) engines with inlet-/outlet cam-phasing (D-VVT), direct injection system (DI) but differ in the charge motion. All engines are equipped with central mounted injector except engine D where a side mounted injector is applied. Mounting position of the injector has some impact on the spark plug position, typically engines equipped with central mounted injectors have a spark plug with a slight offset to the combustion chamber center. Even for tumble charge motion concepts this can affect the flow field in the vicinity of the spark plug. Furthermore, the investigated engines follow different charge motion layouts from medium to high level.

Table 2. Engines used for CEI performance investigation

Engine	Fuel system	Air system	Charge motion
A Full engine	GDI central	T/C, D-VVT + VVLin Int. EGR	High
B Single cyl.	GDI central	T/C, D-VVT Ext. EGR	Medium
C Single cyl.	GDI central	T/C, D-VVT Ext. EGR	Medium - High
D Full engine	GDI side	T/C, D-VVT Ext. EGR	Medium

Typically dilution by EGR respectively air is limited when the combustion stability becomes worse which in general is described by the CoV of the mean effective pressure (CoV_{pmi}). A typical limit of CoV = 5 % for single cylinder engines and CoV = 3 % at full engine application (in view of series application) is chosen. This covers the fact that at full engine CoV is evaluated as a mean value over all cylinders. With the lower CoV limit for a full engine it is ensured, that a single cylinder (e.g. due to cylinder-cylinder variation of charge and flow) not exceeds values of CoV > 5 %. It is widely known [19–22] that by an increase of ignition energy combustion stability limits can be shifted to higher dilution ratios. Current study therefore focuses on the potential of CEI extending the dilution limits compared to conventional TCI Systems.

4.1 Potential Study for EGR Combustion Concepts

First evaluations of the CEI potential have been carried out on single cylinder engine B with external EGR. Since engine B is equipped with a variable charge motion system (tumble flap) it is possible to study also the effect of charge motion on combustion and ignition. An operating point (2000 rpm/2 bar_i) critical for ignition has been chosen. CEI is compared with two transformer coils, providing 90 mJ resp. 140 mJ nominal ignition energy. The 140 mJ coil is the formerly known PowerPlus prototype from Bosch [17]. Furthermore the 140 mJ coil was operated at different energy levels between 96 mJ and 167 mJ.

Figure 14 shows results with active tumble flaps which would be applied for this operating point if a variable tumble flap is installed. The application of gas exchange analysis allowed for estimation of residual gas fraction at inlet valve closing. The x-Axis therefore indicates the total in-cylinder residual gas fraction. Following the rule of CoV limit = 5 % for single cylinder engines it can be observed that CEI enables highest internal EGR rates. Compared to the nominal operation of 140 mJ, the EGR-rate can be increased by roughly $\Delta EGR = 5$ %, whereas compared to the nowadays typically standard energy level of around 90 mJ the shift of combustion limit is even higher $\Delta EGR \sim 8$ %. Only the overloaded 140 mJ coil providing an ignition energy of 167 mJ allows an extension of the combustion stability limit close to the CEI performance. Nevertheless this operation would not be possible under vehicle respectively series conditions with this coil.

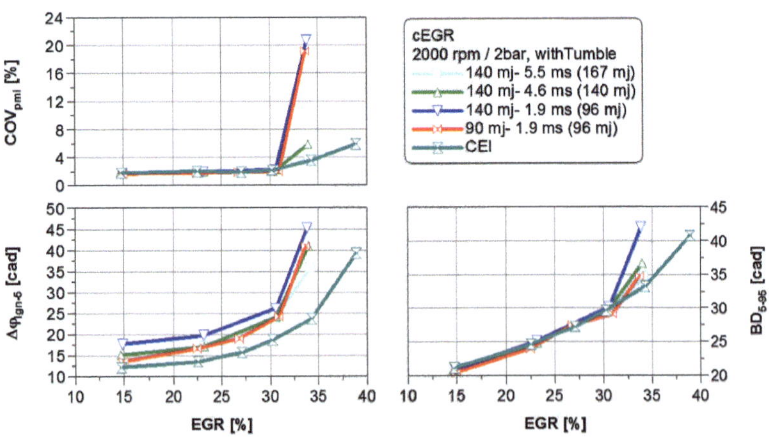

Fig. 14. External EGR variation and stability limits on engine B with activated tumble flap at 2000 rpm/2 bar_i. x-axis indicates total in cylinder residual gas fraction.

Additional to the CoV evaluation ignition delay values ($\Delta\phi_{ign-5}$) are plotted in the lower left plot. Clearly lowest values can be observed with the CEI system, a more detailed discussion will follow below. On the other hand it can be observed that the effect of the investigated ignition systems on the combustion duration ($BD_{5\text{-}95}$) is small until corresponding combustion stability limits are reached. Since the spatial influence

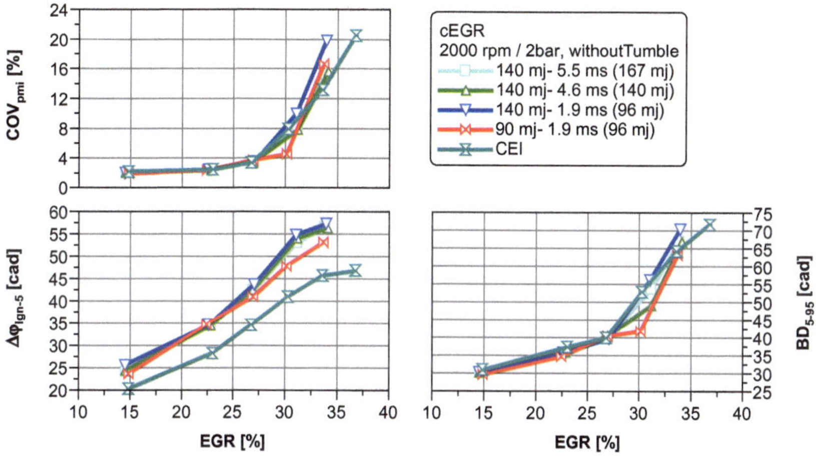

Fig. 15. External EGR variation and stability limits on engine B with not active tumble flap at 2000 rpm/2 bar_i, x-axes indicates total in cylinder residual gas fraction.

of spark plug ignition systems is limited to a small volume (see Sect. 2.1) this behavior is expected. More investigations have been carried out on engine B with deactivated tumble flap. Results for the same operation point are summarized in Fig. 15. Here combustion stability limit is reached at much lower internal EGR values EGR_{Limit} ~ 28 % compared to EGRLimit 32 % … 38 % for the case with active tumble flap. In addition the impact of the ignition system on stability limit is rather small. It is assumed that this behavior relates to the rather low charge motion values in case of deactivated tumble flaps. Spark elongation by the flow field in the near field of the spark plug is very small. At such conditions a large ratio of ignition energy provided to the spark plug is lost via the spark plug electrodes (see Chap. 2.1).

Evaluating ignition delay times ($\Delta\phi_{ign-5}$) also shows on average a slower flame kernel growth compared to the active tumble flap case. Nevertheless also here CEI shows significant faster flame kernel growth. This in general leads to stabilization of the combustion process but due to the very long burn durations in case of the deactivated tumble flaps the CoV value indicates, that the overall stability of the combustion cannot be significantly improved by higher ignition performance.

As mentioned at the presentation of results on engine B with active tumble flaps, a detailed view on the combustion stability will be discussed in terms of combustion statistic plots shown in Fig. 16. The data corresponds to the operation point with EGR-rates ~ 33 % in Fig. 14. Investigations have been carried out using standard ignition transformers (TCI) with nominal energy levels of 96 mJ respectively 140 mJ as well with CEI. The statistic plots show the cycle individual values of indicated mean effective pressure (high pressure loop, imep-hp) and the ignition delay defined as the interval between ignition timing and 5 % heat release (fhr0-5). Increase of ignition energy applied by standard transformers results only in a slight decrease of the ignition delay whereas CEI leads to significant reduction of ignition delay. The continuous feeding of high energy provided by the CEI system enables a faster establishment of

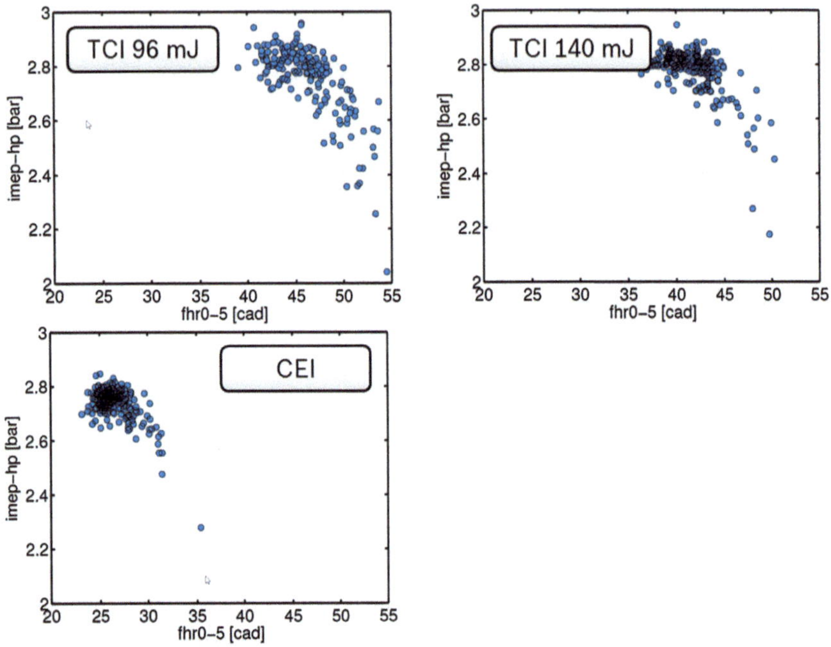

Fig. 16. Combustion statistics on engine B for different ignition systems for operation point 2000 rpm/2 bar_i/EGR ~ 33 %, activated tumble flaps

the flame kernel. Assuming a constant combustion phasing (50 % heat release mfb50) at mfb50 = 8 °CA, this results in later ignition timings for the CEI system. Later ignition timing then profit from improved ignition conditions, clearly indicating a considerably lower scatter band width CEI.

Full engine results are shown for engine A in Fig. 17 for the operation point 1000 rpm/2 bar_i. Fuel consumption is plotted in terms of inlet and exhaust cam timing position, whereby timings are given with respect to the gas exchange TDC (top dead center). Variation of cam timings results in different internal EGR rates. Highest internal EGR ratios are achieved with late outlet and early intake cam timings resp. high valve overlap. Combustion stability limits for different ignition systems are included (solid lines) as well as the series calibration point (dashed circle) as function of valve timings.

Stability limits based on CoV < 3 % are plotted for the series coil with 90 mJ, the 140 mJ PowerPlus coil and the CEI system. In the relevant area stability limits of series coil and PowerPlus coil behave quite similar. However CEI shifts the stability limit to significantly higher valve overlap which means higher internal EGR rates. Series calibration for valve train target positions is somewhat away from the stability limit which reflects the safety margin needed to cover tolerances in the system (engine resp. valve train tolerances, transient behavior …). If an identical safety margin is applied to CEI (red solid circle), the series calibration point (dashed circle) can be shifted towards higher EGR rates respectively larger valve overlaps. Fuel economy benefit accounted for this operating point was around Δb_e = 11 g/kWh which corresponds to roughly 3 %.

It should be mentioned that additional test have been carried out with a reduced electrode gap (Δl = 0.2 mm). In general reduction of electrode gap leads to even worse ignition conditions especially at low engine speed and low load, where charge motion is weak. With CEI it was possible to maintain the stability limit found for the series coil with the larger electrode gap. Smaller electrode gaps in general reduce the requirements of ignition voltage demand.

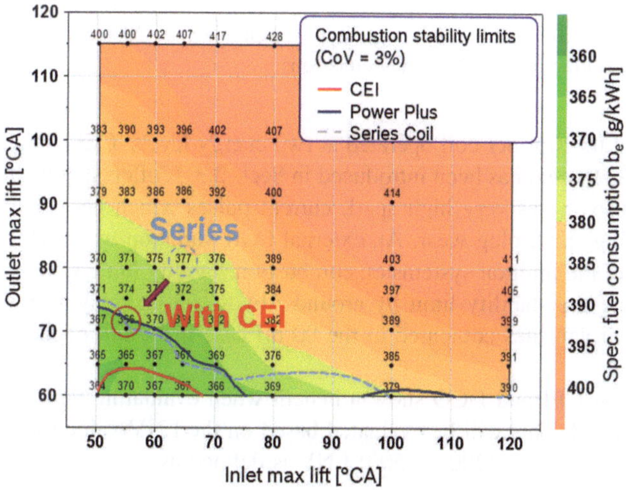

Fig. 17. Cam timing variation and stability limits achieved with different ignition systems on engine A at 1000 rpm/2 bar

Modern and especially future engines tend to have increased charge motion level. Hence experiments have been carried out on single cylinder engine C to investigate also the performance of CEI for higher charge motion levels on representative D-VVT engines. Here for the operating point 2000 rpm/3 bar_i a variation of the external EGR was investigated. In these tests CEI is compared against the MSI-System (Multi Spark Ignition) and the series coil with 90 mJ. Here CEI enables a significant shift of the combustion stability of ΔEGR = 12 % compared to both the MSI and the series system. In this case the stability limit reached with CEI is in the range where fuel consumption already starts to increase again. This is related to due to increasing burn durations. However, also here safety margins need to be considered (shaded areas in left figure). If the target EGR rate is at minimum fuel consumption, applying the safety margin leads to fuel consumption benefit with CEI of roughly 1 % (dashed vertical lines in right figure). Furthermore with the increased EGR limit also NOx raw emissions are reduced but this is not main topic for stoichiometric homogenous operation (Fig. 18).

More benchmark results based on measurements on full engine D are presented for the operation point 1500 rpm/3.2 bari in Fig. 19. Here a comparison between

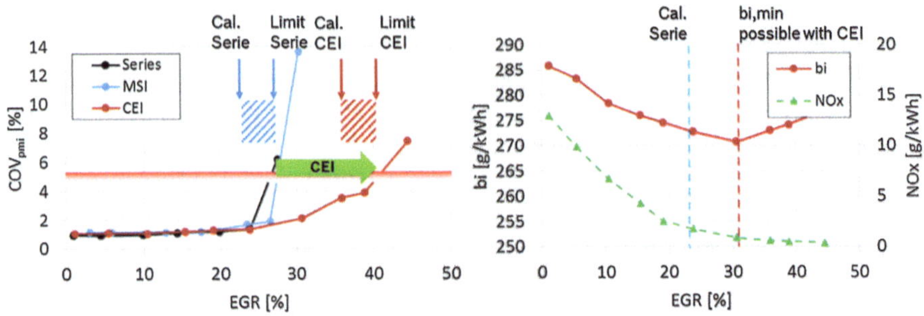

Fig. 18. External EGR variation on single cylinder engine C at 2000 rpm/3 bar$_i$; combustion stability, fuel consumption and NOx raw emissions

a conventional transformer coil operated at two energy levels, CEI and a DCO system is shown. DCO system has been introduced in Sect. 2.3. It offers a quite high energy of 450 mJ but also shows very high spark current pulses which need to be considered when evaluating spark plug wear. An external EGR variation was performed to study the influence of the ignition systems on combustion stability limit. CEI as well as DCO enables shifting the stability limit by around ΔEGR = 5 % applying a COV limit of 5 % but benefit vanishes compared to the 80 mJ case if the full engine COV limit of 3 % is chosen.

Nevertheless CEI and DCO show a benefit when evaluating which safety margin need to be applied. This will be evaluated based on the LNV value [20]. Evaluating a certain number of cycles (200 … 500) LNV is defined as:

$$LNV = \frac{\min(\text{imep})}{\text{mean}(\text{imep})} 100\,\% \tag{1}$$

This property is of great importance for transient operation, since it reflects better the driver feeling during acceleration. During transient operation, increased tolerance of the cylinder charge can be observed since the gas exchange flow varies from stationary conditions. Temperatures as well as pressures relevant for the gas exchange are away from their equilibrium state. This can lead to EGR-rate overshoots nevertheless a continuous evolution of the torque is expected. Typically the LNV value should drop below 80 %.

Above presented results for EGR combustion concepts can be summarized as follows:

- Depending on engine combustion concept and load point EGR can be increased up to ΔEGR = 12 % compared to a nowadays ignition system ranging between 70 … 90 mJ.
- Operating points and engine types have been observed, where the impact of a more powerful ignition system is rather small. However, considering system tolerances higher robustness can be achieved with CEI.

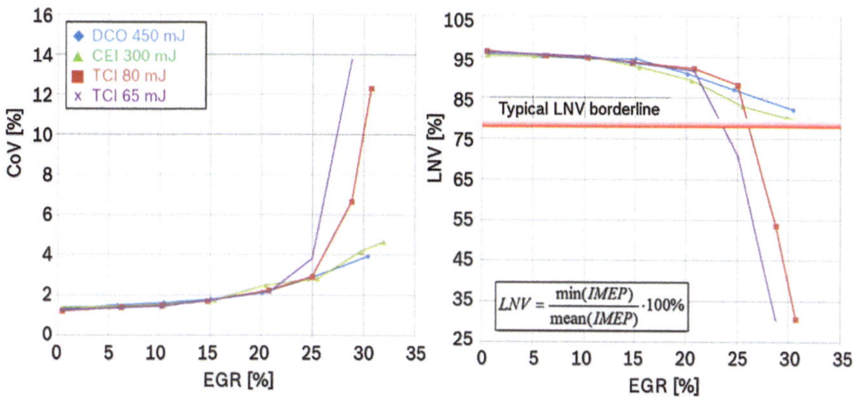

Fig. 19. External EGR variation on engine D at 1500 rpm/3.2 bar_i

- Higher charge motion levels generally lead to higher benefits, since spark is elongated stronger and the ratio of spark energy lost via the spark plug electrodes is reduced.
- Especially at low load conditions, improved combustion stability limits enable increased EGR rates allowing fuel economy benefits typically around 1 %, up to 3 % at specific operating points and engines.

4.2 Potential Study for Lean Combustion Concepts

In the future, homogenous lean combustion concepts will play a much more significant role than today. Therefore investigations have been carried out evaluating the impact of the CEI ignition system on the air dilution limits.

Since conventional 3-way catalyst does not allow sufficient high conversion rates of NOx at lean operation conditions, optimization of NOx raw emissions to lowest values is of great importance. The level of NOx raw emissions defines what type of NOx aftertreatment must be applied to fulfil emission legislation. Hence, additionally to the design of the combustion process itself, a powerful and suited ignition system enabling high dilution rates is a key technology to allow simple exhaust gas treatment e.g. passive systems for lean combustion engines.

Results for lambda variation on engine B are presented in Fig. 20. Here, for CEI different spark durations between 1.5 and 3.5 ms are chosen, a PowerPlus coil operated at 140 mJ is used as reference. It can be found that CEI enables highest lambda (air-fuel equivalence ratio) values. Compared to the PowerPlus coil, lambda can be increased by $\Delta\lambda = 0.1$. It should be mentioned that comparing against todays coils with around 80 mJ the benefit would be even higher.

Additional study of the CEI potential for homogenous lean combustion concept was performed on single cylinder engine C and is presented in Fig. 21. This engine has a higher charge motion level than engine B used for the investigation shown in Fig. 20. On engine C CEI was benchmarked against MSI and the series coil. With CEI it was

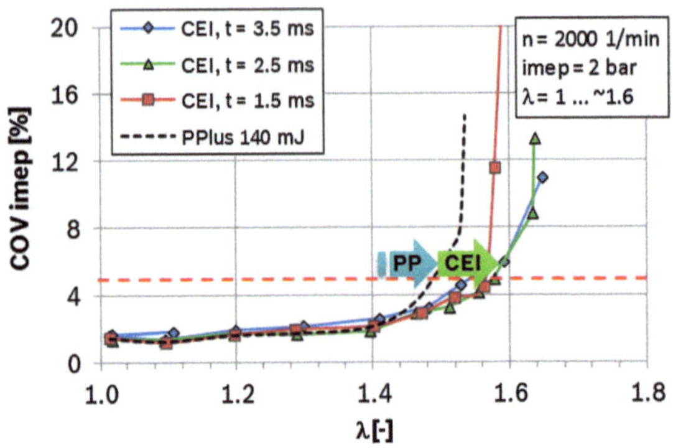

Fig. 20. Lambda variation for engine B at operating point 2000 rpm/2 bar_i

possible to significantly increase the lambda limit. Compared to the other ignition systems, a gain of $\Delta\lambda = 0.17$ was found and a remarkable lambda value around $\lambda = 1.9$ has been achieved for the operating point 2000 rpm/9 bar_i. Indeed, also the other ignition systems enable an operation close to optimal fuel consumption (b_i-best point) since the b_i-curve becomes quite flat in this range. Nevertheless, as described above, NOx raw emissions are of big interest for lean combustion concepts. In this case with CEI it was possible to reduce the NOx raw emissions by 50 % compared to series coil as wells as the MSI system. These results confirm the basic investigations presented in Sect. 2.1 that a continuous energy fed into the spark is beneficial to achieve a long and stable spark elongation which finally leads to improved ignition of lean mixtures.

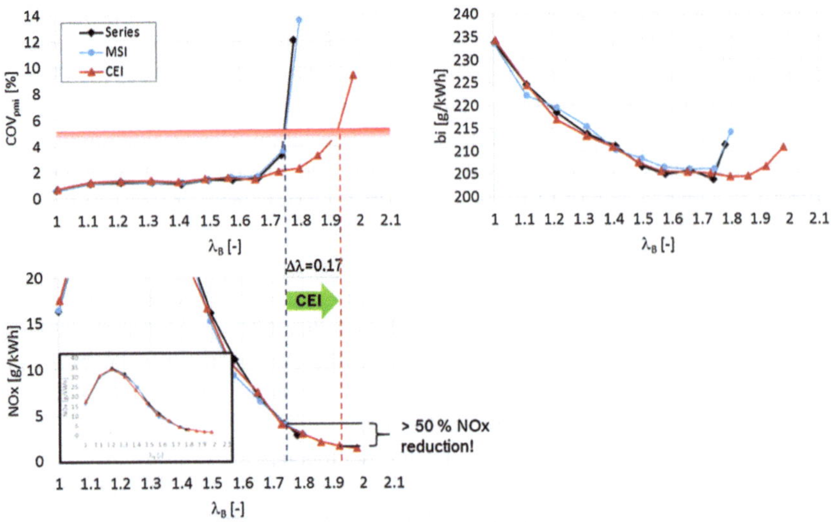

Fig. 21. Homogenous lean combustion lambda variation on engine C at 1250 rpm/9 bar_i

Considering that for series calibration the target values for lambda would be chosen with a certain safety margin to the stability limit, the impact on NOx would be even higher since the gradient dNOx/dλ increase for lower lambda values.

Nearly the same benefit in terms of enleanment extension as well as raw NOx decrease was found at engine C for other operation points e.g. 2000 rpm/3 bar_i.

Above presented results for lean homogenous combustion concepts can be summarized as follows:

- Depending on engine combustion concept and operating point the lean limit can be extended by up to $\Delta\lambda = 0.17$ compared to conventional TCI-Systems ranging between 70 … 90 mJ and MSI.
- A continuous energy fed into the spark is essential for reaching highest air dilution values.
- Even if benefits in terms of fuel economy at some operation points are small, the main focus is put on significant reduction of the NOx raw emissions. Those significantly determine, if a simpler/cheaper NOx aftertreatment can be chosen.

5 Summary

CEI (Controlled Electronic Ignition) as a new approach for high performance ignition is presented; being derived from generic analysis of requirements driven by from market and combustion trends but also considering aspects of industrialization for automotive high volume products. CEI offers high performance up to 300 mJ spark energy and also high flexibility to select a set of secondary performance parameters. Further benefits are high energy efficiency, a balanced package size and weight resulting in an outstanding value of energy density as ratio of spark energy per package volume. Present samples of CEI allow operation at engine test bench and are prepared for vehicle testing as well.

Basic requirements on ignition and combustion system were described, in order to extend the dilution limits. Measurements in a flow combustion vessel have shown that the decreasing spark current profile of TCI-Systems leads to frequently occurring restrikes if the spark current level drops below 50 mA. Restrikes shorten the plasma channel, which impede further deflection. Therefore, the current threshold must be considered when designing high performance spark ignition systems. However, besides electrical spark parameters the combustion system must provide adequate boundary conditions at ignition timing. Optical measurements on a single cylinder engine revealed, that the reduced flow velocity at ignition timing lead to inefficient energy transfer from spark to mixture increasing the probability of misfires and delayed combustion cycles. Furthermore, attention must not only be paid to ignition but also to the entire combustion process when evaluating the lean limits. The ignition system is specifically capable to influence the very early phase of combustion (typically up to 3…5 % heat release point). The acceleration of second half of combustion (mfb50-95) is one of the key factors to extend the lean limits.

Several engine tests have been carried out on different engines to benchmark the CEI system against standard transformer type coils as well as against the MSI system.

Here the main focus was on evaluating the benefit in terms of higher dilution rates for EGR and for homogenous lean combustion concepts.

Variations of internal and external EGR rates show in general an increased combustion robustness enabling higher EGR rates. Depending on engine and reference ignition, EGR rates could be extended with CEI by ΔEGR 5…12 %. This in general enables fuel economy benefits of around 1 % but at some specific points up to 3 %. Compared to MSI a benefit was observed which supports the conclusions made in Sect. 2.1, that a continuous feeding of high spark energy is necessary to create a stable spark providing further elongation in the flow fields. Nevertheless, it must be mentioned that besides the ignition system the layout of the combustion concepts (charge motion) influences the inflammation significantly and is limited for extended EGR rates.

Evaluations carried out to analyze the potential of CEI for homogenous lean combustions concepts also showed promising results. Depending on engine and reference ignition system it was possible to increase the air dilution in the range of $\Delta\lambda = 0.10 \ldots 0.17$. While fuel consumption is already close to optimal values for the investigated operation points with the reference ignition systems, NOx raw emissions are further reduced by 50 % using CEI. This is of big interest since NOx raw emissions determine which type of NOx aftertreatment becomes necessary. Hence, a powerful ignition system as CEI is a key technology to enable homogenous lean combustion concepts with cost optimized NOx aftertreatment.

References

1. Wenig, M.: Simulation der ottomotorischen Zyklenschwankungen. Dissertation, Universität Stuttgart (2013)
2. Ball, J.K.: Cycle-by-cycle variation in spark ignition internal combustion engines. Ph.D. theses, University of Oxford (1999)
3. Kettner, M.: Technologiebewertung Zündsysteme für magere Gemische. Forschungsvereinigung Verbrennungskraftmaschinen (FVV), 834 (2004)
4. Graf, J., Lauer, T.: Innovative Zündung. Forschungsvereinigung Verbrennungskraftmaschinen (FVV), 1008 (2011)
5. Günter, M., Tröger, R., Ignition systems for gasoline engines. In: 1st International Conference on Ignition Systems for Gasoline Engines, Berlin (2012)
6. Günter, M., Tröger, R.: Ignition systems for gasoline engines. In: 2nd International Conference on Ignition Systems for Gasoline Engines, Berlin (2014)
7. Schneider, A., Hettinger A., Kufferath, A., Rottengruber, H.: Requirements on inductive ignitions systems under engine and steady state conditions. In: 2nd International Conference on Ignition Systems for Gasoline Engines, Berlin (2014)
8. Kiefer, A., Lippisch, M., Storch, A., Kufferath, A.: Entwicklung eines strahlgeführten Magerbrennverfahrens mit Mehrloch-Magnetventil-Injektoren für turboaufgeladene Motoren. 9. Tagung Diesel- und Benzindirekteinspritzung, Berlin (2014)
9. Schneider, A., Hettinger, A., Schünemann, E., Rottengruber, H.: Optische Untersuchungen zur Flammenkernbildung im Ottomotor bei Ladungsverdünnung. 12 Internationales Symposium für Verbrennungs-diagnostik, Baden Baden (2016)

10. Schneider, A., Hettinger, A., Leick, P., Rottengruber, H.: experimental studies on spark stability in an optical combustion vessel under flowing conditions. 3 Internationaler Motorenkongress, Baden Baden (2016)
11. Nakata, K., Nogawa, S., Takahashi, D., Yoshihara, Y., et al.: Engine technologies for achieving 45 % thermal efficiency of S.I. engine. SAE Technical Paper 2015-01-1896 (2016)
12. Lückert, P., Doll, G., Merdes, N., Waltner, A., Eder, T.: Die Schwerpunkte der Otto-Motoren-Entwicklung von Mercedes – Benz, 19. Aachener Kolloquium Fahrzeug- und Motorentechnik, Aachen (2010)
13. Sackmann, M., Adolf, M.; Becker, M.: High energy dual coil ignition system for gasoline engines with EGR or lean combustion. In: SIA International Conference and Exhibition: The Spark Ignition Engine of the Future, Strasbourg (2013)
14. Alger, T., Gingrich, J., Mangold, B., Roberts, C.: A continuous discharge ignition system for EGR limit extension in SI engines. SAE Int. J. Engines **4**(1), 677–692 (2011)
15. Weyand, P., Lorenz, F., Schilling, S., Geringer, B., Hofmann, P., Graf, J., Hofherr, T.: Adaptive continuous spark ignition as enabler for high dilution EGR operation. In: 2nd International Conference on Ignition Systems for Gasoline Engines, Berlin, pp. 199–228 (2014)
16. Anderson, R.W.: The effect of ignition system power on fast burn engine combustion. SAE Paper, 870549, Society of Automotive Engineers, Warrendale, PA (1987)
17. Pischinger, S., Geiger, J., Neff, W., Böwing, R., Thiemann, J.: Einfluss von Zündung und Zylinderinnenströmung auf die ottomotorische Verbrennung bei hoher Ladungsverdünnung. MTZ Motortechnische Zeitschrift **63**(5), 388–399 (2003)
18. Kufferath, A., Stroch, A., Hese, M., Tschöke, H.: Potenziale des Zündsystems beim strahlgeführten Brennverfahren mit Piezo-Einspritztechnik. Der Arbeitsprozess des Verbrennungsmotors, Graz (2009)
19. Chiriac, R., Radu, R., Apostolescu, N.: A study of ignition discharge parameters effects on homogeneous mixtures combustion in engine. SAE Paper, 2001-01-1953 (2001)
20. Chen, W., Madison, D., Dice, P., Naber, J., et al.: impact of ignition energy phasing and spark gap on combustion in a homogenous direct injection gasoline SI engine near the EGR limit. SAE Paper, 2013-01-1630 (2013)
21. Gukelberger, R., Alger, T., Gingrich, J., Mangold, B.: Impact of operating parameters on ignition system energy consumption. SAE Paper, 2014-01-1233 (2014)
22. Alger, T., Gingrich, J., Roberts, C., Mangold, B., et al.: A high-energy continuous discharge ignition system for dilute engine applications. SAE Technical Paper 2013-01-1628 (2013)
23. Ayala, F.A., Gerty, M.D., Heywood, J.B.: Effects of combustion phasing, relative air-fuel ratio, compression ratio, and load on SI engine efficiency. SAE Technical Paper 2006-01-0229 (2006)
24. Nakata, K., Shimizu, R.: Toyota's new combustion technology for high engine thermal efficiency and high engine output performance. 37. Internationales Wiener Motorensymposium

Demonstration of Improved Dilution Tolerance Using a Production-Intent Compact Nanosecond Pulse Ignition System

Daniel Singleton[1(✉)], Jason M. Sanders[1], Mark A. Thomas[1], Magnus Sjöberg Jr.[2], James Sevik[3], Michael Pamminger[3], and Thomas Wallner[3]

[1] Transient Plasma Systems, Inc., Torrance, CA, USA
dan@transientplasmasystems.com
[2] Sandia National Laboratories, Livermore, CA, USA
[3] Argonne National Laboratory, Lemont, IL, USA

Abstract. Transient plasma ignition using nanosecond pulses has demonstrated the potential to enable improved fuel economy and reduced emissions by enabling lean and EGR limit extension in dilute burn engines. Existing spark ignition technology is not adequate because the energy transfer mechanisms between the spark and the fuel-air mixture are not efficient enough to guarantee stable ignition for dilute mixtures at high-load conditions. Additionally, long duration sparks and other advanced ignition solutions that require increased energy delivered accelerate spark plug electrode wear. To date, non-thermal plasma ignition with nanosecond pulses have demonstrated a lean ignition limit beyond an air/fuel ratio of 24 [1], demonstrated high-pressure ignition at densities equivalent to over 100 bar at the time of ignition [2], and demonstrated stable (COV <3 %) ignition at EGR dilution levels >20 % [3]. While low-energy nanosecond pulses have demonstrated strong performance compared to existing solutions, they currently only exist on the market in laboratory systems, rather than a production ready system in a single rugged, weather-proof, under-the-hood enclosure. Transient Plasma Systems (TPS) has recently demonstrated the potential for a retroffitable solution similar to coil-on-plug architecture that allows a direct replacement of existing ignition technology without any engine modification. The system was run on a gasoline direct injection engine at Argonne National Laboratory and demonstrated the same trends as previously observed with research grade systems, including lean and EGR limit extension and more stable ignition across a range of loads. The system was capable of delivering 30 kV pulses in bursts of up to 20 pulses at 30 kHz, and demonstrated stable combustion at an air/fuel ratio of 23.5, exhaust gas recirculation of 23 %, and ignition at 19.2 bar with COV <3 % using only 20 kV pulses.

Keywords: Non-equilibrium plasma · Nanosecond pulsed power · Transient plasma ignition

M. Günther and M. Sens (eds.), *Ignition Systems for Gasoline Engines*,
DOI 10.1007/978-3-319-45504-4_3

1 Introduction

With over 16 million new vehicles with internal combustion engines sold each year in the U.S., even small advances in fuel economy will have a significant impact on the health, economy, and environment. Dilute-burn gasoline engines are considered one of the most attractive solutions to developing more energy efficient and environmentally friendly highway vehicles; however, ignition instability associated with dilute mixtures prevents wide-spread application [4]. For this work, the concept for a novel low-energy and cost effective nanosecond pulsed ignition system based on technology that has been shown to reliably ignite dilute mixtures [1, 3] was developed and demonstrated in a plug & play form factor similar to coil-on-plug ignition modules.

Dilute burn gasoline engines, whether lean or EGR, allow for improved fuel efficiency and reduced emissions by limiting thermal losses, but they also reduce the engine's peak power. This lost power can be recovered through turbocharging, but the combination of dilution and intake pressure boost makes ignition increasingly difficult. Traditional ignition methods can ignite this mixture if the ignition energy is significantly increased, but this causes unacceptable spark plug wear. Data collected from numerous engine experiments conducted by Transient Plasma Systems, Inc. with collaborators including Argonne National Laboratory [3], Sandia Combustion Research Facility [1], and the University of California at Berkeley [5] have shown that low-energy, high peak power nanosecond pulses show great potential as an ignition source to enable stable, dilute-burn ignition without accelerated ignition system fatigue.

Low-energy nanosecond pulses have demonstrated performance benefits compared to existing solutions, but they currently only exist in laboratory systems, rather than a production ready system in a single rugged, weather-proof, under-the-hood enclosure. The work reported here as part of a U.S. Department of Energy funded project resulted in a prototype system demonstrating the feasibility of developing such a production ready system.

1.1 Technical Approach

The widely used capacitive and/or inductive coil spark ignition system has remained essentially unchanged for over 100 years. For dilute-burn engines, existing spark ignition technology is not adequate because the energy transfer mechanisms between the spark and the fuel-air mixture are not efficient enough to guarantee stable ignition for dilute mixtures at high-load conditions. Dilute-burn engines require a high-energy conventional spark to transfer enough heat into the combustible mixture for ignition, with the energy required generally increasing with higher pressure and dilution. The current state-of-the-art for ignition systems requires long duration sparks at high-voltage and current, which accelerates spark plug wear and minimizes spark plug wire life. TPS's approach is based on the generation of a non-thermal, low-energy transient plasma with nanosecond pulses that enables extended dilute-burn combustion (Fig. 1).

Low-energy nanosecond pulses can be used to produce a non-thermal plasma (also referred to here as transient plasma) which uses less power and is more effective compared with the thermal or quasi-thermal plasmas that are commonly generated by

Fig. 1. Photo of transient plasma (low-temperature plasma) formed by TPS nanosecond pulsed power technology.

alternative advanced ignition techniques [6]. Traditional thermal ignition systems heat a zone of combustible mixture to a temperature where radicals are thermally generated and produce rapid chemical reactions that generate heat faster than heat transfer can cool the zone, resulting in ignition. This technique has limitations in dilute mixtures because thermally generating radicals is inefficient and slow, and excess air in lean mixtures cools the ignition zone more quickly. In non-thermal plasma, radicals are produced directly, which is more than 50 times more efficient [7], and therefore less energy is needed to generate heat more quickly and the mixture can be leaner (Fig. 2).

To date, non-thermal plasma ignition with nanosecond pulses has:

1. Demonstrated a lean ignition limit lower than a fuel/air equivalence ratio (ϕ) of 0.50 [1];
2. Demonstrated high-pressure ignition in static cells > 60 bar in cold air [2];
3. Enabled operation under high EGR [3];
4. Demonstrated stable (COV <3 %) ignition at EGR dilution levels up to 35 % [8].

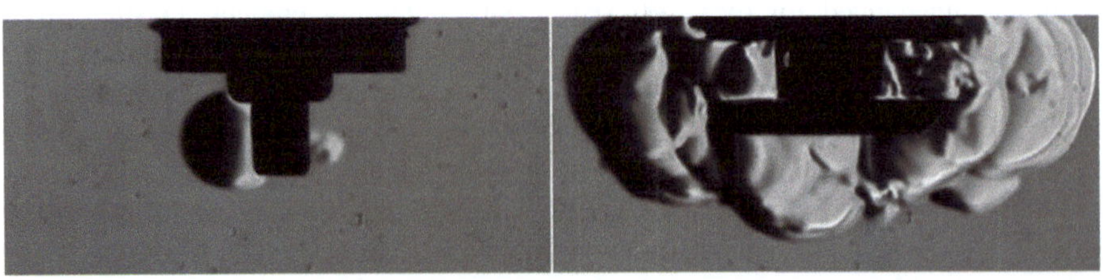

Fig. 2. Ignition of natural as part of EISG grant 13-03T NG demonstrating the feasibility of the TPS technology for enabling stable lean burn combustion in NG SI engines. The work was performed with Prof. Robert Dibble at UC Berkeley. The images show ignition of a lean mixture ($\lambda = 1.67$, $\varphi = 0.599$) of natural gas and air 5 ms after ignition. Left: high-energy spark ignition (300 mJ). Right: low-energy non-thermal plasma ignition (approximately 20 mJ/pulse). This clearly demonstrates a better investment of energy.

There are two distinct advantages to transient plasma ignition via low-energy nanosecond pulses:

- Fundamentally different ignition chemistry using low-energy nanosecond pulses that enables dilute ignition and combustion stability, potentially improving electrode life.
- Multi-point ignition due to the plasma's volumetric properties that improves dilute-burn efficiency and stability by increasing flame wrinkling and propagation speed (Fig. 3).

Non-thermal plasmas are characterized by a high electron temperature T_e and a low gas temperature T_g ($T_e >> T_g$). In the extreme case, the gas temperature can be close to room temperature and the electron temperature can be well above 20,000 K [9]. In non-thermal plasma, the ionization process is dominated by field-driven, energetic electrons impacting with non-excited atoms and molecules. The energy transfer efficiency is high and significant energy goes into creating highly energetic electrons instead of heating of the gas. The energetic electrons collide with the fuel and air molecules, producing highly reactive species such as atomic oxygen [10], reducing ignition delays and improving dilute burn capability [7].

Non-thermal plasmas can be produced using various methodologies, including glow and RF discharges at low pressures, barrier and hollow cathode discharges, and nanosecond pulses. TPS's nanosecond pulse approach produces streamers with energetic electrons filling 60–80 % of the volume with highly reactive species, accelerating combustion reactions and enabling multi-point ignition [11].

The goal of using nanosecond pulses is to apply an electric field as quickly as possible and avoid spark breakdown. Operating in this mode requires the voltage, gap, and electrode configuration to be considered with respect to the temperature and pressure between the electrodes. When primary streamers bridge the gap between the anode and the cathode, secondary streamers form and the plasma channel becomes highly conductive, leading to a spark [12]. When a spark forms, energy delivered increases, but combustion is not enhanced, since the combustion chemistry has already been initiated by the active species formed during the rise of the voltage pulse. Therefore, it is desirable to avoid spark breakdown to extend spark plug life by adjusting the peak amplitude of the voltage pulses. In this work, voltage and current

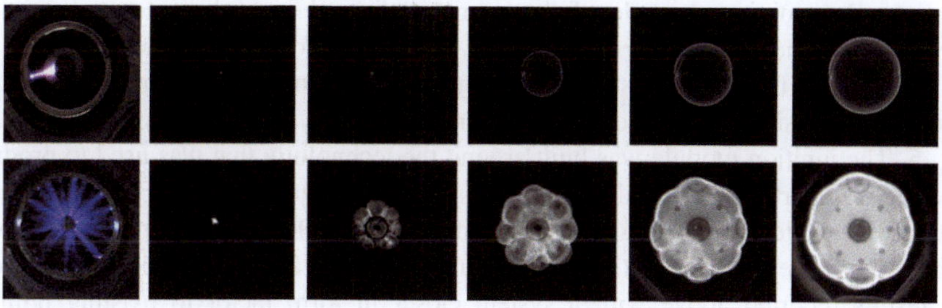

Fig. 3. 105 mJ conventional spark ignition (above) compared to 40 mJ transient plasma ignition (below) in a stoichiometric methane-air mixture. Images taken 1.5 ms apart.

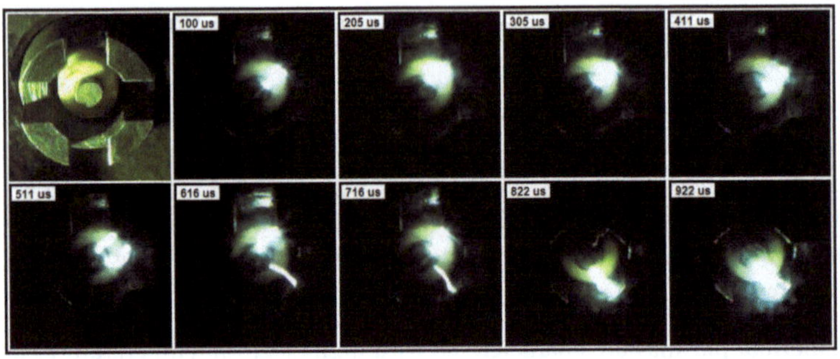

Fig. 4. High-speed imaging of multi-pulse transient plasma in an optically accessible engine, showing an example of breakdown-site procession. Time is given in μs from the first pulse, which did not transition into an arc. Exposure duration = 4.8 μs. All dark or nearly dark images in between pulses have been omitted.

was monitored during the engine operation and the voltage was adjusted to avoid full spark breakdown wherever possible, but could not be avoided entirely. Future work will focus on optical diagnostics along with voltage and current measurement to clarify the role of primary streamers, secondary streamers, and spark breakdown.

The ability to achieve multi-point ignition can be enhanced by operating the ignition system in burst mode, whereby a pulse train of nanosecond pulses is generated, typically at kHz rates. When a spark plug with multiple ground cathodes is used, this can generate a volumetric distribution of primary streamers, secondary streamers, and sparks, as demonstrated in Fig. 4. This example is for a 10-pulse burst with a repetition rate of 9.7 kHz. It was captured by a Phantom v611 color camera, operated at 180 kHz. The spark timing was 30°CA before TDC, at a gas pressure of 12 bar. In this engine cycle, the first nanosecond pulse did not transition into spark, so it could not be detected with this optical setup. At the voltage setting being used for this experiment, pulses 2 through 10 all transitioned into visible streamers and sparks. Pulses 2–5 all engaged with the ground cathode in the 2 o'clock direction, showing only one spark per pulse. However, each of pulses 6–9 showed two sparks or streamers, while a gradual transition to the ground cathode in the 4 o'clock direction occurred. For pulse 10, only one ground cathode was engaged with the two closely situated sparks. This procession of the sparks means that a large volume of fuel-air mixture can be ignited during a burst, shortening the time required for a fully turbulent flame to develop. Determining the role of the energy delivered in each stage, e.g. primary streamers, secondary streamers, and sparks, will be the focus of future work.

The goal of this work was to develop and test a prototype ignition module with a form factor consistent with modern day, commercially available ignition systems, e.g. Fig. 5. The ignition module was built around the technology that was used to produce the results reported in [3], advancing the state of that technology. The system was tested on a single-cylinder gasoline direct injection (GDI) engine at Argonne National Laboratory in Lemont, IL, with the intention of demonstrating the potential to achieve the results reported above with a system closer in form factor to existing ignition technology.

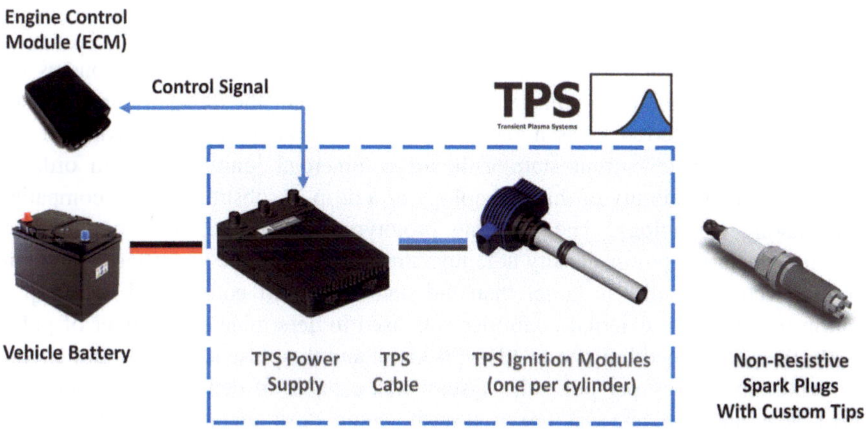

Fig. 5. Schematic of a concept for a low-energy non-thermal plasma ignition system. The system is targeted for OEM use in existing or future gasoline engines.

2 Ignition System

A prototype ignition module based on TPS's solid state, low-energy, high-peak power, nanosecond pulse technology was developed in a form factor similar to existing ignition technologies. The system produces pulses with 25 ns duration. It was bench tested up to 30 kV, 30 kHz burst pulse repetition rate (PRR), 20 pulses per burst with a duty cycle equivalent to 2400 RPM.

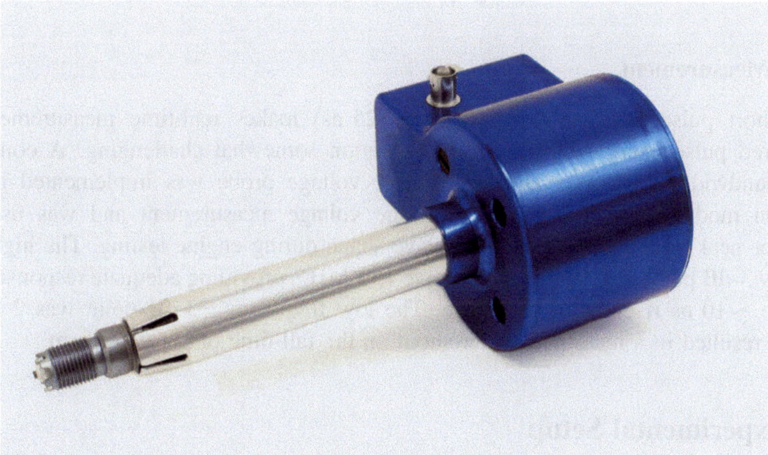

Fig. 6. TPS prototype ignition module TPS to demonstrate that the technology can be made as a retrofit solution in the form factor of existing ignition technology.

2.1 Ignition Module

The module, shown in Fig. 6, was designed to house the switching components that produce the fast rising, high-voltage pulse, which produces the transient plasma inside the engine. It is referred to as a "coil-on-plug" form factor module because it was designed to replicate current state-of-the-art commercial ignition coils in order to demonstrate the feasibility of this technology as a drop-in substitute that is compatible with existing technology. The resulting prototype has not yet been optimized to maximize operating power density at temperatures present in the vicinity of the engine block and consequently, is larger than the state of the art commercial coil-on-plug ignition modules. An external controller was used to determine the number of pulses (1–20), the pulse repetition rate (1 kHz–30 kHz), and the pulse amplitude and energy. The maximum energy per pulse the system was capable of delivering was less than 50 mJ. Maximum voltage was not used in the engine tests reported here. The energy per pulse delivered was less than 10 mJ at the 20 kV setting, resulting in approximately 200 mJ delivered in a 20 pulse burst. The system parameters are shown in Table 1.

Table 1. Ignition module specifications

System parameter	Specification
Max peak amplitude	30 kV
Pulse duration	25 ns
Pulse rise time	10 ns
Max pulse repetition rate	30 kHz
Max pulses per burst	20
Max burst repetition rate	20 Hz
Max average power rating	50 W

2.2 Measurement

The short pulse duration (approximately 25 ns) makes real-time measurement of delivered pulse energy during engine operation somewhat challenging. A compact, high-bandwidth, high-impedance capacitive voltage probe was implemented in the ignition module, which provided real-time voltage measurement and was used to monitor peak voltage amplitude and wave shape during engine testing. The high frequency 3-dB point of the probe was above 150 MHz, providing adequate response time for the $\sim$10 ns risetime of the pulse. The low frequency −3-dB point was 2 MHz, which resulted in a measured under-shoot on the fall-time of the waveform.

3 Experimental Setup

Experiments were carried out on a single-cylinder research engine, set up for steady-state operation. The engine is representative of GDI engines currently available on the market, with specifications listed in Table 2. More details about the engine can be found in [3]. The spark plug used is shown in Fig. 7.

Fig. 7. Modified spark plug. The base plug is a Brisk spark plug, the four-pronged tip was laser welded onto the shell to create four 1.5 mm gaps.

Table 2. Single-cylinder research engine specifications

Displacement	0.6 L
Bore	89.04 mm
Stroke	100.6 mm
Compression ratio	12.1:1
Intake Valve Maximum Opening Position (MOP)	100°CA ATDC
Exhaust valve Open/Close [°aTDC fired]	150/−350
Intake valve Open/Close [°aTDC fired]	350/−140
Exhaust valve MOP	255°CA ATDC
GDI injector	6 hole, solenoid
Injection pressure	150 bar
Spark plug	Custom
Fuel	EPA Tier II EEE Gasoline

4 Results

Figure 8 shows a sweep of EGR dilution at 1500 RPM 5.6 bar IMEP. EGR extension was demonstrated compared to conventional ignition, which in this case was a production Ford EcoBoost coil that nominally delivers 75 mJ to a 0.8 mm gap NGK-R spark plug. The TPS system was set to deliver 20 kV, 20 pulses, delivered at 20 kHz. The EGR limit was extended from around 20 % for spark to 23 % at the 3 % COV_{IMEP} stability limit for the TPS system.

The indicated thermal efficiency was measured to be very similar to conventional spark ignition, as seen in Fig. 9. This is similar to previous results obtained in [3]. There, with respect to the conventional ignition system, the TPS system the low-energy transient plasma system allowed for an increase in EGR dilution, while maintaining ITE levels and marginally increasing losses associated with combustion.

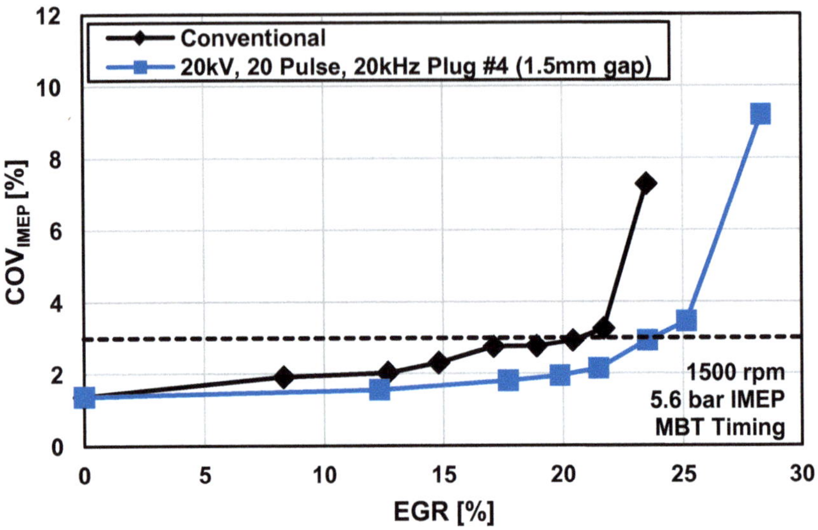

Fig. 8. EGR sweep comparing conventional ignition with compact TPS prototype ignition system at 1500 RPM, 5.6 bar IMEP.

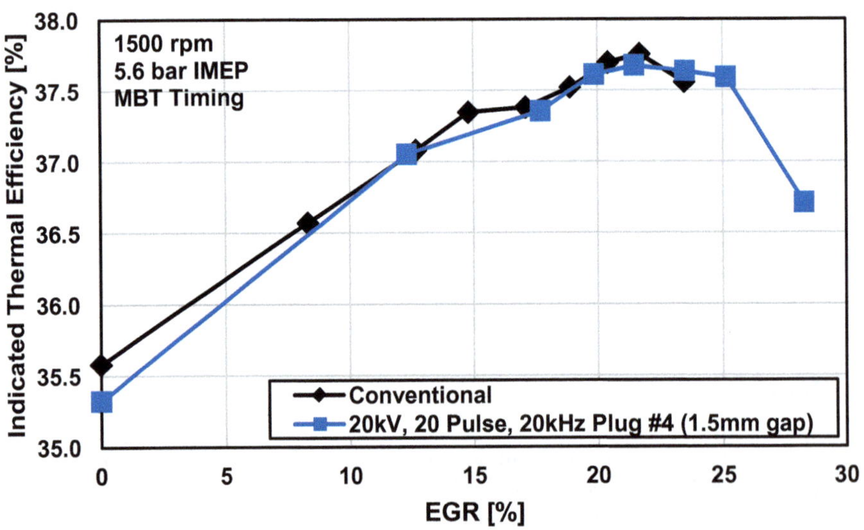

Fig. 9. EGR sweep comparing indicated thermal efficiency with conventional ignition and compact TPS prototype ignition system.

The flame development angle, or ignition delay, was shown to be faster with the TPS system compared to conventional ignition, as shown in Fig. 10. This is expected due to the non-thermal plasma ignition chemistry, which decreases the time duration

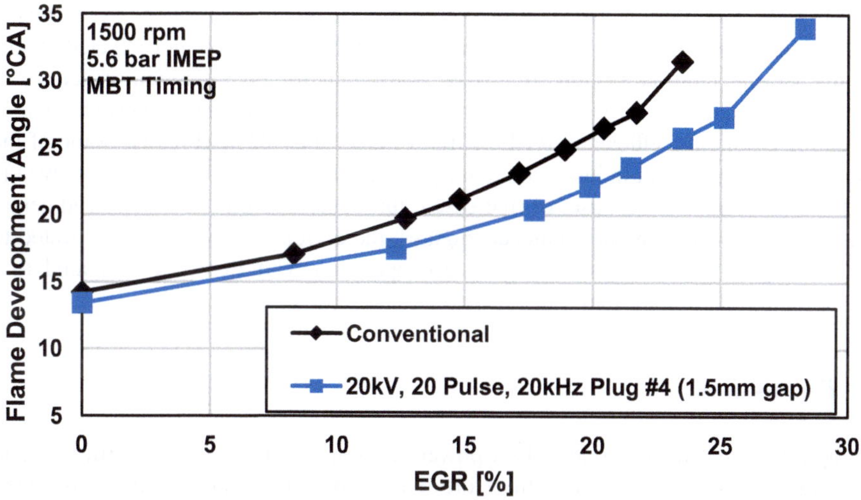

Fig. 10. EGR sweep comparing flame development angle with conventional ignition and compact TPS prototype ignition system.

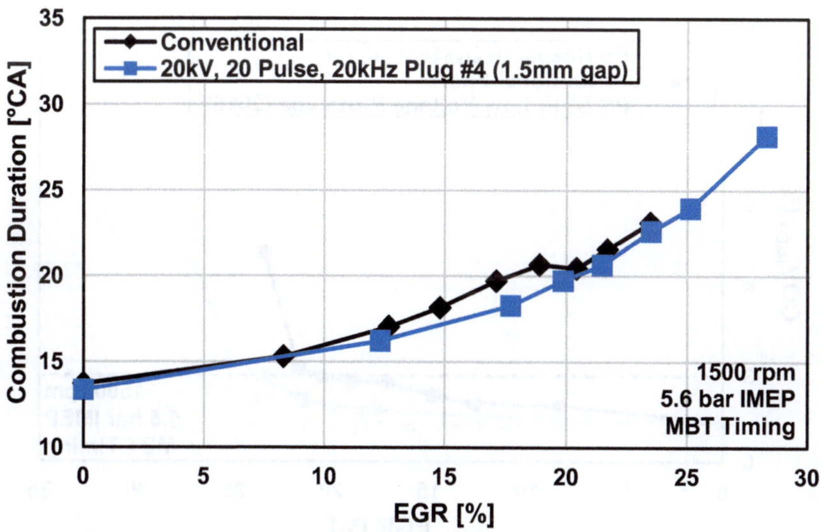

Fig. 11. EGR sweep comparing 10–90 % combustion duration with conventional ignition and compact TPS prototype ignition system.

from early flame kernel to turbulent combustion, as shown by the decrease in flame development angle.

Finally, the combustion duration was shown to be only slightly faster with the TPS system for a given EGR rate, as shown in Fig. 11. This is expected since the heat-release rate during the main combustion phase is controlled by turbulent deflagration throughout the charge, with general turbulence levels not affected significantly by the ignition system. It can be noted that for both ignition systems, the combustion duration increases rapidly with increasing EGR rates, contributing to a reduction of thermal efficiency for EGR rates higher than 22 % (Fig. 8). Some of the current design efforts in industry are aiming at maintaining short burn duration even at high EGR rates by the use of increased turbulence levels near top-dead-center (TDC) [13]. For such high-turbulence engines, the current TPS system is expected to provide higher thermal efficiency gains for operation with EGR, as compared to a conventional ignition system.

5 Discussion

An approach to a nanosecond pulsed power ignition system similar in form factor to existing ignition technology and therefore retrofittable to existing engines was developed and demonstrated on a single-cylinder engine. One way to benchmark this system is to compare the results to previously published engine test results in the same engine with a TPS lab bench research system [3], shown in Fig. 12.

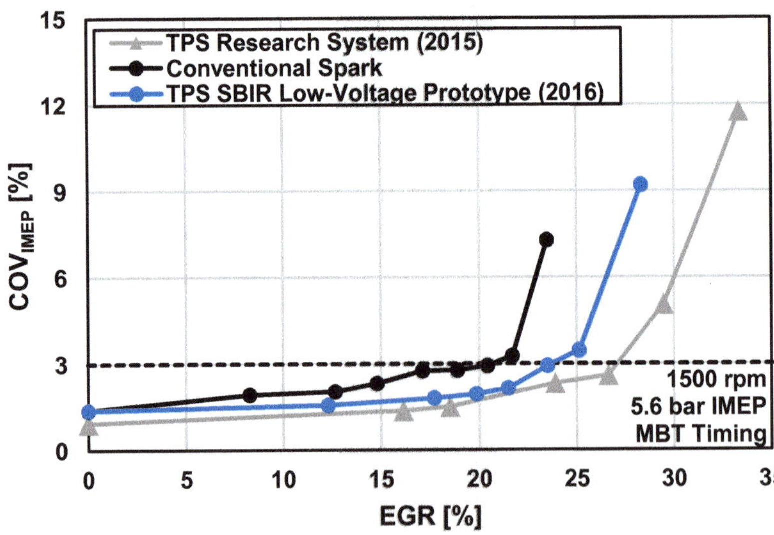

Fig. 12. Comparison of EGR extension using a 20 kV ignition module compared with a with 40 kV research system from [3].

The performance of the new 20 kV prototype system is in between conventional spark, and the 40 kV TPS research system. There were several differences between the two systems, including pulse width, pulse amplitude, electrode gap, and electrode geometry, so it is not possible to point to one feature to explain the difference in

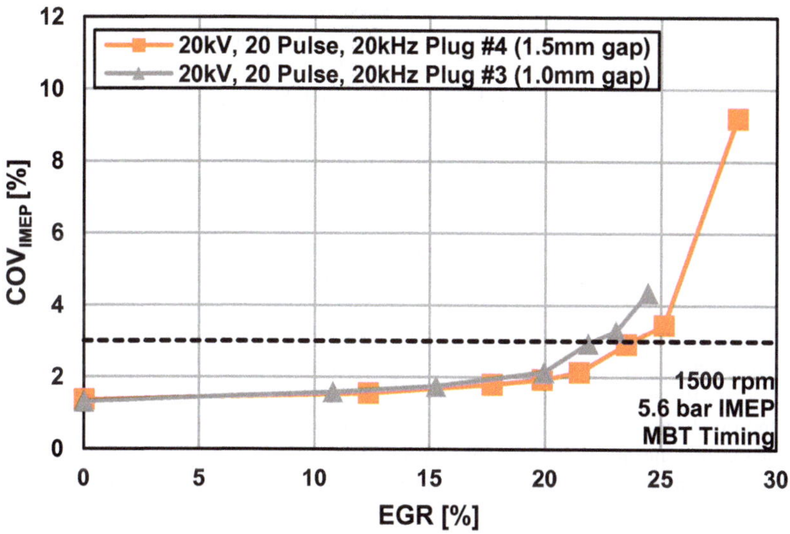

Fig. 13. EGR sweep with increasing spark plug gap.

performance. However, voltage likely played a significant role. It is well known that there is a relationship between the reduced electric field (E/n) and the density of radicals produced in the discharge available to enhance energy release. An increase in E/n beyond 300 Td does not significantly improve combustion, however, prior to that an increase of E/n significantly accelerates ignition development [14]. It is estimated that the peak voltage of the TPS SBIR system was delivering max E/n lower than 100 Td. Follow up work will include increased maximum voltage capability.

Another benefit of increased voltage is improved volumetric ignition. Increasing the gap size increased the system performance, until the voltage could no longer be increased to maintain the minimum reduced electric field necessary for ignition (Fig. 13), which in this case, was 1.5 mm. The improvement in using a larger gap is likely due to the separation of multiple ignition sites.

6 Conclusion

A prototype low-energy nanosecond pulse based ignition system was developed and tested on a single-cylinder gasoline direct injected engine, demonstrating EGR extension compared to conventional spark ignition technology. This milestone marks the first time this nanosecond pulsed power system architecture similar to existing "coil-on-plug" form factor was developed and tested, critically moving the technology toward a "plug & play" solution. The system demonstrated stable combustion at an air/fuel ratio of 23.5, exhaust gas recirculation of 23 %, and ignition at 19.2 bar with COV of IMEP <3 %. Future work will focus on the relationship between primary streamers, secondary streamers, and sparks with ignition, combustion, and spark plug lifetime.

Acknowledgements. This research is funded by SBIR Phase I DOE Grant DE-SC0013824 under Leo Breton. The topic was Advanced Ignition System for Internal Combustion Engines Enabling Lean-Burn and Dilute Gasoline Ignition and the title of the grant was "Application of low-energy nanosecond pulses for enabling lean-burn and dilute gasoline ignition". The authors greatly acknowledge the technical discussions with Isaac Ekoto and Ben Wolk from Sandia National Laboratories, and Riccardo Scarcelli at Argonne National Laboratory.

Part of the submitted manuscript has been created by UChicago Argonne, LLC, Operator of Argonne National Laboratory ("Argonne"). Argonne, a U.S. Department of Energy Office of Science laboratory, is operated under Contract No. DE-AC02-06CH11357. The U.S. Government retains for itself, and others acting on its behalf, a paid-up nonexclusive, irrevocable worldwide license in said article to reproduce, prepare derivative works, distribute copies to the public, and perform publicly and display publicly, by or on behalf of the Government.

This research is also funded by DOE's Vehicle Technologies Program, Office of Energy Efficiency and Renewable Energy. The authors would like to express their gratitude to Gurpreet Singh and Leo Breton, program managers at DOE, for their support. The research engine used to run these experiments at Argonne was provided by Ford Motor Company. Special thanks to Brad Boyer and Steven Wooldridge and their team from Ford Motor Company for their guidance and support.

References

1. Sjöberg, M., Zeng, W., Singleton, D., Sanders, J., Gundersen, M.: Combined effects of multi-pulse transient plasma ignition and intake heating on lean limits of well-mixed E85 DISI engine operation. SAE Int. J. Engines **7**(4), 1781–1801 (2014)
2. Lin, Y., Singleton, D., Sanders, J., Kuthi, A., Gundersen, M.: Experimental study of pulsed corona discharge in air at high pressures. In: Bulletin of the American Physical Society, Austin (2012)
3. Sevik, J., Wallner, T., Pamminger, M., Scarcelli, R., Singleton, D., Sanders, J.: Extending lean and EGR-dilute operating limits of a modern GDI engine using a low-energy transient plasma ignition system. ASME. J. Eng. Gas Turbines Power **138**(11), 112807-1–112807-8 (2016)
4. Danaiah, P., Kumar, R., Kumar, V.: Lean combustion technology for internal combustion engines: a review. Sci. Technol. **2**(1), 47–50 (2012)
5. Singleton, D.: Advanced low-energy ignition system for improved fuel efficiency and reduced emissions. EISG Grant # 57828A/13-03TNG Final report (2015)
6. Briggs, T., Alger, T., Mangold, B.: Advanced ignition systems evaluations for high-dilution SI engines. SAE Int. J. Engines **7**(4), 1802–1807 (2014)
7. Starikovskaia, S.: Plasma assisted ignition and combustion. J. Phys. D Appl. Phys. **39**(16), R265–R299 (2006)
8. Sjoberg, M., Zeng, W.: Combined effects of fuel and dilution type on efficiency gains of lean well-mixed DISI engine operation with enhanced ignition and intake HEating for enabling mixed-mode combustion. SAE Int. J. Engines **9**(2), 750–767 (2016)
9. Becker, K., Kogelschatz, U., Schoenbach, K., Barker, R.: Non-Equilibrium Air Plasmas at Atmospheric Pressure. IOP Publishing LTD, Philadelphia (2005)
10. Pendleton, S., Bowman, S., Carter, C., Gundersen, M., Lempert, W.: The production and evolution of ataomix oxygen in the afterglow of streamer discharge in atmposhpheric pressure fuel/air mixutre. J. Phys. D Appl. Phys. **46**, 305202 (2013)

11. Singleton, D., Pendleton, S., Gundersen, M.: The role of non-thermal transient plasma for enhanced flame ignition in C2H4-air. J. Phys. D Appl. Phys. **44**, 022001 (2011)
12. Winands, G.J.J., Liu, Z., Pemen, A.J.M., Van Heesch, E.J.M., Yan, K.: Analysis of streamer properties in air as function of pulse and reactor parameters by ICCD Photography. J. Phys. D Appl. Phys. **41**, 234001–234010 (2008)
13. Matsuo, S., Ikeda, E., Ito, Y., Nishiura, H.: The New Toyota Inline 4 Cylinder 1.8L ESTEC 2ZR-FXE Gasoline Engine for Hybrid Car. SAE Technical Paper. 2016-01-0684 (2016)
14. Aleksandrov, N., Anikin, N., Bazelyan, E., Zatsepin, D., Starikovskaia, S., Starikovskii, A.: Chemical reactions and ignition initiation in hydrocarbon air mixtures by high voltage nanosecond gas discharge. In: 32nd AIAA Plasma dynamics and Lasers Conference and 4th Weakly Ionized Gases Workshop, Anaheim, CA (2001)

Operating Conditions/Flammability

Study of Ignitability in Strong Flow Field

Kotaro Suzuki[1(✉)], Kazuhiro Uehara[2], Eiji Murase[1], and Shinichiro Nogawa[1]

[1] Engine Management System Development Division, Toyota Motor Corporation, Toyota, Japan
{kotaro_suzuki,eiji_murase_aa, shinichiro_nogawa}@mail.toyota.co.jp
[2] Unit Development Digital Innovation Division, Toyota Motor Corporation, Toyota, Japan
Kazuhiro_uehara@mail.toyota.co.jp

Abstract. To respond to the social requirement such as energy security and climate change, engine thermal efficiency has been improved. The use of charge dilution such as lean combustion and exhaust gas recirculation (EGR) has been considered as effective option to meet these requirements. In diluted combustion, in-cylinder flow-structures such as tumble are used to intensify in-cylinder turbulence and promote flame propagation. However, it is not clarified how it can contribute to the formation of flame kernel and promotion of the flame propagation under highly turbulent and diluted charge conditions. The purpose of this study is to achieve superior thermal efficiency with stable ignitability under high in-cylinder flow and diluted charge conditions. This paper describes the effect of flow velocity and turbulence intensity on the ignitability through analysis of the discharge channel behavior and initial stages of combustion phenomenon such as formation of flame kernel and flame propagation.

Keywords: Diluted combustion · High tumble · Discharge channel · Flame kernel · Discharge specification

1 Introduction

Improving vehicle fuel economy is strongly required for measures to cope with the issues such as energy security and climate change. Enhancing the engine thermal efficiency is one of the effective ways for these. Upper graph of Fig. 1 shows the historical trend of maximum engine thermal efficiency continuously increases in recent years by OEM's technology development [1–3]. Especially, diluted combustion technology such as cooled EGR and lean burn is well known as a method to enhance thermal efficiency, and intensifying turbulence intensity such as tumble expands dilution limit and enhances the effect of diluted combustion. Bottom graph of Fig. 1 shows the trend of tumble ratio. It is also continuously increasing same as maximum thermal efficiency.

This part describes the effect of intensifying turbulence flow on dilution limit and thermal efficiency. Figure 2 shows the comparison of dilution limit and fuel consumption improvement with low and high tumble cases. In the case of low tumble, improvement of

M. Günther and M. Sens (eds.), *Ignition Systems for Gasoline Engines*,
DOI 10.1007/978-3-319-45504-4_4

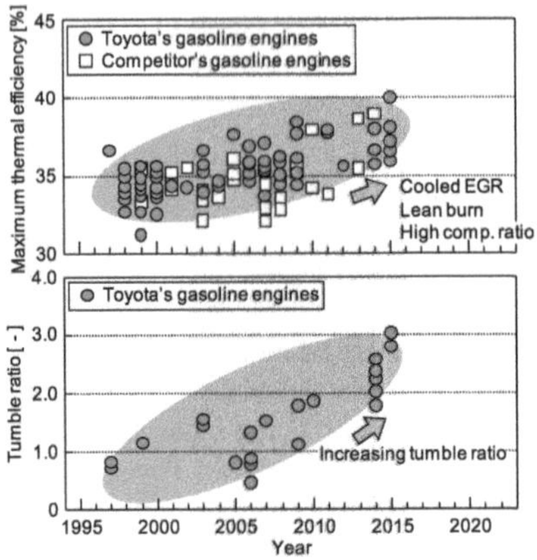

Fig. 1. Trend of thermal efficiency and tumble ratio

fuel consumption is limited at excessive dilution by lowering laminar burning velocity and finally it reaches dilution limit due to the failure of flame propagation. For preventing that situation, in-cylinder flow-structures such as tumble are used to intensify in-cylinder turbulence intensity and promote flame propagation. Figure 2 shows also that high tumble realizes stable combustion duration and improves fuel economy by the intensified in-cylinder turbulence intensity. Although for further more improvement of fuel economy, expanding dilution limit cause the issue of misfire. Therefore, clarifying the root cause of misfire is essential to realize superior thermal efficiency.

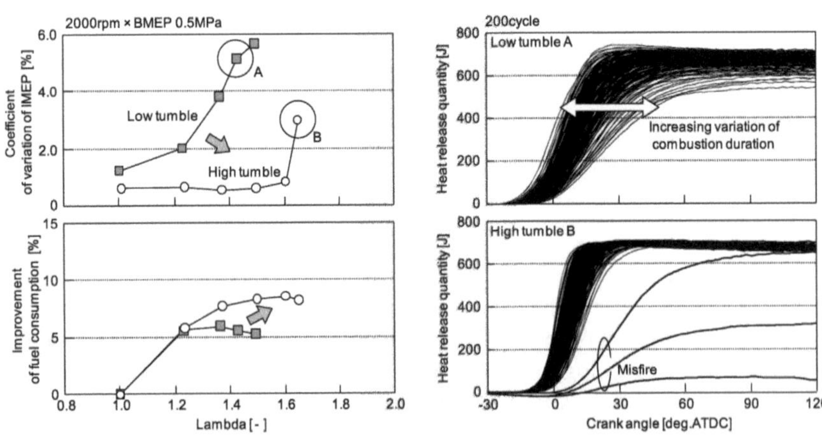

Fig. 2. The effect of tumble on dilution limit and fuel economy

Ignition system has been improved for enhancing ignitability. Figure 3 shows the history of combustion concept and ignition technology which Toyota has adopted. To improve the fuel economy, stratified lean burn, variable valve train system and cooled

EGR have been adapted. For stratified lean burn, discharge duration has been extended to improve ignition probability of stratified mixture [4]. For weak in-cylinder turbulence combustion concept such as Atkinson cycle, spark plug electrode has been thinned to reduce energy loss [5]. And for recent high tumble concept, intensifying discharge energy and aligning ground electrode position to stretch discharge channel have improved ignitability [6]. It is necessary to clarify the required function of ignition system for realizing superior ignitability for higher tumble concept.

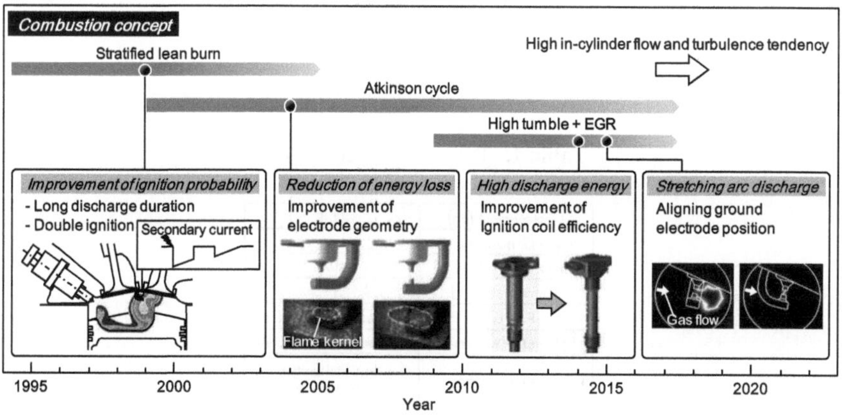

Fig. 3. History of Toyota's combustion concept and ignition system development

2 Direction of the Study

The objective of this study is to achieve superior thermal efficiency with stable ignitability under high in-cylinder flow and diluted mixture conditions. This paper describes the effect of in-cylinder gas flow and turbulence on ignitability through analysis of initial stages of ignition phenomenon such as the discharge channel behavior, formation of flame kernel and flame propagation. Homogeneous air lean burn is used as the way to create diluted conditions in this study. Firstly, the mechanism of misfire in strong flow fields is investigated with numerical simulation and an optical engine. Secondary, the effect of discharge specifications on physical phenomenon is clarified so that the most effective discharge specifications in strong flow fields are proposed.

3 Analysis of Discharge Channel Behavior and Initial Flame Propagation in Strong Flow Fields

3.1 Effect of Strong Flow Fields on Ignitability

Firstly, the mechanism of misfire in strong flow fields is investigated with an optical engine. Table 1 shows its specifications. The spark plug with a mass production specification is mounted so that the ground electrode orientation is perpendicular to gas flow direction. Figure 4 shows the outline of optical measurement. To observe in-cylinder phenomenon, it has a sapphire glass window on the rear side of a fourth cylinder. Flow velocity at the spark gap is calculated as shown in Eq. 1 that is developed based on a past study in consideration of cylinder pressure, lambda and discharge current [7]. u is velocity at the spark gap. V is secondary voltage. Ia is decreasing rate of secondary current. ρ is discharge resistivity. d is spark gap. P is cylinder pressure at ignition timing. α and β is calibrated to adapt the flow velocity measured in advance with LDV (Laser Doppler Velocimeter).

$$u = \frac{\frac{dV}{dt} + Ia}{2\rho d(\alpha P^2 + \beta)} \tag{1}$$

Table 1. Test engine specifications

Displacement [cc]	2489
Cylinder number [-]	4
Compression ratio [-]	10.0
Tumble ratio [-]	3.5
Fuel type	Gasoline 95RON
Injection system	Port and direct injection
Ignition system	Conventional ignition coil (120mAx 1.2ms)
Boost	Turbocharger

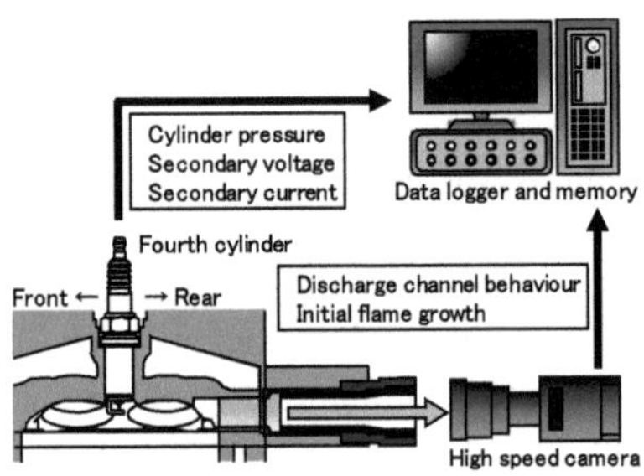

Fig. 4. Outline of the optical measurement

Figure 5 shows the effect of flow velocity on initial combustion duration, discharge channel holding period and discharge channel length. Here, a period from discharge initiation to first blow-off is defined as discharge channel holding period, and discharge channel displacement just before first blow-off is defined as discharge channel length. Flow velocity and turbulence intensity are controlled by changing engine speed. Initial combustion duration is improved by increasing flow velocity as shown in upper graph. Although, it becomes worse above 30 m/s and finally misfire is occurs. Middle graph of Fig. 5 shows discharge channel holding period is shorten with flow velocity, and bottom graph shows discharge channel length is not influenced by flow velocity.

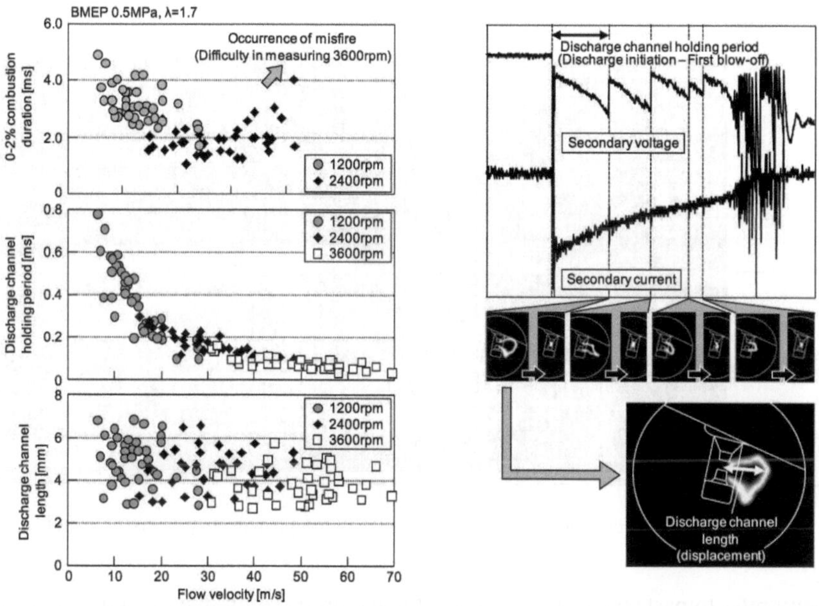

Fig. 5. The relationship between gas flow velocity and ignitability

Figure 6 shows the discharge channel behavior and the growth process of flame kernel formation in various cases of flow velocity. In the case of 10 m/s, the area of flame kernel is small due to short extension of discharge channel per millisecond. In the case of 30 m/s, flame area becomes large with extension of discharge channel. Furthermore, occurrence of flame fusion by repeating blow-off and re-discharge is confirmed. Referring to this phenomenon, a small flame is formed at the blow-off place after first blow-off occurrence, and then second flame kernel is formed by re-discharge. While repeating blow-off and re-discharge, these flames are formed and fused with propagation. In the case of 50 m/s, flame kernel becomes small with repeating frequently blow-off and re-discharge. In addition, flame kernel disappearance is confirmed.

As described above, ignitability improvement with large flame area by the extension of discharge channel is confirmed under 30 m/s in these test conditions. On the other hand, in the case of more than 30 m/s of flow velocity, reduction of flame area due to shorten discharge channel holding period and flame disappearance is confirmed. To understand this phenomenon better, further analysis is to be continued.

3.2 Analysis of Misfire Mechanism

In this section, to clarify the mechanism of misfire, discharge behavior and initial flame propagation in strong flow fields is investigated with numerical simulation. STAR-CD of CD-adapco corporation is used to calculate. G-equation method is used to calculate flame propagation. The calculation of laminar burning velocity and turbulence burning

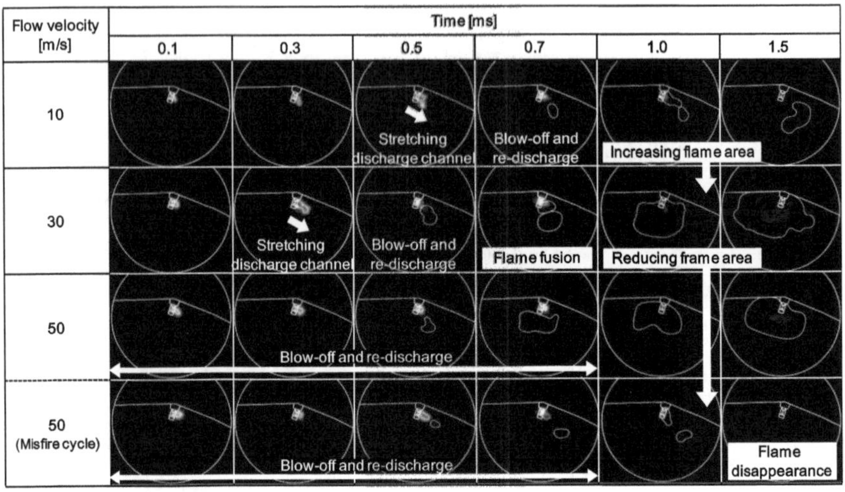

Fig. 6. Discharge channel behavior and initial flame growth in various flow fields

velocity refer to past studies [8, 9]. Start of flame propagation in each cell is defined at laminar burning velocity exceeding proper value as Eq. 2. S_L is laminar burning velocity. In this equation, S_L is calculated by one-dimensional calculation of detailed chemical reaction with gas temperature and lambda. The efficiency in converting discharge energy to heat energy is 30 %. a is calibrated to adapt the result of other experiments.

$$S_L > a \tag{2}$$

Judgment of discharge blow-off refers to discharge channel length as Eq. 3. $L_{channel}$ is discharge channel length. Threshold value is adapted to the previous experimental results.

$$L_{channel} > 10\,[mm] \tag{3}$$

And as shown in Eq. 4, flame disappearance in each cell is defined in the case of Kalrovitz number above 1 [10]. Ka is Kalrovitz number.

$$Ka > 1 \tag{4}$$

Table 2 shows the conditions of calculation. The spark plug is located in the duct and the calculation is conducted from discharge initiation to flame propagation.

Firstly, flow velocity and turbulence intensity at the ignition position of the test engine are measured by LDV as the conditions of numerical simulation. As shown in Fig. 7, turbulence intensity is increasing with flow velocity. As the conditions of measuring result, initial flame growth is calculated. This result shows in Fig. 8. As same as the experimental result, flame growth speed is slow in the case of low flow

Table 2. Calculation conditions

Excess air ratio [–]	1.8
Temperature [K]	600
Pressure [kPa]	1000
Flow velocity [m/s]	2 – 60
Turbulence intensity [m/s]	2 – 20
Discharge specifications Current [mA]× Duration [ms]	100 × 1.0 (Triangle shape)
Spark gap [mm]	0.8

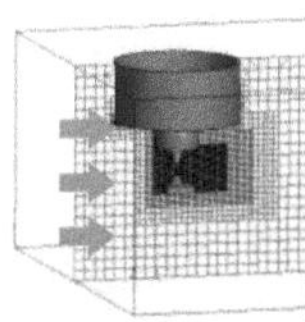

	Cell size [mm]
Base	1
Region 1	0.5
Region 2	0.25
Region 3	0.125

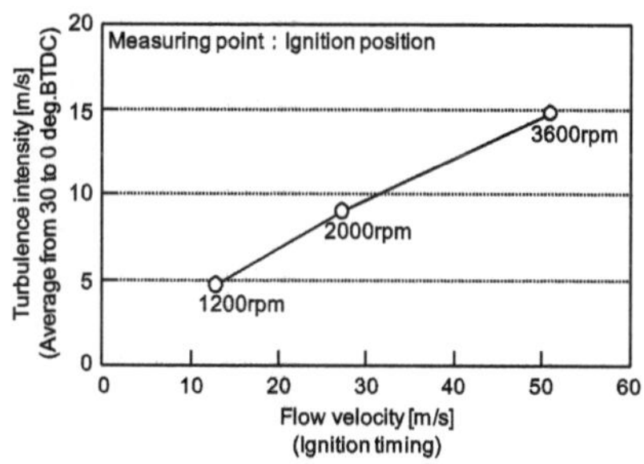

Fig. 7. The measuring result of flow velocity and turbulence intensity in the test engine

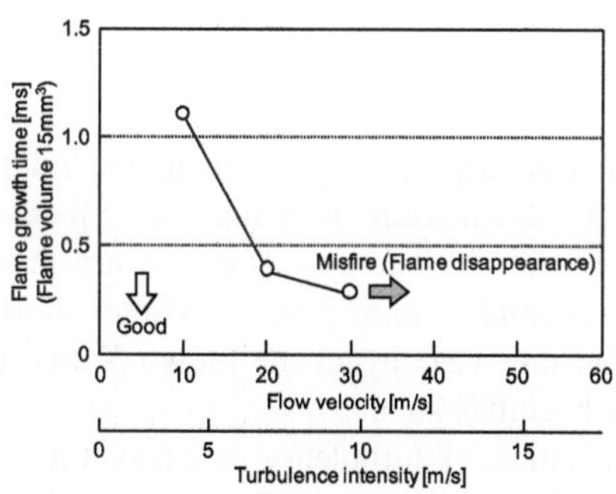

Fig. 8. The calculation result of flame growth speed

velocity. It becomes faster with increase in flow velocity and turbulence intensity. Although in the case of more than 30 m/s, flame kernel disappears.

Here, to clarify the mechanism of misfire, the effect of flow velocity and turbulence intensity at ignition position as individual parameter is investigated. Figure 9 shows the effect of flow velocity on growth process of flame kernel. Flame volume increases with high flow velocity by the stretching discharge channel. Although, it becomes small in the case of more than 30 m/s as shown in condition D, due to the lack of energy caused by repeating blow-off and re-discharge.

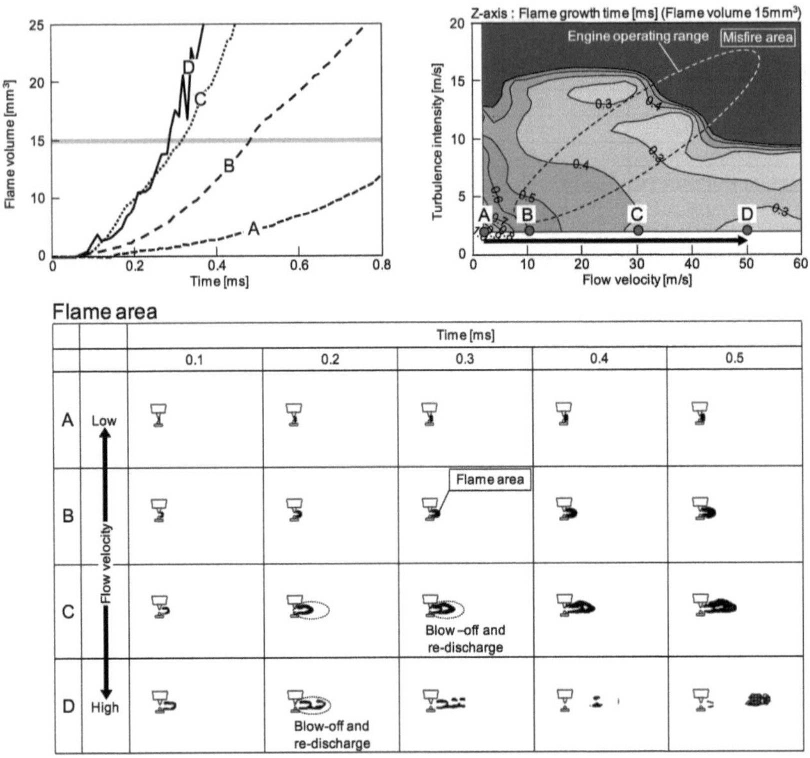

Fig. 9. The effect of flow velocity on initial flame growth

Figure 10 shows the discharge energy given to the mixture. In this graph, the discharge energy given to the mixture is defined as effective ignition energy. The higher flow velocity gives smaller energy to the mixture in same discharge channel length because discharge channel holding period becomes shorter by discharge channel blow-off. The laminar burning velocity in the blow-off area is low. As a result, the flame kernel in this area becomes smaller.

Figure 11 shows the effect of turbulence intensity on growth process of flame kernel at the high flow velocity that causes frequent blow-off of discharge channel. Flame kernel becomes small due to blow-off and finally it disappears completely when flame propagation in the strongest turbulence intensity fails. Figure 12 shows the distribution of Karlovits number. In high flow velocity, it increases and flame disappearance area is large compared to slow flow velocity.

As described above, the mechanism of misfire in strong flow fields is clarified. In the high flow conditions, laminar burning velocity in the discharge channel blow-off area is low and flame volume becomes smaller, because shorten discharge channel holding period causes the lack of energy for the mixture. Moreover, flame disappearance occurs due to strengthening turbulence intensity with increasing Karlovits number of the small flame area.

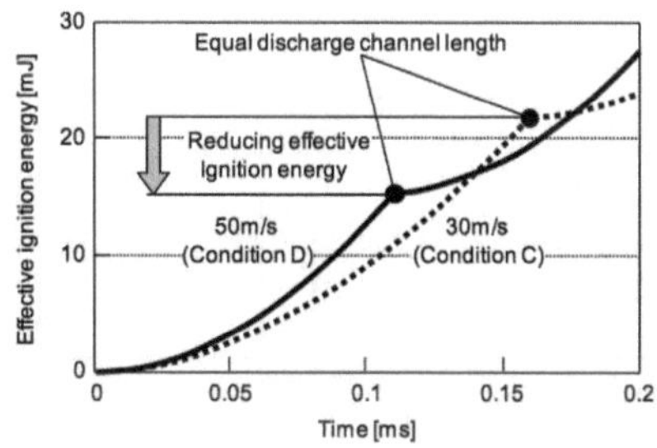

Fig. 10. The effect of flow velocity on effective ignition energy

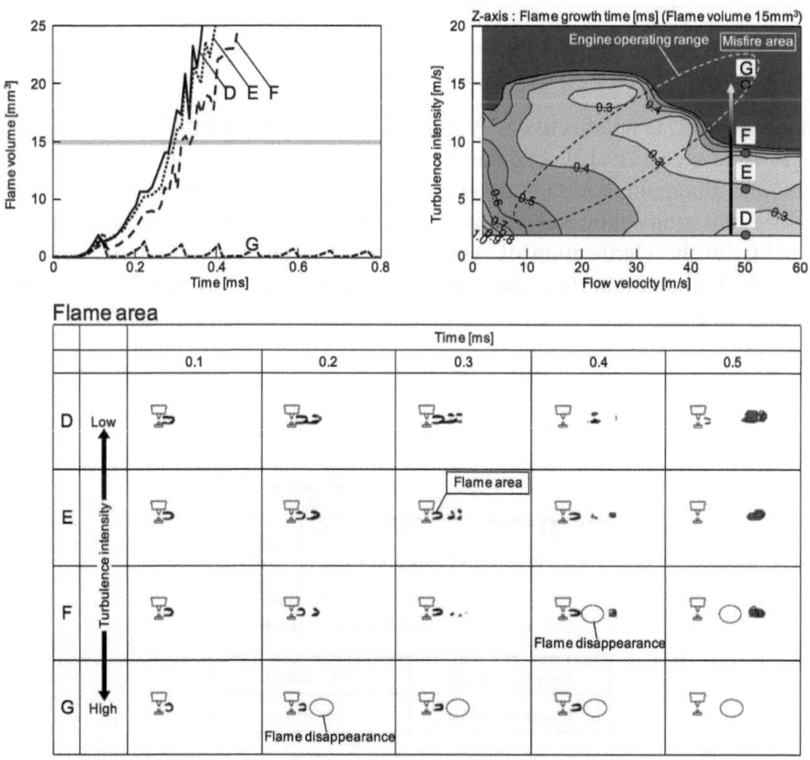

Fig. 11. The effect of turbulence intensity on initial flame growth

4 Effect of Discharge Specifications on Discharge Channel Behavior and Initial Flame Propagation

4.1 Effect of Discharge Current and Duration

This section describes the effect of ignition system on discharge channel behavior and initial flame growth. It is well known that intensifying ignition energy, especially

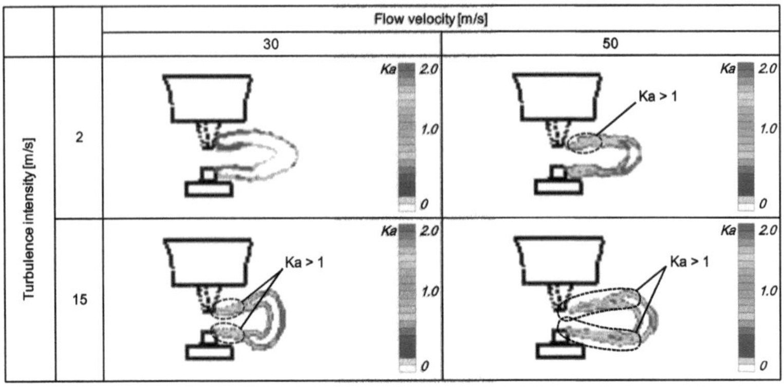

Fig. 12. The distribution of Karlovits number

discharge current, is an effective way to prevent discharge channel from blow-off and enlarge flame area [11, 12]. Discharge specifications are constructed with discharge current and duration, but the individual effects of each have not been clarified. Therefore, this study proposes the guideline to set discharge specifications in strong flow fields by the clarification of these effects. The effect of discharge current and duration on initial combustion phenomenon is individually investigated with the test ignition system which can create a rectangular shape secondary current. Table 3 shows the test discharge specifications.

Table 3. Test discharge specifications

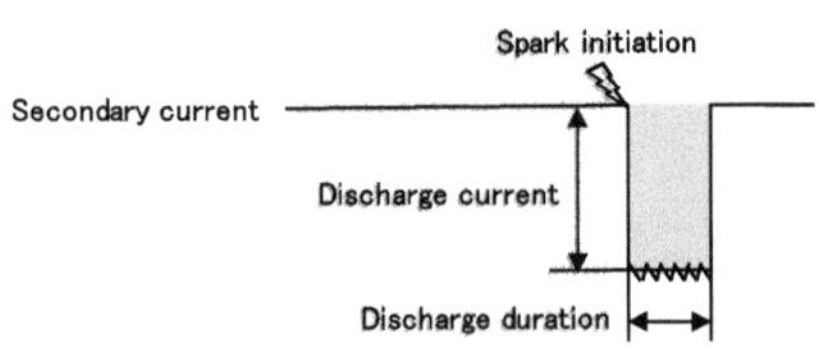

Discharge current	100 – 250 mA
Discharge duration	0.5– 1.5 ms

Figure 13 shows discharge channel behavior and growth process of flame kernel and Fig. 14 shows flame area, discharge channel holding period and discharge channel length with increasing discharge current. Flame kernel becomes large by extending the discharge channel holding period from 100 mA to 200 mA. In addition, the extension of discharge channel length is also observed. Although, the growing tendency of flame kernel area is saturated when discharge current is more than 200 mA. Discharge channel holding period and discharge channel length are also same tendency as shown in the middle and bottom of Fig. 14.

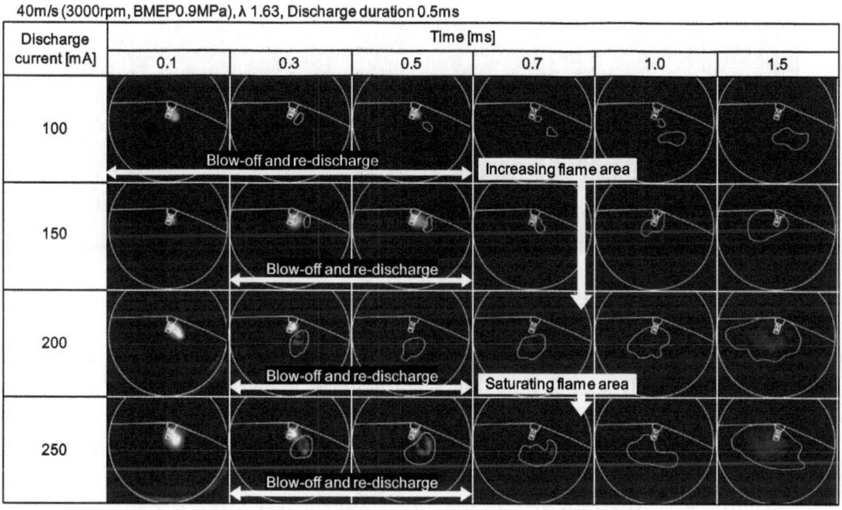

Fig. 13. The effect of discharge current on initial flame growth

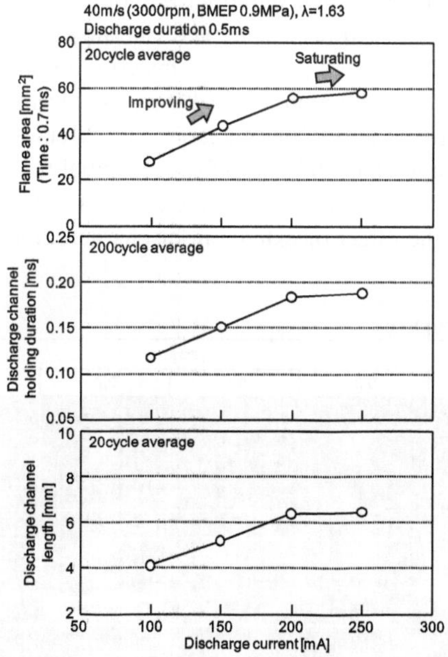

Fig. 14. The relationship between discharge current and flame area

Figure 15 shows the discharge behavior of 250 mA, short-circuited discharge channel on a different path before discharge channel blow-off is confirmed. It is estimated that the phenomenon occurs by the electronic field strength around the discharge channel and the high discharge resistance due to extended discharge channel length. In

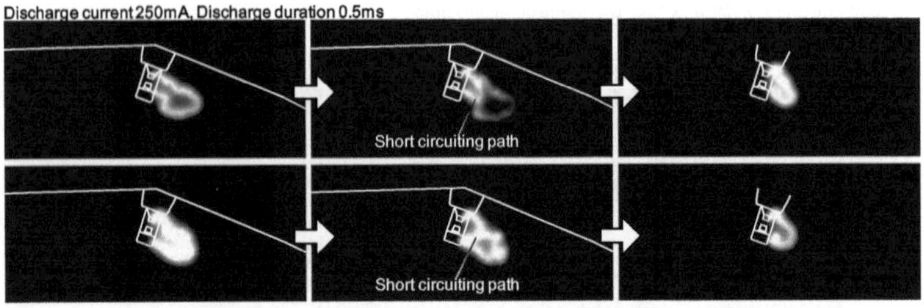

Fig. 15. Short-circuiting of discharge channel

this study, when discharge channel length achieves about 6 mm, the short-circuiting phenomenon occurs. It seems that this phenomenon leads the effect of intensifying discharge current to a saturation.

Figure 16 shows discharge channel behavior and growth process of flame kernel with extending discharge duration. Flame kernel becomes larger with extending discharge duration from 0.5 to 0.75 ms. It is also confirmed that flame fusion leads flame kernel to enlargement with repeating blow-off and re-discharge. Figure 17 shows flame area and the frequency of flame fusion. The increasing tendency of flame fusion is confirmed with extending discharge duration. Although the effect of extending discharge duration from 0.75 to 1.0 ms is saturated. As shown in Fig. 18, it is considered that flame area reaches the ignition point at 1.0 ms after discharge initiation so that the effect of extending discharge duration above 1.0 ms is extremely small. In the case of 100 mA, initial flame area does not become large with discharge duration of more than 1.0 ms. Frequent discharge channel blow-off causes initial flame area to reduce so that the frequency of flame fusion is low. It is indicated that holding discharge channel is necessary to maximize the effect of extending discharge duration.

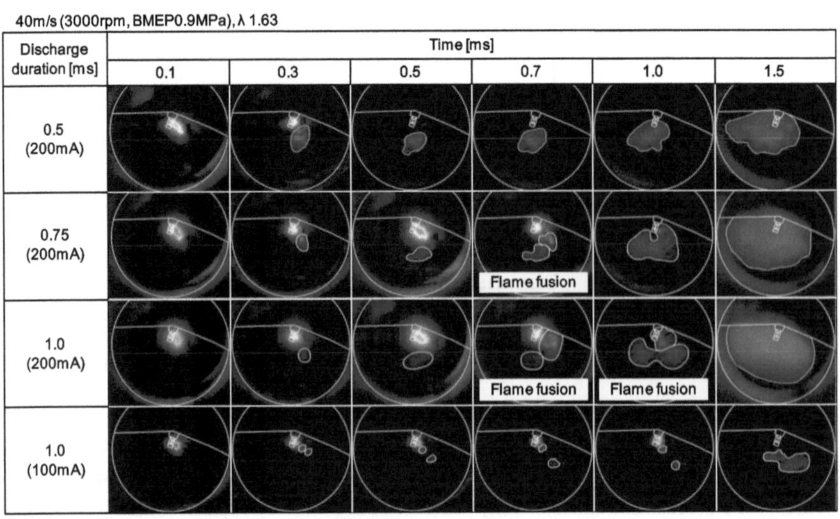

Fig. 16. Effect of discharge duration on initial flame growth

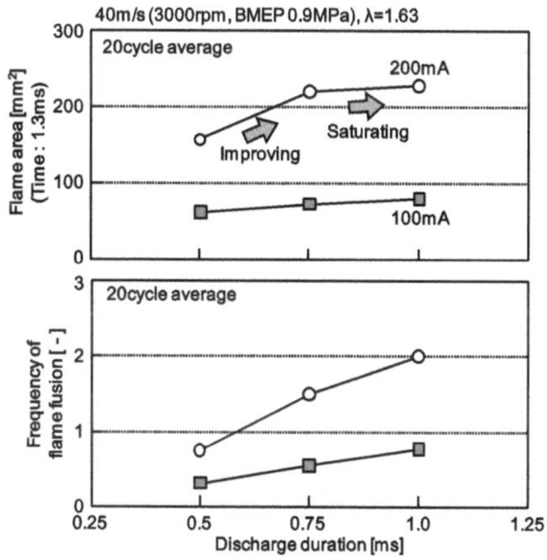

Fig. 17. The relationship between discharge duration and flame area

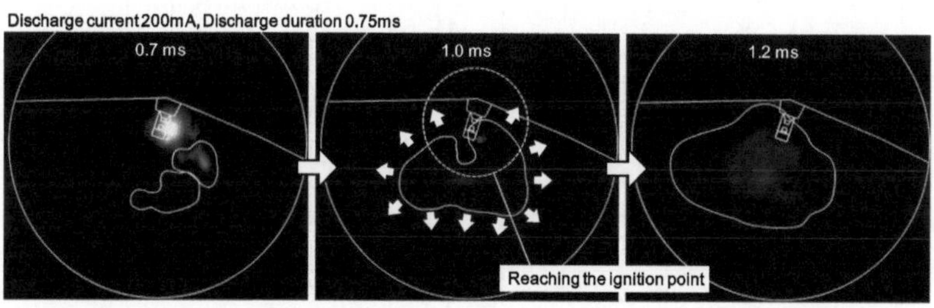

Fig. 18. The relationship between flame area and ignition point

4.2 Optimizing Discharge Specifications

In the final part of this paper, discharge specifications to realize superior ignitability in strong flow fields is mentioned. Through the previous study, the functions of discharge specifications to form initial flame are clarified. The function of discharge current is to hold discharge channel for forming large flame kernel. The function of discharge duration is to lead flame fusion with repeating blow-off and re-discharge. Therefore required specifications for ignitability are proposed such as rectangular shape shown Fig. 19.

Figure 20 shows the result of optimizing discharge specifications and its effect on dilution limit and fuel economy. The benefit of intensifying discharge current above 200 mA is quite small. Extending discharge duration under the conditions below 200 mA is not effective to improve dilution limit. Although extending discharge duration under conditions above 200 mA improved dilution limit. As it was shown in the previous section, the discharge current to hold discharge channel is essential for maximizing the effect of extending discharge duration. As shown in Fig. 20, in the result of optimizing discharge specifications, dilution limit can be extended 0.2 point of

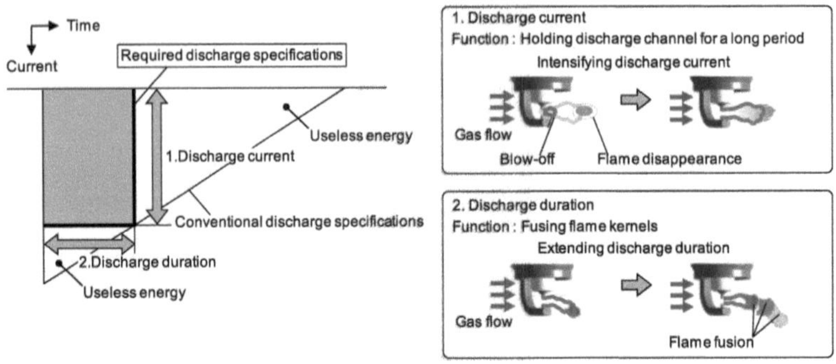

Fig. 19. Required discharge specifications

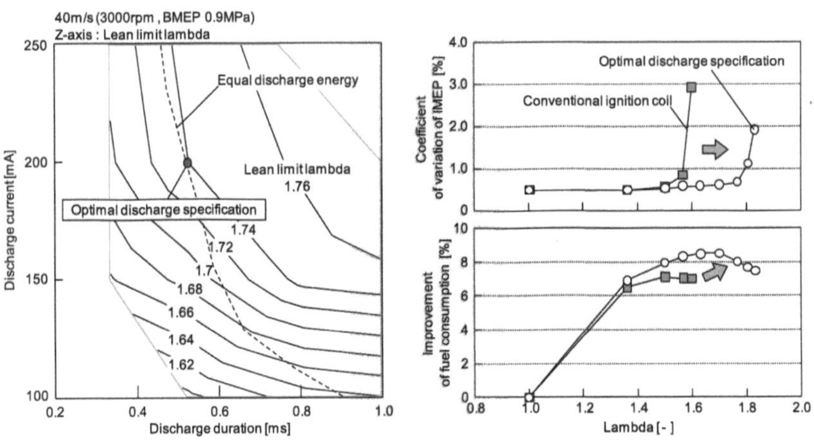

Fig. 20. The effect of optimizing discharge specifications on dilution limit and fuel economy

lambda. As a result, it is confirmed that improving fuel economy is extended by 2% compared to conventional ignition coil.

5 Conclusion

In this study, the underlying mechanism of misfire in strong flow fields is clarified and superior ignitability can be realized by optimizing discharge specifications to suppress the occurrence of misfire. By means of numerical calculation and optical engine experiment, the following knowledge is obtained.

1. It is clarified that the misfire in strong flow fields is caused by following discharge channel behavior and flame propagation.

- Blow-off of discharge channel by strong gas flow reduces the period for supplying discharge energy to the mixture, and the laminar burning velocity in the blow-off area is low. The flame kernel in this area becomes smaller.
- Strong turbulence prevents initial flame propagation. Especially, under the above condition where laminar burning velocity is low, the misfire is caused by flame kernel disappearance.

From these results, the most effective way for suppressing the occurrence of misfire is keeping a sufficient period for supplying effective energy from discharge channel to the mixture and forming large initial flame immediately after discharge initiation.

2. Intensifying discharge current prevents discharge channel blow-off and improves initial flame formation. Although when discharge channel length achieves about 6 mm, a short-circuiting phenomenon occurs and the effect of intensifying discharge current saturates. Furthermore flame kernel formed by repeating blow-off and re-discharge or short-circuiting is fused and this phenomenon leads initial flame to enlargement. It is confirmed that extending discharge duration increases the frequency of flame fusion and improves ignitability.

By applying above knowledge to optimize discharge current and duration, dilution limit can be expanded 0.2 point of lambda. As a result, it is confirmed that improving fuel economy is extended by 2% compared to conventional ignition coil.

Acknowledgment. The author would like to thank NIPPON SOKEN for assisting the analysis of optical tests and DENSO for supplying the test ignition system.

References

1. Yamada, T., Adachi, S., Nakata, K., Kurauchi, T., Takagi, I.: Economy with superior thermal efficient combustion (ESTEC), SAE Technical Paper 2014-01-1192
2. Nakata, K., Shimizu, R.: Toyota's new combustion technology for high engine thermal efficiency and high engine output performance. In: 37th International Vienna Motor Symposium (2016)
3. Murase, E., Shimizu, R.: Innovative gasoline combustion concepts for toyota new global architecture. In: 25th Aachen Colloquium and Engine Technology (2016)
4. Miwa, T., Okabe, S.: Study of the ignition specifications in gasoline direct injection engine, SAE Technical Paper 2001-08-0213
5. Hori, T., Shibata, M., Okabe, S., Hashizume, K.: Super ignition spark plug with fine center & ground electrodes, SAE Technical Paper 2003-01-0404
6. Ota, Y., Fushiki, S., Matsuo, S.: The new PRIUS powertrain: the new 1.8L ESTEC engine with the new generation hybrid system. In: 37th International Vienna Motor Symposium (2016)
7. Kim, J., Anderson, R.W.: Spark Anemometry of bulk gas velocity at the plug gap of a firing engine, SAE Technical Paper 952459
8. Rhodes D., Keck J.C.: Laminar burning speed measurements of indolene-air-diluent mixtures at high pressures and temperatures, SAE Paper 8500047

9. Herweg, R., Maly, R.R.: A fundamental model for flame kernel formation in S.I. engines, SAE Technical Paper 922243
10. Peters, N.: The turbulent burning velocity for large scale and small-scale turbulence. J. Fluid Mech. **384**, 107–132 (1999)
11. Nogawa, S., Koichi, N., Mohammadi, A.: Effect of ignition system on combustion. In: 1st International Conference: Advanced Ignition Systems for Gasoline Engines, Berlin (2012)
12. Shiraishi, T., Teraji, A., Moriyoshi, Y.: The effects of ignition environment and discharge waveform characteristics on spark channel formation and relationship between the discharge parameters and the EGR combustion limit, SAE Technical Paper 2015-01-1895

Simulation of Ignition

Simulating Extreme Lean Gasoline Combustion – Flow Effects on Ignition

B. Morcinkowski[1(✉)], P. Hoppe[2], F. Hoppe[2], M. Mally[2], P. Adomeit[1], T. Uhlmann[1], M. Thewes[1], J. Scharf[1], and H. Baumgarten[1]

[1] FEV GmbH, Aachen, Germany
morcinkowski@vka.rwth-aachen.de
[2] Institute of Combustion Engines, RWTH Aachen University, Aachen, Germany

Abstract. Engines with extreme lean combustion exhibit a significant potential to improve efficiency and to reduce NO_X raw emissions. The development of advanced combustion systems is based on the extensive usage of simulation tools to reduce the number of testing variants and development time. This paper presents a comprehensive simulation methodology for the layout of combustion systems. The presented models are able to predict the ignitibility limits and the cycle-to-cycle resolved combustion with a high degree of accuracy. An interlink approach of 1D gas exchange and 3D in-cylinder flow simulations is being used. The presented models are based on physical considerations and correlation approaches. The physical models are used to predict the ignition limits and to describe the early combustion. These models are validated by test bench measurements. The prediction of late combustion and cycle-to-cycle fluctuations is based on correlation approaches which includes a large number of different engine concepts. The methodology enables a standardized work flow for complex development projects such as required for extreme lean combustion systems.

Keywords: Ignition · Lean combustion · Computational fluid dynamics · Combustion modeling

1 Introduction

Enleanment in gasoline engine has proven to significantly increase the efficiency [1]. On the contrary, expensive exhaust aftertreatment is necessary with moderate lean combustion of air/fuel ratios (AFR) below 1.8 to reduce tail pipe NO_X emissions, as the NO_X conversion rate of commonly used three-way catalyst is strongly limited at lean AFR [2]. Operating the engine under extreme lean conditions (rel. AFR > 1.8) not only improves the efficiency further, but also drastically reduces engine out NO_X emissions. This allows to utilize the efficiency benefits of lean combustion and fullfill future NO_X emission limits with reasonable exhaust aftertreatment system costs at the same time. Figure 1 shows exemplary the operating modes of an extreme lean concept in the engine map including the operating range of the WLTC.

M. Günther and M. Sens (eds.), *Ignition Systems for Gasoline Engines*,
DOI 10.1007/978-3-319-45504-4_5

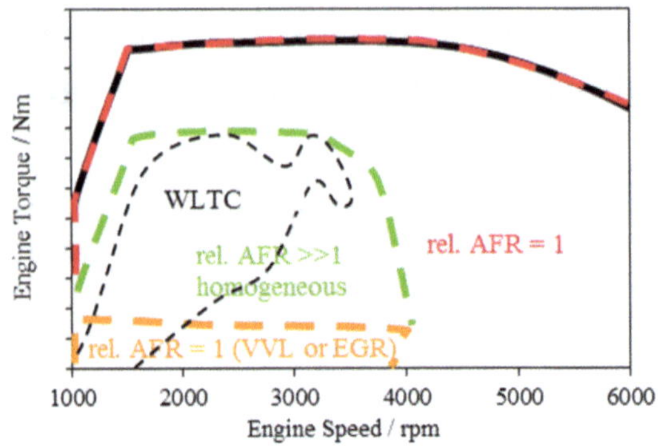

Fig. 1. Extreme lean concept operating modes in engine map

The shift to higher loads is favorable especially for extreme lean concepts, as a wide operating range lays in the area where extreme lean combustion is feasible to be applied.

Besides the increased demands on the charging system, further challenges arise with the extreme lean operation. The ignition of the diluted mixture is stronger hindered by the poor thermal boundary conditions that might lead to misfiring. An additional limitation under lean conditions are the increased Cycle-to-Cycle Variations (CCV) due to the slower combustion. CCV can be minimized with higher turbulent combustion velocity by enhancing the in-cylinder charge motion. Hence, the port and combustion chamber design together with the spray pattern layout play a key role in the development of an extreme lean combustion system. Finally, the transient operation including the switching of operating modes from stoichiometric to lean air/fuel ratios is a challenging topic.

Flow effects will significantly influence the ignition and early combustion process. The heat delivered by the ignition system will activate a specific volume which has to withstand quenching. The deflection of this volume will be affected by the flow. In case ignition is stable a flame kernel will growth. At the beginning the flame is a laminar one and therefore comparatively slow. After a certain time and size of flame kernel the flame will become more and more turbulent, thus the burn rate will increase.

All these effects are important to consider for reduction of CCV respectively stability limit.

2 Challenges of Lean Burn Combustion

The processes from ignition to early combustion can be divided into different phases, see Fig. 2. The arc-over is mainly influenced by the ignition system performance and the thermodynamic state. The heat transfer from the spark has to be larger than the heat losses to the environment. In this phase chemical reactions lead to an early flame kernel development. Approaching flow will deflect such a flame structure by convectional processes, but the flame kernel is too small to be affected by small scale turbulence.

However, turbulence will influence deflection indirectly by having influence on large scale flow. Deflection will influence the volume activated by energy from the ignition system. The combustion speed is rather slow since laminar flame speed is significantly lower than turbulent flame speed. This early flame kernel develops to a shape which could be approximated as a sphere. During this phase the transition from a laminar to a turbulent flame takes place.

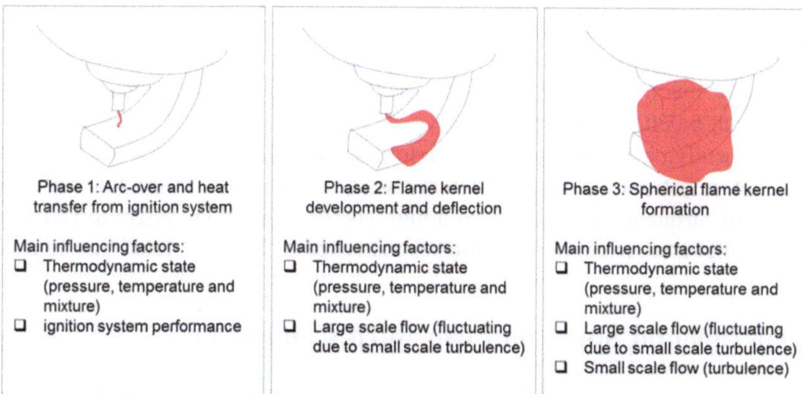

Fig. 2. Phases of flame kernel development

The enleanment capability of homogenous combustion in gasoline engines is mainly limited by two phenomena. On the one hand, the high charge dilution results in longer burn delays leading to increased CCV [3]. On the other hand, sufficient spark energy has to be provided to ensure the ignition of the mixture.

The decreasing laminar flame speed with stronger enleanment requires a more advanced spark timing to maintain the optimal combustion phasing for maximum brake efficiency. This leads to earlier spark timings where the ignition of the mixture is stronger hindered by the higher thermal diffusivity of the mixture [4]. The disadvantageous ignition conditions will lead to misfiring when the applied energy amount during ignition and the according heat release is insufficient to compensate for the thermal losses of the flame kernel.

However, this restriction can be overcome by introducing a higher energy amount to the initial flame kernel. Reports about ignition systems that provide high-energy content during the ignition of the mixture like laser, corona or microwave ignition can be found in literature [5–10]. Reliability issues and/or costs as well as the limited benefit under stoichiometric operation of these systems prevailed an application in series production engines of passenger cars so far. Nevertheless, these systems have the potential to shift the border of the ignition limit of earlier spark timings and hence, enable further enleanment.

Furthermore, introducing a high energy amount to the initial flame kernel allows for the activation of a bigger volume of gas mixture by the plasma. This results in a self-exciting effect, as the burn delay itself is shortened and later spark timings at more favorable thermodynamic ignition conditions are applied for the same combustion phasing. This leads to further enleanment capability.

Beside the activation of a bigger volume during ignition, the burn delay can be shortened by an increased turbulence level. The higher turbulence leads to a faster turbulent flame propagation, which drastically increases the flame speed. The higher turbulence level also works on the other limitation mentioned above, the CCV. Increasing charge dilution due to enleanment significantly reduces the laminar flame speed of the mixture. Hence, the total flame speed during combustion is lowered and the longer combustion duration causes higher CCV. Increasing the turbulence greatly affects the turbulent flame speed, which is characteristic for the speed of combustion after the conversion from laminar to turbulent flame. Thus, the combustion duration and CCV can be effectively reduced.

Consequently, a combustion system for extreme lean homogenous combustion benefits not only from a port, combustion chamber and piston design that generates a high level of turbulence but also from a high energetic ignition system to address the limitations that arise from the decreased flame speed of highly diluted gas mixtures.

3 Experimental Study on Lean Burn Combustion

The tools used for an advanced development process - that is suitable for lean combustion system layout - are developed and validated with the help of experimental data.

3.1 Test Bench Setup and Instrumentation

The single cylinder research engine used for experimental investigations uses the same basis as already described in previous publications [11–13]. In order to achieve a high reasonable compression ratio of 13.5 the stroke of the engine is extended to a stroke/bore-ratio of 1.21 resulting in a cylinder displacement of 400 cm^3. More technical data can be derived from Table 1.

Table 1. Technical data of the research engine

Displacement	cm^3	400
Bore	mm	75
Stroke	mm	90.5
Stroke /bore	1	1.21
Compression ratio	1	13.5
Valves per cylinder		4
Maximum fuel pressure	bar	200
Maximum peak firing pressure	bar	170
Intake valve angle	°	17.5
Exhaust valve angle	°	17.5
Event length intake (1 mm)	° CA	186
Event length exhaust (1 mm)	° CA	186

The central DI injector and the spark plug are placed in central cross position in the combustion chamber roof which means that the spark plug is installed between the exhaust valves while the injector is installed between the intake valves. The injector has an inclination of 6° and the spark plug of 11.5°. The injector features a 6 hole spray targeting which was found to be optimal for the engine in terms of mixture homogenization, oil dilution behavior and pre-ignition probability. A volume flow controlled 200 bar high pressure fuel pump is used to pressurize the conventional RON98 pump fuel used for the presented investigations. The engine is equipped with a multi charge ignition (MCI) coil from Delphi. The ignition system is operated in single charge operation mode with an operating voltage of 12 V and a charge time of 1.0 ms. A series production spark plug with a heat value of 8 according to NGK and an electrode gap of 1 mm was used.

For the thermodynamic measurements, the cylinder pressure is measured with two Kistler 6045B pressure transducers, which are flush-mounted in the combustion chamber roof between each intake and exhaust valve seat ring. Sampling is per-formed via Kistler 5064 charge amplifiers and an FEV Combustion Analysis System (FEVIS) in a resolution of 0.1° CA. Dynamic intake and exhaust gas pressures are measured with Kistler 4045 A5 pressure transducers and sampled via Kistler 4665 and Kistler 4603 charge amplifiers in 0.1° CA resolution. In total, 1000 consecutive cycles are measured. The measurements of static pressures and temperatures are performed with conventional pressure transducers and thermocouples during an averaging interval of 30 s. Oil and water conditioning systems allow steady-state operation. The intake air is conditioned to 50°C downstream of the electronically con-trolled throttle flap. The pressure upstream of the throttle flap and in the exhaust manifold is controlled to 1013 mbar during throttled operation. For charged operation, the cycle-average pressure in the exhaust system is set identical to the pressure in the intake manifold. The engine is coupled to an eddy-current brake and an electric dynamometer to maintain the desired engine speed with an accuracy of ±1 1/min, regardless of the engine load. The intake air mass flow is measured with an ultrasonic air mass meter. Fuel is measured via Coriolis-type mass flow sensor. The rel. air/fuel-ratio (λ) was derived via exhaust gas composition according to the formula of Spindt [14] with the extension for oxygenated fuels of Bresenham [15]. The gaseous exhaust gas composition is determined from a partial mass flow of exhaust gases, which is sampled 60 cm downstream of the cylinder head flange. The exhaust gas sample is fed to the emission analyzers via a transfer line heated to 193°C depicted in Table 2.

Table 2. Technical data of the research engine

HC	Flame ionization detector (Rosemount NGA 2000)
O_2	Paramagnetic oxygen analyzer (Rosemount NGA 2000)
CO	Infrared gas analyzer (Rosemount NGA 2000)
CO_2	Infrared gas analyzer (Rosemount NGA 2000)
NO_X	Chemiluminescence analyzer (Eco Physics 700 EL ht)
PM	Smoke meter (AVL 415 s)

3.2 Thermodynamic Engine Testing

The experimental investigations were performed at an engine speed of 2000 1/min and two loads 6 and 10 bar indicated mean effective pressure (IMEP). In both cases variations of the AFR were performed. Table 3 depicts the engine parameters that were applied for the subsequent measurements. The results of these investigations are displayed in Fig. 3.

Further investigations were performed at engine speeds of 1500 1/min and 2280 1/min at mid-range loads. The results of these investigations give further operating points for verification of the models introduced in Chap. 4 (Figs. 7, 8 and 13).

Table 3. Engine parameters

Start of injection	° CA BTDC	300
Exhaust valve closing (1 mm)	° CA BTDC	10
Intake valve opening (1 mm)	° CA ATDC	10
Intake temperature	°C	50
Oil & coolant temperature	°C	105

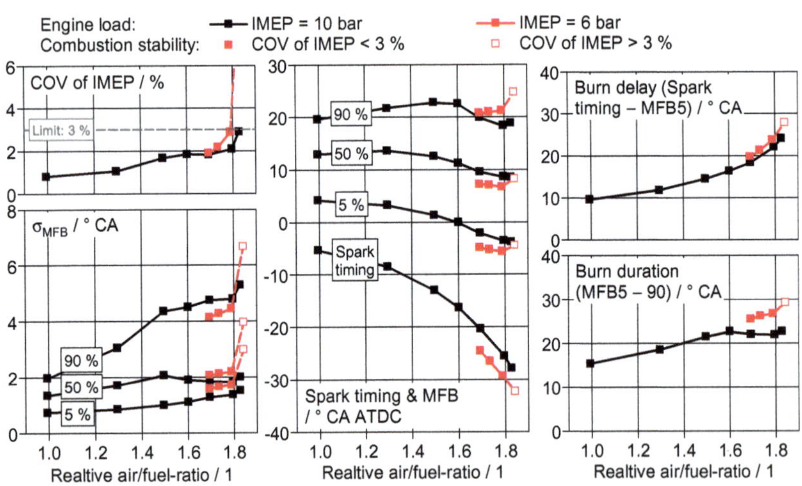

Fig. 3. Variation of rel. air/fuel-ratio at n = 2000 1/min

In case of the IMEP = 10 bar load point (Fig. 3) the enleanment capability starting from stoichiometric combustion was evaluated. The limit for the combustion stability was defined by the coefficient of variation of the IMEP < 3 %. Coming from stoichiometric combustion the spark timing can be advanced with increasing AFR due to lower knock restriction. However, as the burn delay increases a later mass fraction burned of 50 % is present for rel. AFRs below 1.5. At the same time a strong increase of the standard deviation (σ) of the mass fraction burned 90 % is visible whereas only slight increase for the lower MFBs and combustion fluctuations (COV of IMEP) is measured. With further increasing of the AFR earlier combustion is possible resulting in an optimal MFB 50 % at the maximal rel. air/fuel-ratio of 1.83. At the lower load of IMEP = 6 bar (Fig. 3) the enleanment capability is reduced to a rel. AFR of 1.79. In general the lower pressure and temperature lead here to more challenging ignition conditions. Further enleanment results in a decrease of combustion stability and partial misfiring. The reason therefore can be seen in the exponential increase of the burn delay with higher AFR which shifts the spark timing to unfavorable pressure and temperature conditions. For both loads a burn delay greater than 25° CA was found to be critical. Further advancing of the spark timing results in more fluctuating combustion and later MFBs with increased standard deviations (σ) of the mass fraction burned (MFB).

All results were not only basis for the development of models that predict extreme lean combustion, but were also used for an extreme lean concept study in which this engine is the base design for an virtually optimized variant with high indicated efficiency.

4 Lean Mixtures Ignition and Combustion Model

Based on experimental data FEV developed an ignition model to evaluate the ignitability of the mixture towards lean operation of advanced as well as conventional ignition systems. It is based on the proportionality between minimal required ignition energy ($E_{ign,min}$) and laminar flame thickness (l_f) [3], which itself is a function of pressure (p), Temperature (T), air/fuel ratio (AFR) and residual gas fraction (y_{Res}), as Eq. 1 shows:

$$E_{ign,min} \sim l_f(p,\ T,\ AFR,\ y_{Res}) \tag{1}$$

The model allows the determination of earliest possible spark timing resulting in an inflammation of the mixture throughout the complete engine operation map. Figure 4 shows exemplary results under lean operation for different amounts of ignition energy provided by the ignition coil represented by different dwell times.

At late spark timing, where increasing burn delay and hence, the combustion stability is limited by the turbulence a gradually increase in CoV of IMEP is visible. Whereas with advanced spark timings the trend of the CoV of IMEP indicates a more stable combustion until the ignition limit is reached and misfiring occurs.

FEV has furthermore developed models to predict the combustion performance of the engine. On the one hand the early phase of combustion (0–5 % mass fraction burned) is evaluated with a burn delay model [16]. On the other hand the second phase of combustion (5–50 % mass fraction burned) is evaluated with a burn duration correlation, which correlates a burn duration with a turbulent characteristic number [17].

Initially, the burn delay model is used during the post-processing of CFD in-cylinder flow field simulations. It bases on a spherical burn delay correlation introduced by Ewald [18]. Additionally, it considers the displacement of the spark

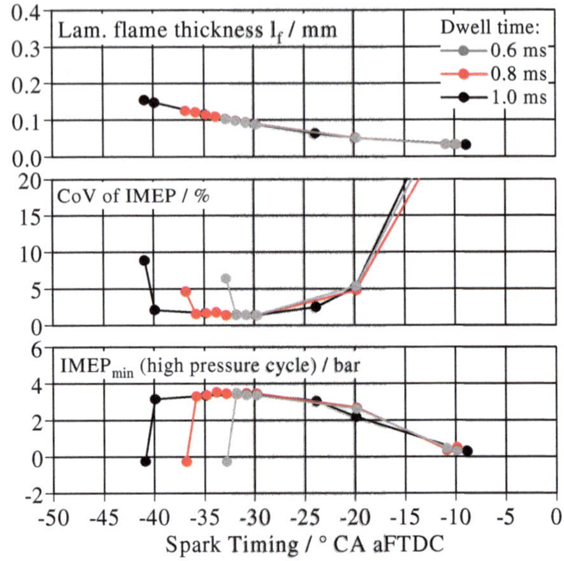

Fig. 4. Simulative and experimental investigation of ignition limits

plasma channel during the glow phase of the ignition. The break-through phase and the arc phase are considered only energetically as their combined duration of 10^{-6} s is too short for the post-processing time scales.

For the determination of the spark plasma channel's displacement, a defined number of tracer particles is introduced into the gap between the ground electrode and the middle electrode in a constant distance. The movement of those particles due to the flow field is calculated and according to the length of the plasma channel new particles are introduced to keep the distances constant. Between the tracer particles, cylinder segments are analyzed representing the plasma channel and the burned mass. The radii of those cylinder segments grow at the position of the tracer particle which leads to conical enlargement. When the diameter of one channel cone becomes larger than the total length of the plasma channel, the development of a spherical structure is assumed. From this point on, the burned mass of all channel segments is transferred into the sphere and the flame propagation is assumed to be purely spherical. Figure 5 illustrates this burn delay model.

Equation 2 describes the flamelet core temperature $T_{Z,i}$ of each channel segment i where N is the number of all channel segments:

$$\frac{dT_{Z,i}}{dt} = \frac{\dot{m}_{Z,i}}{m_{Z,i}}(T_{b,i} - T_{Z,i}) + \frac{\dot{Q}_{spk} - \dot{Q}_{lass}}{N\, m_{Z,i}\, c_{p,i}} + \frac{1}{\rho_{Z,i}\, c_{p,i}} \frac{dp}{dt} \tag{2}$$

The initial values for burned mass $m_{Z,i}$, the channel radius $r_{Z,i}$ and the channel temperature T_Z are estimated.

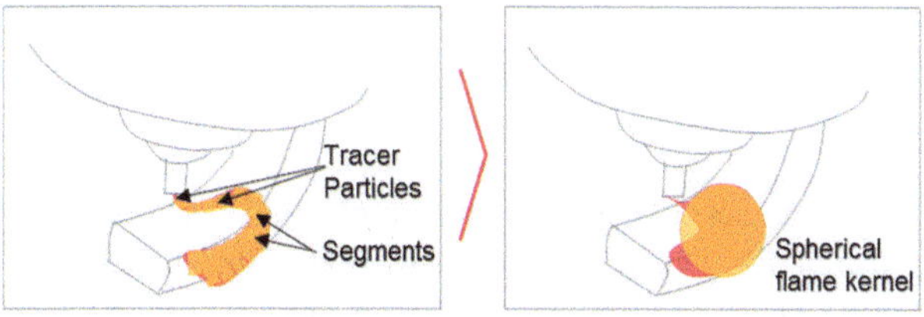

Fig. 5. Burn delay model with tracer particles, consideration of spark channel displacement and transition to spherical flame propagation

Heat losses to the spark plug $\dot{Q}_{loss}$ are calculated according to Pischinger [19] and Willems and Sierens [20] as heat transfer with a heat transfer coefficient h_c defined by the laminar burning velocity s_L, the unburned density ρ_u and the heat capacity c_p averaged between the unburned temperature and the burned temperature.

The electric energy transferred from the spark to the ignition volume $\dot{Q}_{spk}$ is calculated by the voltage loss over the spark gap u_{spk} and the current through the spark plasma. According to Kim and Anderson [21] the voltage loss u_{spk} depends on the length of the spark l_{spk}, the current through the spark plasma i_{spk} and the pressure p:

$$u_{spk} = 40.46 l_{spk} i_{spk}^{-0.32} p^{0.51} \tag{3}$$

The current through the spark plasma depends mainly on the ohmic resistance of the secondary circuit R and the inductance of the firing coil L. It decreases during the glow phase with:

$$i_{spk} = i_0 \exp\left(\frac{-t}{L/R}\right) \tag{4}$$

where i_0 is the start value. According to Duclos and Colin [22], a certain amount of the saved energy in the primary circuit is used in the secondary circuit and considered in the model.

The burned volume during the ignition is calculated from the N cylinder segment volumes. The increase in mass of one cylinder segment is defined by:

$$\frac{dm_{Z,i}}{dt} = 2\pi \cdot r_{Z,i} h_{Z,i} \rho_{u,i} s_{t,\kappa,i} \tag{5}$$

Due to the cylinders curvature $\kappa_{Z,i}$, the turbulent burning velocity is influenced by the flamelet diffusivity D and the turbulent diffusivity D_t:

$$s_{t,\kappa,i} = s_t - \kappa_{Z,i}(D + D_t) \tag{6}$$

The flame kernel growth is determined by the turbulent combustion speed s_t according to Peters [23]:

$$\frac{s_t - s_l}{v'} = -\frac{a_4 b_3^2}{2b_1} Da + \sqrt{\left(\frac{a_4 b_3^2}{2b_1} Da\right)^2 + a_4 b_3^2 Da}$$

The turbulent flame speed depends on laminar flame speed resp. the thermodynamic state and turbulence. Due to limited interaction of turbulence with the flame front at early combustion (when the turbulent eddy sizes are larger than the flame kernel) the turbulent quantities are scaled according to the size of the plasma segment and the turbulent length scale l_t:

$$k_{spk,i} = k_i \left(\frac{\min(2r_{Z,i}; h_{Z,i})}{l_t}\right)^{2/3} \tag{7}$$

The second phase of combustion is calculated based on a correlation (Fig. 6). This correlation has been derived from a database of test bench and CFD results. Since this phase of combustion is dominated by the turbulent properties, a turbulent characteristic number is calculated.

The characteristic number is calculated based on turbulent quantities like turbulent velocity v' and turbulent time τ_t, the laminar burning velocity s_l and the Damköhler-number Da. The engine geometry is considered by a correlation based on bore (D), stroke (s) and compression ratio (ε):

$$\Phi = \int \frac{\tau_t}{\frac{s_l}{v'} + f(Da)} g(D, s, \varepsilon) \tag{8}$$

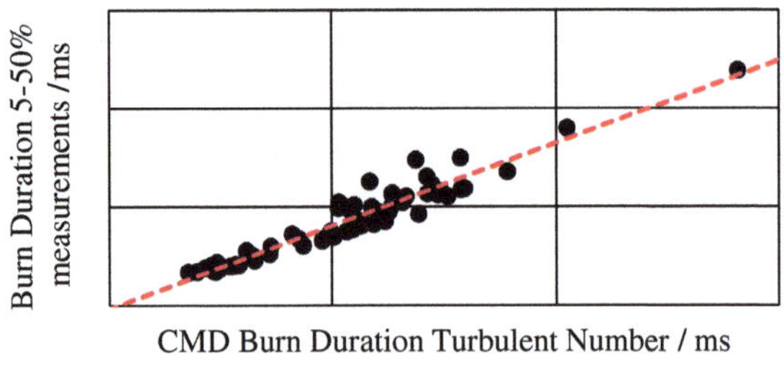

Fig. 6. Burn duration correlation shown for 5–50 % conversion

In Figs. 7 and 8, the comparison between test bench data and combustion prediction is shown. Even for high enleanment both models predict the combustion phases with sufficient accuracy. Thus a reliable prediction of combustion can be derived from the early stage concept results.

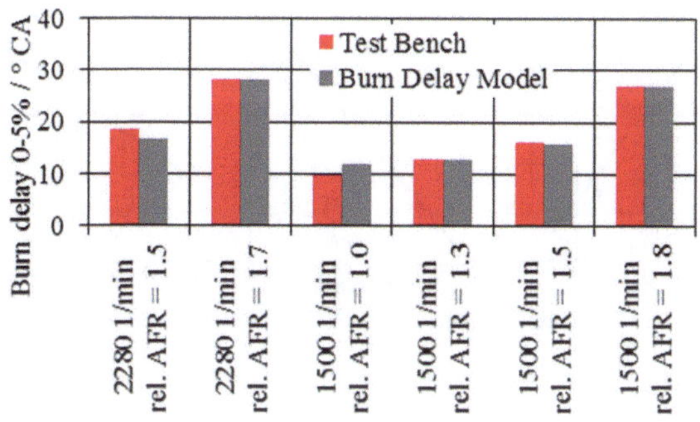

Fig. 7. Burn delay predictions in different operating points compared to test bench data

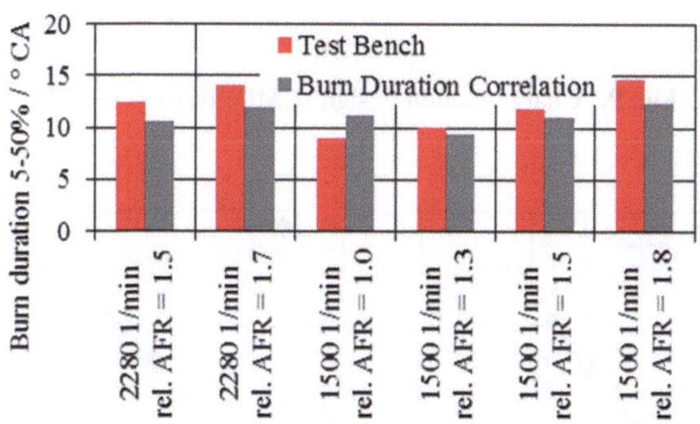

Fig. 8. Burn duration predictions in different operating points compared to test bench data

5 Extreme Lean Concept Study

To face the challenges of extreme lean combustion an effective and predictive virtual development process is needed. FEV has developed such a process, which is called Charge Motion Design (CMD) Process. In this paragraph an exemplary lean concept study is carried out with the CMD Process.

5.1 Charge Motion Design

The target of the CMD process is the optimization of flow guiding surfaces of the combustion system, injector layout and combustion properties. This process is strongly focusing on 3D-CFD simulations in order to minimize testing effort. Backbone of the CMD process are FEV scatterbands on initial charge motion, conserved charge motion and converted charge motion shown in Fig. 9. The scatterbands allow the evaluation of a given design quickly. Furthermore, the underlying data base is used to directly identify required geometry changes.

For extreme lean operation a high tumble level is required to enable stable engine operation for high AFR (e.g., up to rel. AFR of 2). Thus the target of the CMD process

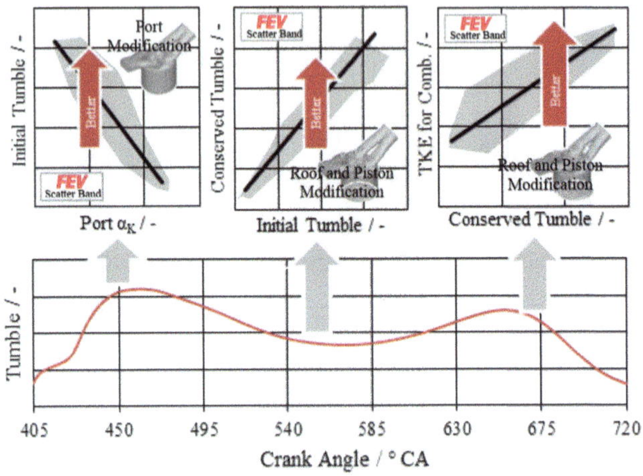

Fig. 9. Charge motion design (CMD) process at FEV

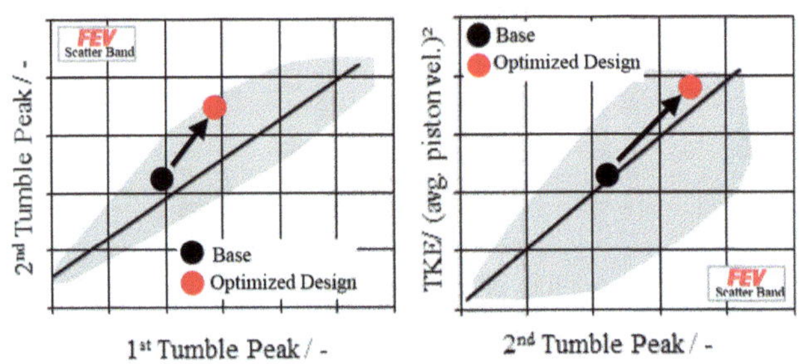

Fig. 10. CMD optimization results

is not only generating a high initial charge motion due to combustion chamber and port layout, but to ensure a good conservation of tumble motion and conversion into turbulent kinetic energy (TKE) near TDC as well.

In Fig. 10 the results of the design optimization are shown. Starting from the base design of the engine described in Chap. 3.1 the geometry was modified to enable a high level of enleanment. Firstly, the design of the intake port was modified. Secondly, the cylinder head was modified and intake valve masking was added to generate a higher initial charge motion. With the initial charge motion optimized, the piston design was changed in order to maximize the conversion of the generated charge motion. A shallow bowl has been identified to be beneficial in terms of conserved tumble motion. The conversion into TKE is following the trend line as shown in the scatterband in Fig. 10. Compared to the base design the TKE near TDC was increased by nearly 70 %.

Since TKE does not directly correlate with the combustion properties FEV has developed models to predict the combustion performance of the engine. These models already have been introduced in Chap. 4.

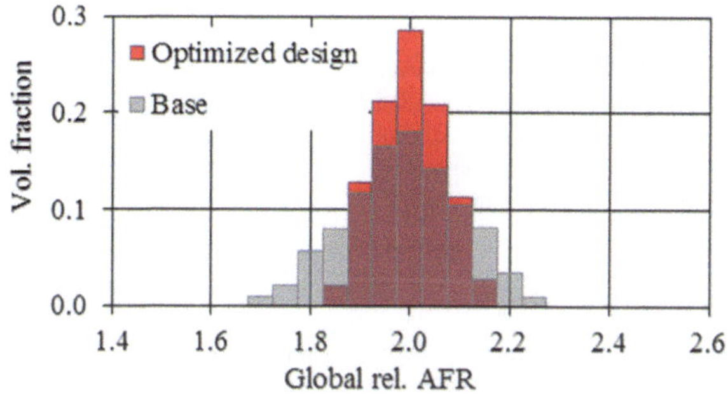

Fig. 11. Exemplary injection optimization results

Another target for the optimization within the CMD process is the injector layout. A high fidelity multi-hole injector model has been developed at FEV. The target for injector layout is the reduction of wall impingement and the overall improvement of mixture homogeneity (Fig. 11). For lean operations the stratification within the cylinder is of particular importance because NO_X emissions have to be minimized. Thus zones with low rel. AFR have to be minimized when operating with global rel. AFR up to 2 due to local higher temperatures and excess of oxygen. Since the spray will interact with the charge motion the injector layout is performed with respect to the flow/injection interaction.

The optimization of the flow guiding surfaces is performed at up to five different characteristic operating points, like the low-end-torque, rated power and the lean operated sweet spot (best efficiency point).

5.2 Predictive 1D-Model

FEV has adopted their combustion prediction and ignitability models for application within 1D simulation. A newly developed turbulence model that uses pre-calculated, interpolated maps from 3D-CFD simulation provides the information about the turbulent quantities. The accuracy of the results is similar to the one of the 3D model shown in Figs. 7 and 8.

Furthermore, in the 1D simulation environment, an experimentally developed combustion stability model is coupled with the FEV predictive combustion model. This model is divided in two sub-models: a CCV model and an ignition model. The CCV sub-model is based on the evaluation of standard deviations of combustion quantities and it depends on the average combustion prediction. These values are used to define normally distributed fluctuations to apply to the mean values. In Fig. 12 a scheme of the CCV model is provided. The CCV model was developed and validated against the extensive FEV measurements database.

In this way, 1D simulation offers the possibility to verify not only the performance of a certain combustion system inserted in a defined air management system, but also the combustion stability which are both very critical issues for lean concepts.

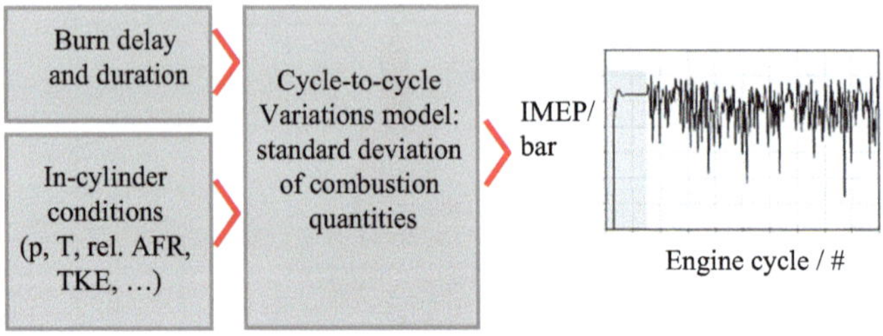

Fig. 12. CCV model scheme

Results of the CCV model are presented in Fig. 13. The accuracy of the CCV prediction is high with deviations on average lower than 0.5 %. The only outlier is the point at 2280 1/min, rel. AFR = 1.7: in this operating point the predicted CCV is sensitively lower than on the test bench. Nevertheless, the reason why the CoV is so high on the test bench can be explained with the ignition model: the operating point is at ignitability limits.

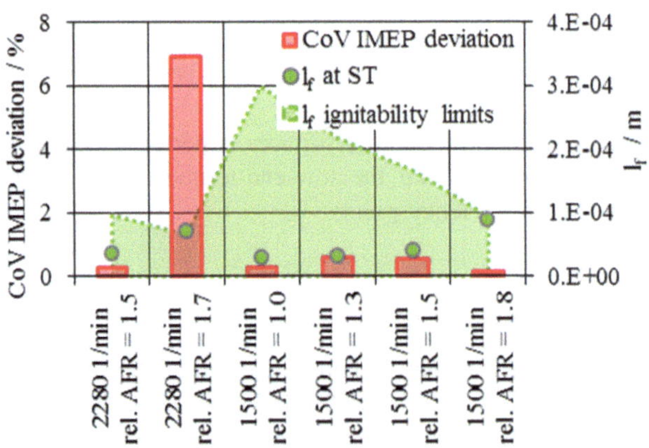

Fig. 13. CCV model and ignition model results

Finally, in order to estimate the influence of the enleanment on the engine NO_X emissions, a NO_X model based on the extended Zeldovich mechanism was calibrated and included into the 1D simulation tool.

5.3 Extreme Lean Concept Study

Once that the prediction capability of the model in lean operations is validated, this can be used for the setup of an extreme lean engine concept. The availability of such a model for 1D simulations is of great importance for new engine concepts. It can be used on the one hand to define the targets in terms of turbulence level for the selected enleanment degree and on the other hand to define requirements for the air management

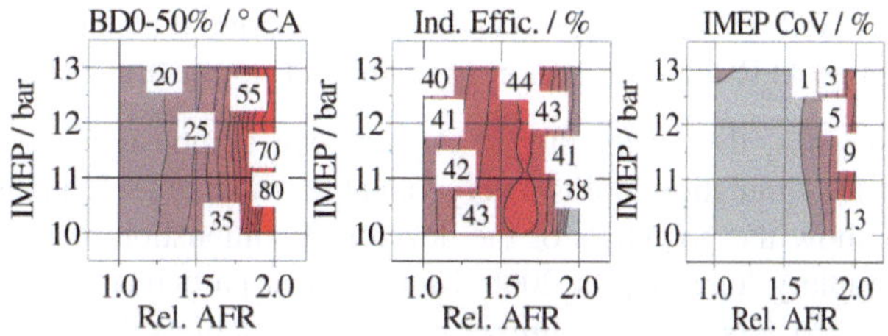

Fig. 14. Simulation results of base design combustion performance

system. In this way, the development process is efficient and it is possible to optimize the entire system layout with lower effort compared to an extensive CFD analysis or to new engine concept prototype construction.

The so-developed model is used in this paper for an exemplary combustion system development and layout of new lean engine concept. The base design turbulence characteristics are used into the 1D simulation tool to check combustion performance and enleanment capability for a certain speed (2000 1/min) in the mid-load area, where the sweet spot is supposed to be. The results (Figs. 14 and 3) show that the base design is not able to have stable and sufficiently short combustion for rel. AFR more than 1.7, with a high worsening of indicated efficiency.

After charge motion design optimization according to Chap. 5.1, the same kind of analysis is repeated with the new turbulence characteristics. The results show that the new design is able to have stable and reasonably fast combustion up to rel. AFR of 2. The maximum indicated efficiency (Fig. 15) is higher than 46 %. That corresponds to an indicated specific fuel consumption (ISFC) below 185 g/kWh.

The coefficient of variance (CoV) of the IMEP is lower than 2 % which means still stable combustions. It is visible that the further enleanment from rel. AFR of 1.5 to 2 brings only an additional increase of approximately 2 % of indicated efficiency, but a reduction of more than 90 % for NO_X raw emissions.

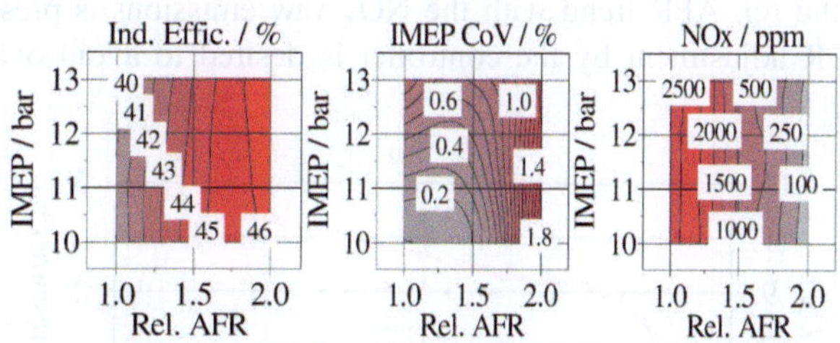

Fig. 15. Simulation results of optimized design combustion performance

Another important aspect for the lean concept layout is linked to transient behavior. Two critical points are:

- Air management system: necessity to build boost pressure in a very fast way in order to be able to increase rel. AFR;

- Combustion stability: Importance to understand if during the transient operation and with increasing AFR the combustion is stable and no misfires occur.

For these reasons, the availability of a predictive combustion model with CCV simulation in a 1D simulation tool is a very valuable resource for lean engine concept.

In order to show the potentials of the developed combustion model for 1D simulations, an exemplary load step at 2000 1/min with a particular rel. AFR switching strategy is presented. The strategy utilizes a 2-stage variable valve lift and an electric charger in addition to the turbocharger. With this setup the lift switch and the fast boost pressure build-up allow to change the AFR without decrease of torque. The strategy is presented in Fig. 16.

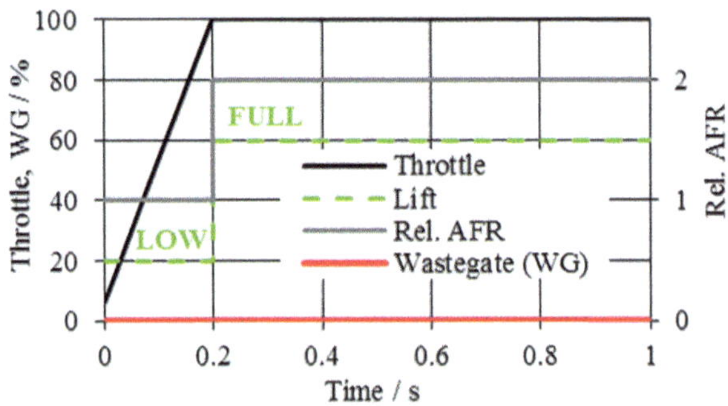

Fig. 16. Transient AFR switching strategy

The performance at transient operation in terms of brake mean effective pressure (BMEP) and the intake manifold pressure build-up is shown in Fig. 17. The initial fast increase of intake pressure is due to throttle opening (first 0.1 s) and the e-charger (from 0.1 to 0.4 s approximately). Further increase of boost pressure is slower due to the inertia of the turbocharger.

In Fig. 18, the rel. AFR trend with the NO_X raw emissions is presented. A short time for the AFR adjustment by the controller is desired to avoid overshooting and

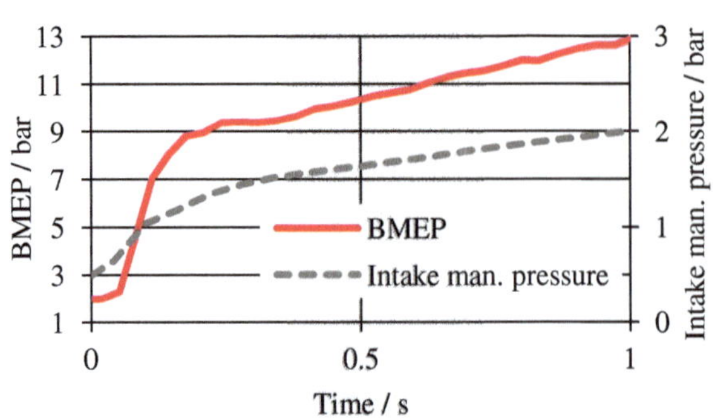

Fig. 17. Transient operation performance

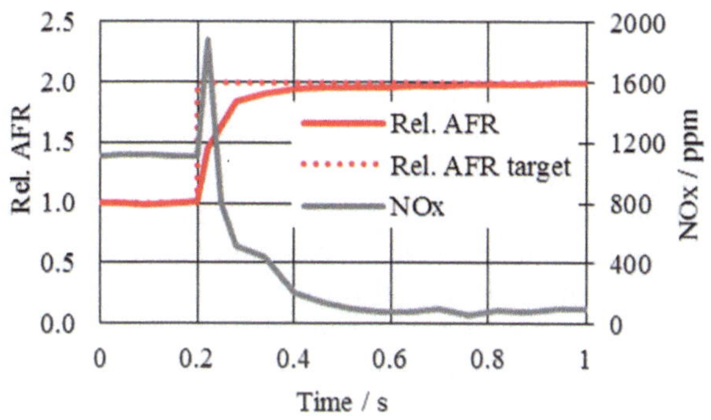

Fig. 18. Transient AFR and NO_X raw emissions

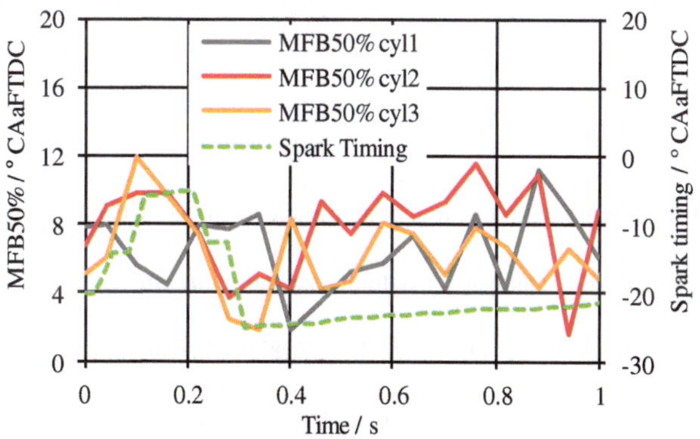

Fig. 19. Cylinders fluctuating MFB50 % and imposed spark timing

torque fluctuations. With the proposed concept the rel. AFR reaches the target value after 0.2 s. The NO_X raw emission are increasing only for a very limited time and in less than 0.3 s values lower than 100 ppm are achieved.

Finally, in Fig. 19 the fluctuation of the MFB 50 % for the different cylinders is presented. In the same figure, the actuated spark timing during the transient is displayed: with increasing rel. AFR an advance of spark timing is needed in order to keep the MFB 50 % at the maximum thermodynamic efficiency.

6 Conclusion and Outlook

This paper presents a comprehensive and powerful simulation methodology for the layout of extreme lean combustion systems within 1D simulation supplied by 3D-CFD simulations. The presented experimental results form the basis of the developed models. These models are based on physical considerations and correlation approaches.

The physical models are used to predict the ignition limits and to describe the early combustion accounting for flow effects. Thermodynamic measurements were used to validate these models for a large number of different engine concepts. The prediction of late combustion and cycle-to-cycle fluctuations is based on correlation approaches which again includes a large number of engine concepts. The same models are used for both, 1D gas exchange simulations to optimize the engine processes globally and 3D in-cylinder flow simulations to optimize the design of flow leading surfaces.

It is shown that this interlink approach will lead to efficient and robust interfaces which enables a standardized work flow for complex development projects. This work flow is exemplary described for a lean burn combustion system. The resulting engine shows an indicated efficiency of more than 46 % in the best point with a rel. AFR of 2.0. Cycle-to-cycle fluctuations are on a low level proved by a CoV of IMEP of less than 2 %.

Finally, the potential of the developed toolchain is presented to evaluate the transient behavior in terms of air management system and cycle-to-cycle fluctuations.

References

1. Heywood, J.B.: Internal Combustion Engine Fundamentals. McGraw-Hill Inc., New York (1988)
2. Matsumoto, S.: DeNOX catalyst for automotive lean-burn engine. Catal. Today **29**, 43–45 (1996)
3. Wenig, M.: Simulation der ottomotrischen Zyklenschwankungen. University Stuttgart, Stuttgart (2013)
4. Böwing, R.: Der Einfluß von Zündung und Zylinderinnenströmung auf Verbrennung in Ottomotoren mit hoher Ladungsverdünnung. Dissertation RWTH Aachen (2000)
5. Geringer, B.: Demands on gasoline engines with focus on the stringent fuel economy and emission limits until 2020. In: Advanced Ignition Systems for Gasoline Engines, No. 1, pp. 1–18 (2012)
6. Ruan, G., Kienzle, D., Roth, D., Becker, M.: Dual coil ignition for gasoline EGR engines. MTZ **74**(03), 19–21 (2013)
7. Ikeda, Y.: Development of microwave plasma ignition system with semi-conductor microwave oscillator and combustion improvement in practical SI engine. In: Ignition Systems for Gasoline Engines, No. 2, 229–250 (2014)
8. Burrows, J., Mixell, K., Reinicke, P.-B., Riess, M., Sens, M.: Corona ignition – assessment of physical effects by pressure chamber, rapid compression machine, and single cylinder engine testing. In: Ignition Systems for Gasoline Engines, No. 2, pp. 87–107 (2014)
9. Graf, J., Lauer, T., Geringer, B., Rottengruber, H.: Innovative ignition systems for highly-supercharged downsized engines. In: Advanced Ignition Systems for Gasoline Engines, No. 1, pp. 205–225 (2012)
10. Schenk, M., Fessler, M., Wolf, T., Klaus, B., Fischer, H.: Comparison of the thermodynamic potential of alternative ignition systems for SI-Engines. In: Advanced Ignition Systems for Gasoline Engines, No. 1, pp. 238–260 (2012)
11. Thewes, M., Baumgarten, H., Dohmen, J., Uhlmann, T., Seibel, J., Balazs, A., Hoppe, F., Krieck, M., Hoppe, P.: Gasoline combustion systems beyond 2020. In: 23rd Aachen Colloquium Automobile and Engine Technology Aachen (2014)

12. Thewes, M., Hoppe, F., Baumgarten, H., Seibel, J.: Water Injection for Gasoline Combustion Systems. MTZ worldwide, February 2015
13. Hoppe, F., Thewes, M., Baumgarten, H., Dohmen, J.: Water injection for gasoline engines: potentials, challenges, and solutions. Int. J. Engine Res. (2015). doi:10.1177/1468087415599867
14. Spindt, R.S.: Air fuel ratios from exhaust gas analysis. SAE Technical Paper 650507 http://dx.doi.org/10.4271/650507
15. Bresenham, D., Reisel, J., Neusen, K.: Spindt air-fuel ratio method generalization for oxygenated fuels. SAE Technical Paper 982054 http://dx.doi.org/10.4271/982054
16. Budde, M.: Untersuchung zyklischer Schwankungen der Entflammung an der Klopfgrenze von magerbetriebenen Ottomotoren. Dissertation RWTH Aachen (2012)
17. Wiese, W., Pischinger, S., Adomeit, P., Ewald, J.: Prediction of combustion delay and duration of homogeneous charge gasoline engines based on in-cylinder flow simulation. SAE Technical Paper, 2009-01-1796 (2009)
18. Ewald, J., Peters, N.: On unsteady premixed turbulent burning velocity prediction in internal combustion engines. Proc. Combust. Inst. **31**, 3051–3058 (2007)
19. Pischinger, S.: Effects of Spark Plug Design Parameters on Ignition and Flame Development in an SI Engine. Dissertation Massachusetts Institute of Technology (1989)
20. Willems, H., Sierens, R.: Modeling the initial growth of the plasma and flame kernel in SI engines. J. Eng. Gas Turbines Power **125**, 479–484 (2003)
21. Kim, J., Anderson, R.W.: Spark anemometry of bulk gas velocity at the plug gap of a firing engine. SAE Technical Paper 952459 (1995)
22. Duclos, J.-M., Colin, O.: Arc and kernel tracking ignition model for 3D spark-ignition engine calculations. In: 5th International Symposium on Diagnostics and Modeling of Combustion in ICE, Nagoya, pp. 343–350 (2001)
23. Peters, N.: Turbulent Combustion. Cambridge University Press, Cambridge (2000)

New Ignition Systems 1

High Energy Multipole Distribution Spark Ignition System

Ming Zheng(✉), Shui Yu, and Jimi Tjong

Department of Mechanical Automotive and Materials Engineering, University of Windsor, 401 Sunset Avenue, Windsor, ON N9E 2W9, Canada
mzheng@uwindsor.ca

Abstract. This work presents a high energy multipole distribution spark ignition system that utilizes a three-pole spark igniter to create spatially distributed sparks within the igniter perimeter. The current prototype can fit in a M14 standard spark plug mounting thread, evenly distributing three spark gaps in a triangular pattern with a circumradius of 2.3 mm. The spark gaps can be individually energized by the attached ignition coils, thereby discharging in either simultaneous or alternating mode. The experimental results from both the single-cylinder engine and constant-volume combustion vessels indicate a clear trend that, on the basis of similar total spark energy levels, the three-pole ignition system can shorten the ignition delay, stabilize the combustion phasing, and extend the ignitable lean limits, compared to a conventional fine-electrode spark plug. Moreover, as found in the combustion vessel tests, the effectiveness of spark discharge enhancement using direct-capacitor discharge can be augmented through simultaneously discharging through the three poles, hence resulting in significantly shortened ignition delay.

Keywords: Multi-pole ignition · Spark distribution · High energy ignition · Diluted combustion

1 Introduction

Charge dilution is employed in advanced clean combustion gasoline engines to improve the fuel efficiency, normally through excess air admission and exhaust gas recirculation (EGR). The improvement, especially for low to intermediate loads, is at least partially attributed to the reduction of the pumping loss, lowered heat transfer loss, and the increase of the ratio of specific heats (γ) [1–3]. Theoretically, the higher the dilution level is applied, the higher the potential of engine efficiency improvement is possibly achievable. Nevertheless, in practice, the operable range of the diluted combustion is often restricted by the deteriorated ignitability and the sluggish combustion processes of the reactivity-suppressed cylinder charge. In order to attain an optimal thermal efficiency, it is suggested to locate one of the representative combustion phasing parameters, i.e. the crank angle of 50 % mass fraction burnt (CA50), at around 6–8 °CA after the top dead center (ATDC) [4]. However, a desired CA50 is not always achievable for a highly diluted mixture. A previous report indicates that in some cases under high EGR, even with an extremely advanced spark timing at earlier than 60 °CA

M. Günther and M. Sens (eds.), *Ignition Systems for Gasoline Engines*,
DOI 10.1007/978-3-319-45504-4_6

before the top dead center (BTDC), a substantially delayed CA50 can still occur because of the slow combustion processes [5]. The over-stretched combustion duration under highly diluted conditions can lower the thermal efficiency, in addition to deferring the CA50. Therefore, to hasten the combustion processes, a diluted combustion system is often incorporated with charge motion intensifying approaches, such as using a tumble intake port, increasing piston velocity, and adopting a squish combustion chamber shape, etc. [3]. To improve the combustion phasing control of diluted mixtures, the shortening of the ignition delay is of particular importance. The burning during early combustion stage, where the cylinder pressure and temperature are relatively low, could be sped up via ignition enhancement, hence decreasing the ignition delay. The other critical aspect for improving the diluted combustion is to secure the ignitability that is directly associated with the improvement of the ignition method.

Fundamentally, the ignition of an air/fuel mixture is initiated from a flame kernel that is generated by an electrical plasma discharge from an igniter. The flame kernel succeeds to self-sustaining propagation by the chemical energy release of combustion, which in many cases is assisted by the concurrent plasma expansion. The improvements on the ignition source are made to either intensify the plasm breakdown e.g. by seeking a large ignition volume [6, 7] or a high-power break down process [8, 9], or to enhance the follow-up energy delivery e.g. a higher quantity [10, 11] or a longer duration [12, 13]. A typical sparking process of a conventional inductive ignition system involves two types of electrical discharge processes, i.e. capacitive discharge and resistive discharge. Although the resistive discharge during the glow phase dominates the overall energy delivery, the capacitive discharge, mainly being effective during the high power breakdown and arc phases, may play avital role in extending the inflammation limits for the air/fuel mixtures that are highly diluted and/or under strong charge motion [14]. The process of capacitive discharge can be enhanced through deploying capacitors directly on the secondary loop of the ignition system. The techniques are categorized as direct-capacitor discharge ignition in this paper, in order to distinguish from the so-called capacitive discharge ignition (CDI) that employs capacitors in the primary loop of the ignition system. A traditional high energy spark ignition system employs auxiliary capacitors that are externally charged to enhance the spark energy, such as the so-called plasma jet [10], and rail plug [15] techniques. The transient power of spark discharge, during and/or in a short period right after break down, can be boosted by using a small shunt capacitor parallel connected to the spark gap, also known as the peaking capacitor technique [9, 16]. With regard to the resistive discharge, the improvement is mainly made by managing the temporal profile of the spark discharge process, such as achieving longer spark duration and multiple events. A multiple-charge ignition technique [12] can energize the spark plug multiple times within a combustion cycle, often with a boosted charging voltage supplied to the primary coil. An innovative dual-coil offset multiple-charge ignition system [13] can produce a long duration continuous spark discharge though alternating the energization of two coils over one spark plug. The engine test results show that the prolonged spark duration can extend the EGR tolerance for a homogenous gasoline engine combustion system.

The afore-mentioned techniques are based on the conventional spark plug architecture that comprises of counter-electrodes forming spark gaps. The ignition plasma effect is typically constrained within the vicinity of a single spark gap. It is always attractive to open up the blockage of the electrodes for achieving a volume-type of ignition with multiple ignition sites created in the engine cylinder. The corona ignition systems based on transient plasma [7] or radio-frequency plasma discharge [17] can produce multiple spatially distributed ignition sites along the discharge streamers. It has been broadly reported in terms of extending the ignitable limits for diluted engine combustion and improving fuel economy. The quarter-wave microwave induced ignition concept [18] has shown the ability of producing a volume ignition process. According to the literatures, those advanced ignition techniques can significantly improve the ignition process of highly diluted combustion, however, in the cost of complicated hardware and control implementations. The present paper introduces a high energy multipole distribution spark ignition system which utilizes a three-pole spark igniter to create three spatially distributed sparks within the igniter perimeter. The current prototype can fit in a M14 standard spark plug mounting thread, evenly distributing three spark gaps in a triangular pattern with a circumradius of 2.3 mm. The spark gaps can be individually energized by the attached ignition coils, thereby discharging in either simultaneous or alternating mode. The three-pole arrangement unfolds a layer of possibilities to implement advanced high energy spark discharge strategies, in addition to applying any proven technologies, such as direct-capacitor, multi-charge, dual-coil offset ignition, etc., onto an individual spark gap. The three-pole igniter contains the ignition sites within the perimeter of the igniter mounting thread; therefore the expected effective ignition volume could be smaller than that of a corona igniter, but would be more tolerant to the ever-changing in-cylinder conditions.

In this work, the three-pole igniter is evaluated on constant-volume combustion vessels and a modified single-cylinder engine running ultra-lean gasoline mixtures. The effects of spark distribution on the ignition delay, combustion phasing control, and the extension of lean operable limit are examined. High-speed shadowgraph imaging of the ignition flame kernel developing process is performed on one optically accessible combustion vessel. Repetitive ignition tests are carried out on a small combustion vessel to evaluate the effect of the spark energy and discharge pattern on the ignition delay and the ignition success rate at the lean inflammable boundaries. Moreover, on the combustion vessel, the effectiveness of spark discharge enhancement with direct-capacitors can be augmented through simultaneously discharging on the three-pole igniter, hence resulting in significantly shortened ignition delay. The engine test results indicate a clear trend that, on the basis of similar total spark energy levels, the three-pole ignition can shorten the ignition delay, stabilize the combustion phasing, and extend the operable lean limits, compared to a conventional fine-electrode spark plug. Taking into account the best efficiency point for the baseline ignition strategy, the operable lean limit is extended by the three-pole ignition from $\lambda 1.61$ to $\lambda 1.84$, which results in a 5 % drop in indicated specific fuel consumption.

2 Experimental Setups

2.1 Three-pole Spark Plug and Ignition System Configurations

The three-pole prototype spark plug is made in production-level processes with conventional sintered aluminum oxide processing methods and steel shell fabrication. It contains three electrically isolated spark gaps equally distributed in a 4.5 mm diameter circle within the perimeter of the standard M14 mounting thread. All the ground electrodes are typical J-type shape. The conductor used to connect each of the core electrodes to spark wires consists of 20AWG (American Wire Gauge) high-voltage cable with silicone insulation material. The conductor cable joints are potted with electrical epoxy to isolate the conductors from each other. The spark gap size is adjusted to 0.86 mm for all three poles. Figure 1 gives the front and the top views of the three-pole electrode configurations, with comparison of a stock NGK fine-electrode spark plug as benchmark. The visual spark discharge channels in the middle picture are generated simultaneously by three identical ignition coils.

Stock spark plug (Fine electrode) | 3pole prototype (Front view) | 3pole prototype (Top view)

Fig. 1. Electrode configurations for the tested spark plugs

Figure 2 gives an example of the three-pole spark breakdown waveforms that are obtained from a discharge event in nitrogen at 20 bar and 25 °C. The breakdown voltage of the three-pole is at similar level of about 25 kV because of the similar spark gap size. While the breakdown timings of the three poles are different, there is no fixed sequence observable. As shown in Fig. 2, the breakdown of an individual pole will drag down the voltage at the other poles, which is possibly due to the charging process into the parasite capacitance that is formed between the poles. The sudden decrease of the voltage thus delays the breakdown process.

The well-referred definitions divide a typical spark discharge process into three stages i.e. the breakdown, arc, and glow phase [14]. In this paper, direct-capacitors are employed to enhance the breakdown and arc phases to form a stronger ignition plasma. Figure 3 gives a comparison of the basic concept of two direct-capacitor discharge methods. The high power discharge employs a small shunt capacitor (~100–200 pF) to boost the capacitive discharge mainly during the breakdown process. The high energy discharge utilizes a relatively larger capacitor (~5 μF) charged by an auxiliary power source at 600 V, storing approximately 900 mJ of energy in the capacitor.

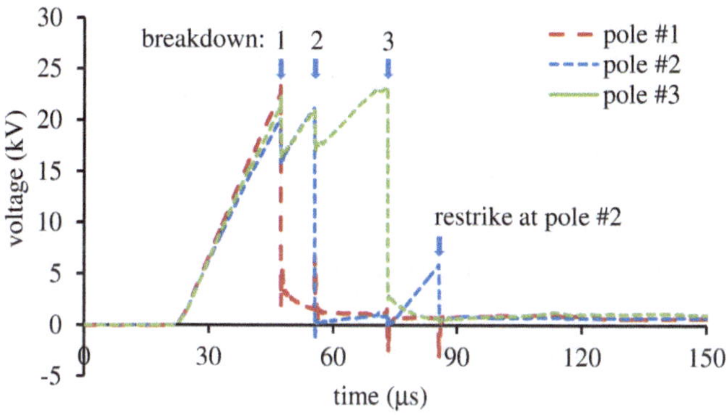

Fig. 2. Three-pole spark breakdown waveforms (N_2 at 20 bar pressure and 25 °C)

The energy stored in the capacitor will be discharged, mainly during the arc period, after a conductive plasma channel within the spark gap is established by spark break down. It is still unclear to the authors how much residual energy will be left in the capacitor after the discharge. However it is obvious that the overall energy is boosted dramatically compared to the conventional spark process. The energy value for the high energy direct-capacitor discharge in this paper is referred to the sum of the stored capacitor energy and the base spark energy. Figure 4 gives a comparison of the capacitive discharge waveforms for the conventional and direct-capacitor discharge processes. The transient power of the discharge process is also calculated based on the voltage and current. Detailed interpretations can be found elsewhere [19]. It can be seen that the transient discharge power can be boosted dramatically by the direct-capacitor. For the high energy direct-capacitor discharge, although it can produce a much higher peak current than that of the high power direct-capacitor discharge, the peak power is lower (~30 kW), because the voltage during the capacitive discharge process is relatively low. In combination with the three-pole igniter, three independent direct-capacitor discharge loops can be implemented to achieve optimal ignition performance.

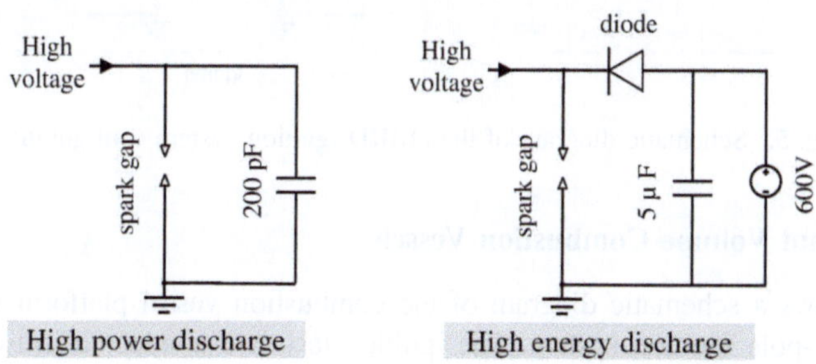

Fig. 3. Electrical circuit arrangements for the direct-capacitor discharge methods

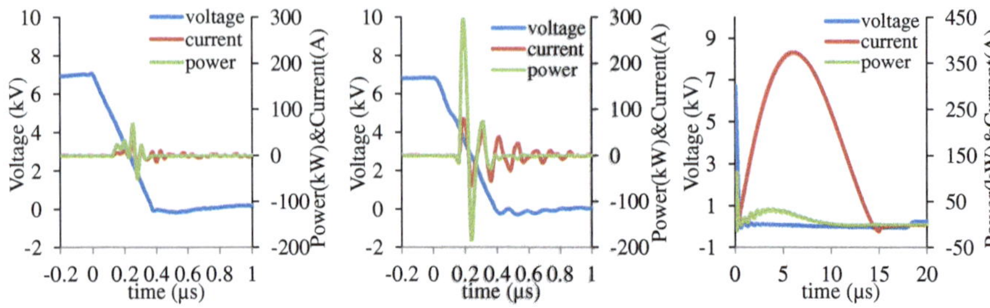

Fig. 4. Direct-capacitor discharge waveforms (one-pole in air ambient condition) [19]

With the three-pole configuration, a novel patent-pending ignition strategy based on the three-pole igniter and the direct deployment of capacitors, named elastic breakdown ignition in responsive distribution (eBIRD), can be realized as depicted in Fig. 5. In each spark gap loop a high-voltage capacitor is connected in-line, hence damping the electrical interaction between the ignition coil and the spark gap. Figure 6 shows the voltage and current waveforms acquired from one discharge event of a selected pole. Detailed descriptions can be found in [20]. It should be noted that the spark current (I1) is more than 10 times higher than the current provided from the coil. One of the advantages of the eBIRD strategy is that only one ignition coil is required for a multipole igniter. It is also feasible to use a resonant ignition coil to drive the eBIRD system, which can produce continuous spark discharges in the multiple spark gaps.

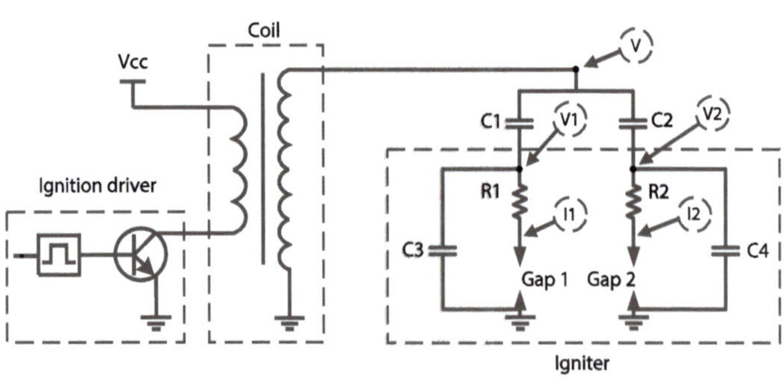

Fig. 5. Schematic diagram of the eBIRD ignition system configuration

2.2 Constant Volume Combustion Vessels

Figure 7 shows a schematic diagram of the combustion vessel platform with implemented three-pole ignition system. The ignition tests are conducted in two optically accessible constant volume combustion vessels. One chamber has a working volume of about 1.2 L with an Ø80 mm view port. The other small chamber has a working volume only about 30 mL with inner dimensions of 25 mm diameter and 50 mm length. In this paper, high-speed Shadowgraph imaging tests are conducted on the

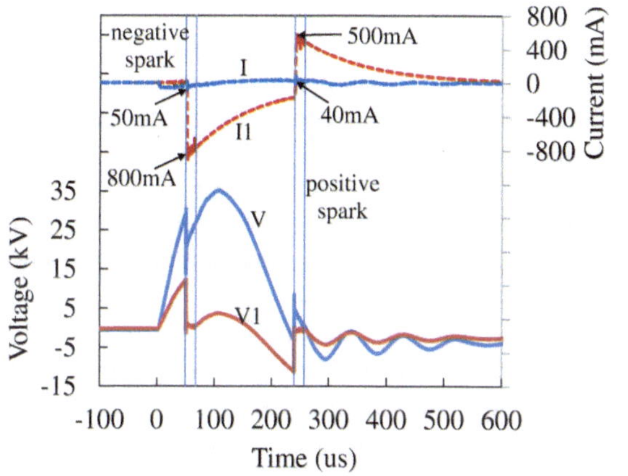

Fig. 6. Voltage and current profiles of the eBIRD ignition system (V, I: voltage and current at upstream to the in-line capacitor, V1, I1: voltage and current at downstream to the in-line capacitor, air at 6 bar pressure and 25 °C) [20]

bigger chamber to visualize the ignition flame kernel development process in a better view. Extensive repetitive ignition tests are performed in the small vessel for safety considerations. The two combustion vessel systems share the same hardware for gas

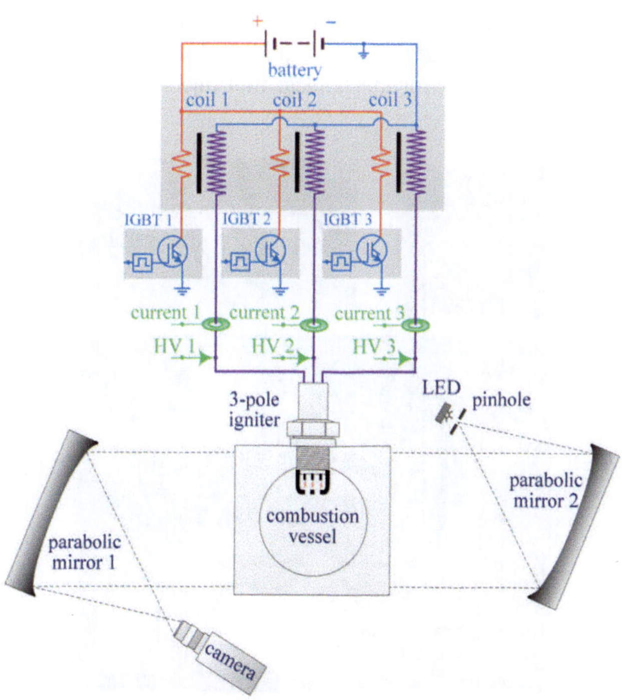

Fig. 7. Schematic diagram of the combustion vessel test platform

preparation. An Environics 4040 gas divider is employed to provide accurate concentration control of the methane (CH_4), carbon dioxide (CO_2) and air. The combustion vessels are fitted with dynamic pressure transducers for combustion pressure measurement. The combustion pressure is recorded by a data acquisition system which is externally triggered by the spark energizing command signal.

For the ignition power drive, three identical ignition coils are connected in parallel to energize the three spark gaps of the igniter. Each ignition coil is independently controlled by an individual insulated gate bipolar transistor (IGBT) chip which is driven by a specifically designed gate-drive circuit. The electronics driver circuit is built in-house. The ignition command signal for each spark event is generated from a National Instruments real-time computer and field programmable gate array setup. In order to characterize the spark discharge process, three identical wide-band current probes (Pearson 411) are employed to measure the discharge current. For the spark discharge high voltage measurement, two Tektronix P6015a probes and one Northstar PVM-6 are used. Good measurement fidelity between three high voltage probes is observed during the validation tests. The discharge voltage and current waveforms are recorded on a digital oscilloscope.

2.3 Single-Cylinder Engine Dynamometer Test

Preliminary engine tests are conducted to examine the three-pole ignition technique. The three-pole igniter is installed on the cylinder head of a modified Yanmar NFD-170 stationary engine. Figure 8 shows the pictures of the engine cylinder head and spark plug adaption. The base engine is a liquid-cooled, two valve single-cylinder direct-injection diesel engine with mechanical fuel injection. The engine is modified

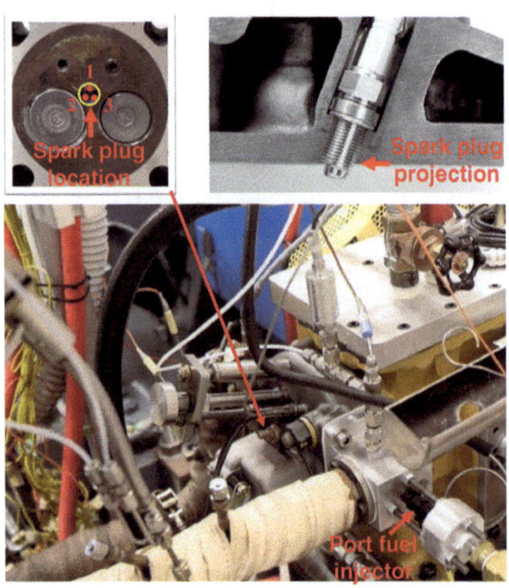

Fig. 8. Single-cylinder engine setup and spark plug adaption

to locate a centrally-mounted spark plug in place of the diesel injector. A diagram of the spark plug location, the orientation of the three electrodes, and key engine parameters are shown in Table 1. This engine is used for the experiment primarily because of its availability and not for any intrinsic advantages of the diesel engine.

Table 1. Engine specifications

Engine parameter	Value
Displacement	0.84 L
Bore	101 mm
Stroke	105 mm
Compression ratio	13:1
Port injection pressure	7 bar
Fuel	#87 gasoline

The piston shape and the intake system are modified to operate in the spark ignition mode. The omega shape combustion bowl of the original diesel combustion system is retained, but the compression ratio was reduced to 13:1 by removing material to form a shallow bowl across approximately 80 % of the piston top surface. An intake manifold is fabricated to integrate a low-pressure port fuel injector. The engine is coupled with a direct current dynamometer. The in-cylinder pressure is acquired with a pressure transducer at 0.1°CA, in synchronization with a crank-mounted optical encoder. The manifold pressure is acquired with a piezo-resistive absolute pressure transducer. Data recording of the pressures at each condition consists of 200 consecutive engine cycles. The fresh engine air intake is provided by a dry, clean compressed-air compressor system and controlled with electronic pressure regulation. A manual gate valve is used to throttle the engine intake. The fresh air flow rate is measured by a mass air flow rate meter (Roots Meter 5M175). No exhaust gas recirculation is used for these tests. The fuel flow rate is measured by a piston-type flow meter (Ono Sokki FP-213). The engine coolant circulation and temperature control is carried out by an external coolant conditioning unit and maintained to at 80 °C. The fuel used in the engine test is pump-octane #87 gasoline, with estimated research octane and motoring octane number of 91, and 83, respectively.

3 Results and Discussions

3.1 Evaluation on the Constant Volume Combustion Vessels

Optical Imaging of Ignition Flame Kernel. Figure 9 shows the high-speed shadowgraph images of the three-pole spark ignition flame kernel taken in the large constant volume combustion vessel, comparing the ignition modes including the convention allow energy, high power direct-capacitor, and high energy direct-capacitor spark discharges. The initial charge in the combustion vessel consists of quiescent λ1.6

methane-air mixtures at a pressure of 4 bar absolute and 25 °C. Images are taken of the full extent of the 80 mm optical window at 67 μs intervals and shown here in cropped view of the spark plug region at the instance of breakdown, 0.66 ms, 3.3 ms, and 6.6 ms following breakdown. The upper three rows show the cases for the one-pole ignition, with the base spark energy of approximately 50 mJ. The second and the third rows are the images for director-capacitor discharge on the same pole. The images suggest that the direct-capacitor discharge hastens the flame propagation at the early combustion stage, resulting in significant expansion of the flame kernel. Although the overall energy of the high-power direct-capacitor discharge is much lower than that of the high energy one, the resulting flame kernels are similar in size. The bottom three rows are the ignition cases with three-pole in simultaneous discharge. When the three poles, in close proximity to each other, are discharged, the resulting flame kernel is measurably larger than that of the one-pole cases. The simultaneous direct-capacitor discharge on three poles shows more prominent effect on expanding the flame kernel, compared to the one-pole cases. The three-pole high energy direct-capacitor discharge could deliver an overall energy of about 3 J in one discharge event, which results in the largest flame kernel of all cases tested.

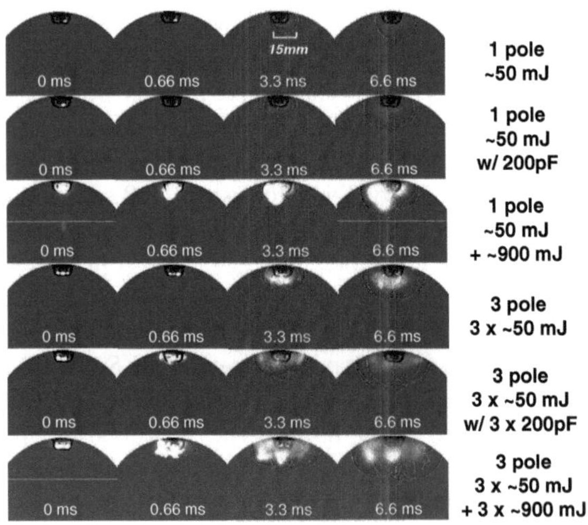

Fig. 9. High-speed shadowgraph images of ignition flame kernel (4 bar and 25 °C, λ 1.6 Methane/air)

Variations of Combustion Phasing Parameters. Combustion analysis is conducted for the repetitive ignition tests with various lambda and CO_2 concentrations. Figure 10 gives an exemplary curve of the pressure-derived MFB for a three-pole ignition case at λ 1.0. The time to the 5 % MFB (denoted by t_5 %) is employed to represent the ignition delay, while the time to the 10 %, 50 %, and 90 % MFB (denoted by t_10 %, t_50 %, and 90 % respectively) are used to indicate the combustion phasing characteristics. Combustion processes, by nature, are variable especially for diluted mixtures.

Figure 11 shows three selected cases of the combustion phasing variations for mixtures with lambda 1.0, 1.6, and 1.8. For each case, repeating ignition tests are performed with a fixed lambda but different ignition strategies including the conventional low energy discharge and direct-capacitor enhanced discharge over both one-pole and three-pole configurations. The errors of lambda, as a result of the variations during the mixture preparation process, and the slight change in combustion vessel conditions between each test are all contributors to the variations of the combustion phasing parameters. However, the stochastic nature of the combustion process under diluted condition is the dominating cause for the combustion phasing variations. According to Fig. 11, the t_10 %, t_50 %, and t_90 % exhibit good linearity to the t_5 % at λ 1.0 and λ 1.6 cases, this means that the post-ignition combustion processes are less variable at those mixture strengths. For the mixtures at λ 1.8, the linearity of t_50 % and t_90 % versus t_5 % degrades, indicating an increased variation of the post-ignition combustion process. For all three lambda cases, in general, the combustion phasing parameters consistently follow the change of t_5 % (ignition delay) that is highly varying for the diluted mixtures. The shortening of the t_5 % can be achieved by the enhancement of ignition process.

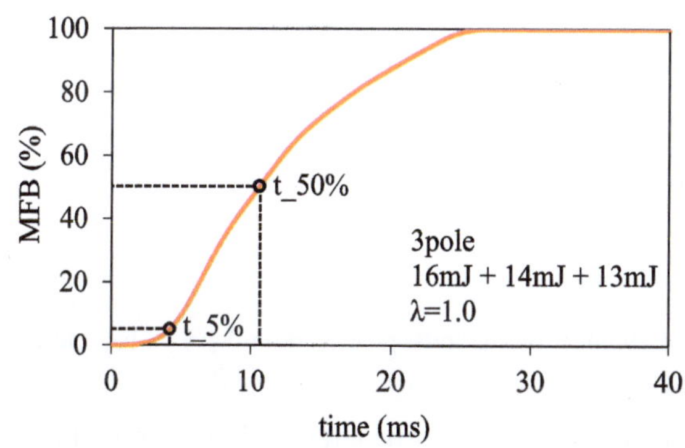

Fig. 10. Definition of time to the x% MFB based on the combustion pressure

Three-pole Ignition Delay

Spark Energy Effect on the Three-pole Ignition Delay. A commercially available spark plug (NGK zfr6fgp), with a ~0.7 mm platinum tip on center electrode and a standard J-type ground electrode, is used as a baseline for comparison with the three-pole igniter. The spark gap is adjusted to 0.86 mm the same to the gaps size of the three-pole igniter. The spark energy is varied by changing the charging duration of the primary coil and the matching of resistance in the ignition cable. In the combustion vessel tests, a non-resistance high voltage cable is used; thus the total resistance in the high voltage loop is about 10 kΩ which is mainly from the secondary winding of the ignition coil. Figure 12 compares the t_5 % time with different spark energy levels delivered into the spark gap. The data presented are averaged from 5 repeated tests for each condition, with

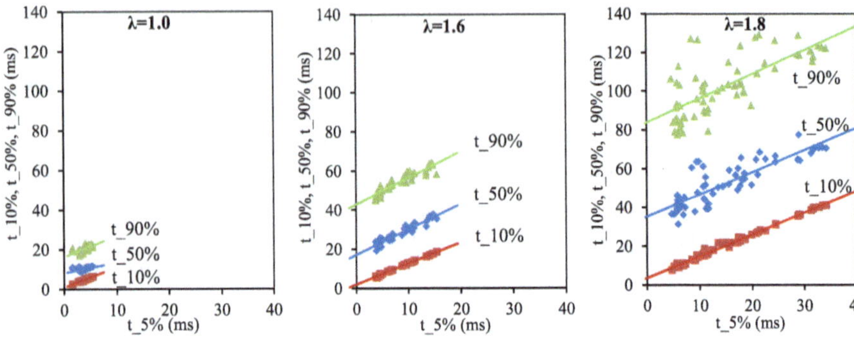

Fig. 11. Variations of combustion phasing parameters with different ignition strategies (2 bar and 25 °C, Methane/air)

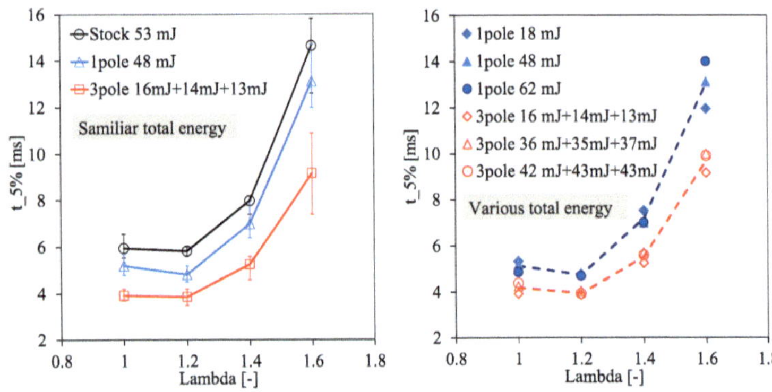

Fig. 12. Spark energy effect on the three-pole ignition delay (2 bar and 25 °C, Methane/air)

the error bar indicating the deviation. The left side figure shows the results with similar total spark energy. Although overlaps in the error bars occur at some points, the one-pole mode consistently reduces the ignition delay compared with the stock baseline spark plug, on the basis of similar total spark energy. Furthermore, the three-pole distribution of spark energy exhibits clear reduction of ignition delay throughout the lambda sweep. The right side figure compares the cases with various spark energy levels over the three-pole igniter. With the charging duration of the ignition coil set at 2 ms, 4 ms, and 5 ms, the spark energy delivered under the one-pole mode is 18 mJ, 48 mJ, and 62 mJ. Correspondingly, the total energy delivered under three-pole spark mode is 43 mJ, 108 mJ, and 128 mJ respectively. The results indicate that, within the tested spark energy range, the three-pole spark distribution clearly reduces the ignition delay throughout the lambda sweep tests, while the spark energy level has marginal effect.

Direct-Capacitor Effect on Three-pole Ignition Delay. As evidenced in Fig. 12, the increase of spark energy through prolonging the charging duration of primary coil has very marginal effect on the reduction of the ignition delay, because the basic spark discharge pattern has not been changed. The direct-capacitor discharge, in principle, enhances the spark plasma intensity, thus it is expected to be more effective in ignition improvement. Figure 13 compares the effect of direct-capacitor discharge in combination with the three-pole configuration during the lambda sweep and CO_2 dilution tests. The baseline spark energy is set at about 45 mJ, a level that would sufficiently charge the direct-capacitor for the high-power case. The results from both the lambda sweep and CO_2 dilution tests indicate agreeable trends with that observed in the optical imaging tests. The reduction of ignition delay by means of three-pole spark distribution is more prominent at leaner and higher CO_2 concentration conditions.

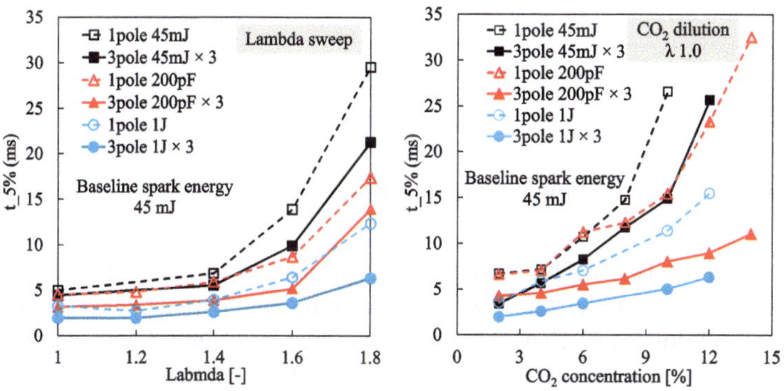

Fig. 13. Direct-capacitor effect on the three-pole ignition delay (2 bar and 25 °C, Methane/air)

Ignition Success Rate in the Combustion Vessel. Repeating tests are carried out to evaluate the ignition success rate of different ignition strategies. The statistical success rate is obtained from a series of 20 repetitive tests for each condition. Figure 14 gives the ignition success rate results for the three-pole igniter at ignitable boundary levels, for both low energy and direct-capacitor discharge. The spark energy for an individual spark gap of the three-pole igniter is set at about 62 mJ with a 5 ms charging duration in the primary coil. With the same charging duration, a simultaneous three-pole discharge delivers about 45 mJ spark energy to an individual pole and thus a total energy of about 135 mJ. The results evidence that the duplication of spark energy onto three spark gaps cannot extend the inflammable dilution limit under the stationary test condition in this combustion vessel. When the three-pole igniter is coupled with direct-capacitor, increased ignition success rate at the inflammable boundaries is observed. For the lambda sweep test, the ignition success rate is increased but the extension of the absolute inflammable limit, i.e. the boundaries to 0 % ignition success, is marginal. For the CO_2 dilution tests, in addition to the increase in ignition success rate, the inflammable limit is extended to a higher CO_2 dilution level by the combination of direct-capacitor and three-pole.

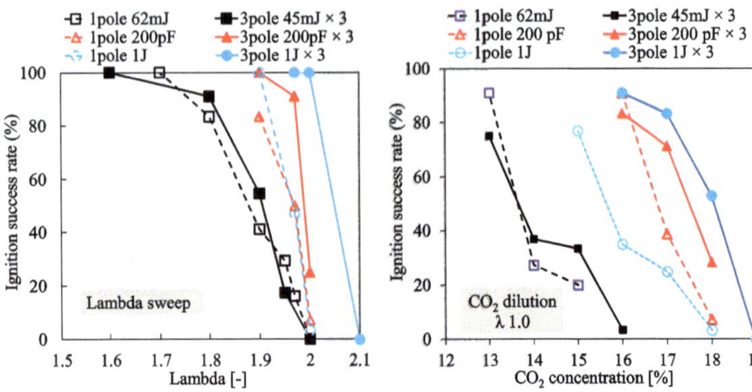

Fig. 14. Three-pole ignition success rate in the combustion vessel (2 bar and 25 °C, Methane/air)

3.2 Evaluation on the Single Cylinder Engine

Baseline Engine Tests. The baseline engine test is conducted with the stock fine-electrode spark plug, examining the spark timing and lambda effects on the engine combustion and emissions characteristics. The three-pole igniter and the tested stock spark park plug have a similar standard 19 mm thread reach that is commonly used for commercial spark plugs. The distance from the spark gap to the flat surface of the cylinder head is adjusted by the thickness of the sealing washer for the spark plug. The projection depth of the spark gap into the cylinder is set at about 8 mm for all the tests, which sufficiently avoids the quenching effect of the boundary layer on the initiation of a flame kernel. The spark plug is mounted through the original diesel injector hole that has been enlarged to fit the M14 spark plug diameter. The orientation angle of the spark gap is fixed for all tests, keeping the same as one of the spark gaps of the three-pole igniter. A few sets of spark timing sweep tests at a fixed lambda of 1.35 are conducted to examine the effect of the orientation angle of the stock spark gap, but there is no noticeable effect observed. The three poles are also examined in turn under one-pole mode; the results indicate marginal difference between the poles. This can be found in the other paper published by the authors' lab members [21]. The spark plug projection depth and the orientation angle chosen in the tests might not be optimal for this specific combustion system, but the baseline can be set, as the engine runs smoothly in a wide range of spark timings and lambda sweeps. The tests are all running at a constant fuel metering of about 19.6 mg/cycle to produce a similar load level, while the lambda is changed by increasing the intake air admission.

Figure 15 shows the spark timing adjustment effect of the engine at a fixed lambda of 1.35, by using the stock spark plug and the one-pole mode of the three-pole igniter. The advance of the spark timing is typically not limited by knocking at the experiment load and dilution levels. The thermal efficiency, however suffers with excessive heat-release in advance of TDC. The retardation is stopped when the IMEP has a substantial drop or an unacceptable coefficient of variation (COV) of the IMEP is observed. Responsive

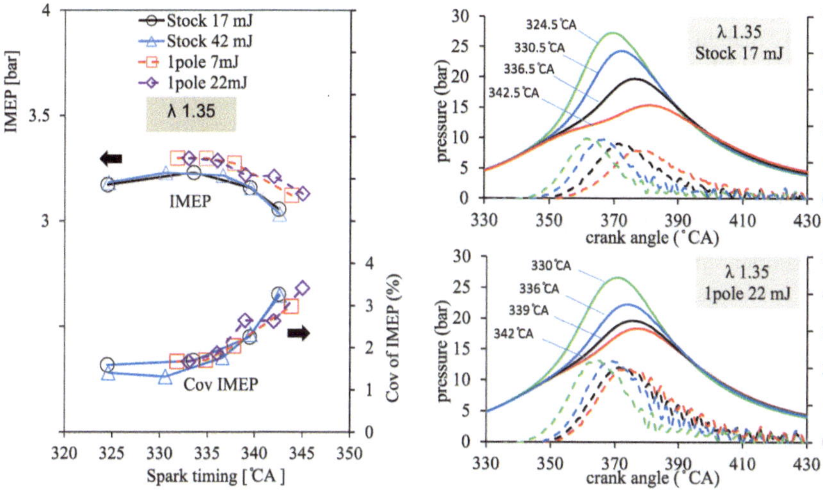

Fig. 15. Combustion phasing adjustment with spark timing

combustion phasing change is observed for both the stock spark plug and the one-pole cases. The cylinder pressure and the heat release rate (HRR) for the stock spark plug and the one-pole mode with close spark energy levels, are shown in the right side of Fig. 15. Different spark energy levels are also tested, showing no significant effect on the power output and engine stability at this mixture strength, as shown in Fig. 15. Compared with the baseline stock spark plug, the one-pole ignition provides at least a similar level in terms of the IMEP and the COV of IMEP. The comparisons made hereafter will mainly be based on the results acquired from the three-pole igniter.

Effect of Three-pole Spark Distribution on Combustion Phasing. The three-pole igniter offers a possibility to create three ignition sites simultaneously in the igniter proximity, in addition to being capable of delivering larger amount of total spark energy. In Fig. 16, comparisons are made on the basis of delivering a similar amount of total spark energy to examine the effect of spark distribution. The tests are performed at two mixture strengths at lambda 1.35 and 1.5 respectively. The spark timings are swept to vary the CA50. The parameters examined include the IMEP, COV of IMEP, standard deviation (STD) of CA5, STD of CA50, ignition delay (crank angle between spark timing and CA5), and the combustion duration (crank angle between CA5 and CA95). The curves are plotted with those combustion parameters versus CA50. A minimal spark energy of 7 mJ per pole is applied for lambda 1.35, while an increased energy of 12 mJ per pole is used for lambda 1.5. For the 1pole ignition mode at lambda 1.35, ignition tests with 7 mJ per pole and the 22 mJ per pole show marginal difference in terms of the engine power output and combustion stability, except at a late CA50, the ignition delay and combustion duration are both slightly shortened. The three-pole simultaneous ignition, with 7 mJ per pole and a total energy of about 21 mJ, shortened the ignition delay significantly. The STD of CA5 and CA50 are substantially decreased by the three-pole spark distribution. For the tests at lambda 1.5, similar trends between one-pole and three-pole ignition tests are observed. The one-pole mode with 12 mJ spark energy can barely maintain the engine operation under the test conditions, with much deteriorated engine stability and late CA50. If look at the exact value of the STD of CA5 and CA50, the reduction by the three-pole ignition at lambda 1.5 is more prominent than that at lambda 1.35. Although the IMEP and the COV of IMEP are insignificantly affected by the ignition strategies over a wide range of CA50 for both 1ambda 1.35 and 1.5, the combustion phasing stability is substantially improved by the spark distribution. The reason for the more stable combustion phasing yielded by the three-pole ignition may lie in the shorter ignition delay that allows closer TDC spark timings offering higher cylinder pressures and temperatures for early stage combustion. This is of critical importance for combustion control under lean operation, especially when the mixture strength is push to extremely lean for achieving the possibly best fuel efficiency.

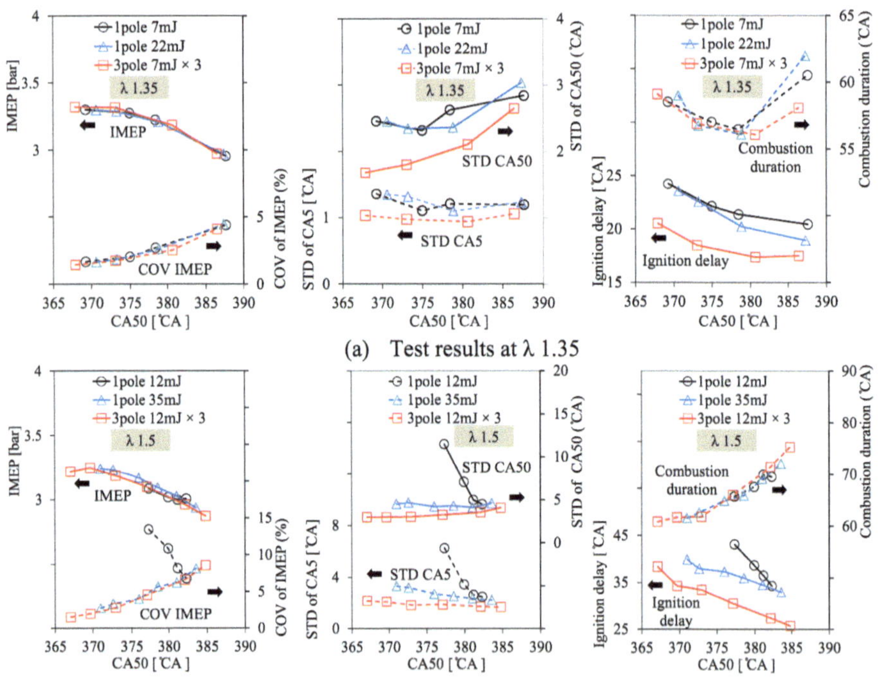

(a) Test results at λ 1.35

(b) Test results at λ 1.5

Fig. 16. Effect of spark energy distribution on engine combustion parameters

Combustion Phasing Control under Ultra-lean Conditions. When pushing the mixture strength towards ultra-lean, e.g. targeting about λ 2.0, the CA50 becomes increasingly delayed from the desired operating window. Figure 17 shows the combustion phasing control during the transition of mixture strength from λ 1.6 to λ 2.12. The one-pole ignition mode uses a spark energy of about 45 mJ that requires almost saturation of the tested ignition coil. Although the coil can store a much larger amount of energy, the discharge energy delivered onto the spark gap is limited due to the energy dissipation on the resistance in the coil winding and along the ignition cable, about 20 kΩ in total in this case. The three-pole ignition mode in this set of tests utilizes the total energy about three times higher than that of the one-pole ignition mode, i.e. about 45 mJ per pole and a total energy of about 135 mJ. This is at the maximum energy level provided by the ignition system setup. As indicated by Fig. 17, the one-pole and three-pole modes perform similarly at λ 1.6 in terms of the CA50 control. When pushing to leaner mixtures at λ 1.84, the one-pole mode exhibits significantly delayed CA50 compared to the three-pole mode. Beyond λ 1.84, the engine is not operable with one-pole mode. The three-pole mode can still operate the engine at λ 2.06, but with excessively late CA50 and poor stability. It should be noted that at λ 2.12 the engine is still ignitable with the three-pole mode despite very poor combustion efficiency and stability.

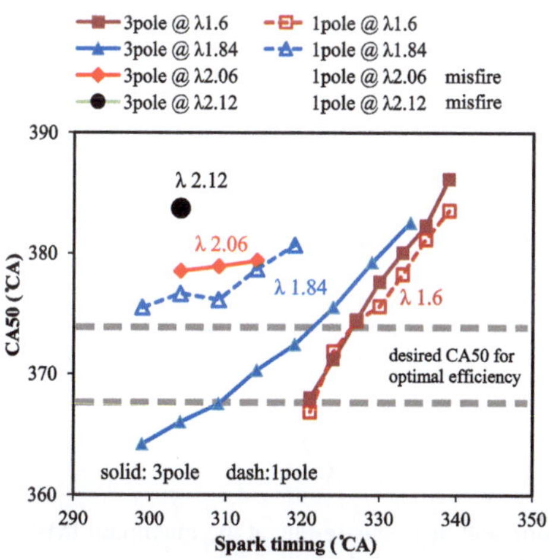

Fig. 17. Combustion phasing controllability of three-pole ignition under ultra-lean conditions

Figure 18 compares the engine combustion parameters with one-pole and three-pole ignition modes at λ 1.84. The CA50 of the three-pole ignition mode is quite responsive to the spark timing, but the one-pole ignition mode is not able to advance the CA50 to the desired optimal efficiency window. The three-pole mode can provide stable combustion phasing, indicated by the low STD of CA5 and CA50 across the spark timing sweep. The lowest COV of IMEP under three-pole mode is 1.9 % at a CA50 of 366.5 °CA where a maximum IMEP is produced. Figure 19 compares the cylinder pressure and heat release rate curves at the maximum IMEP points for both one-pole and three-pole modes. The spark timings are at 309 °CA, the same for the two cases, while the resultant CA50 for the one-pole and three-pole modes are 367.5 °CA and 376.2 °CA respectively. There is no noticeable difference in either the pressure or the HRR curve between the two ignition modes at before about 335 °CA, which is about 2 ms after the initiation of spark discharge. Starting from about 335 °CA, the heat

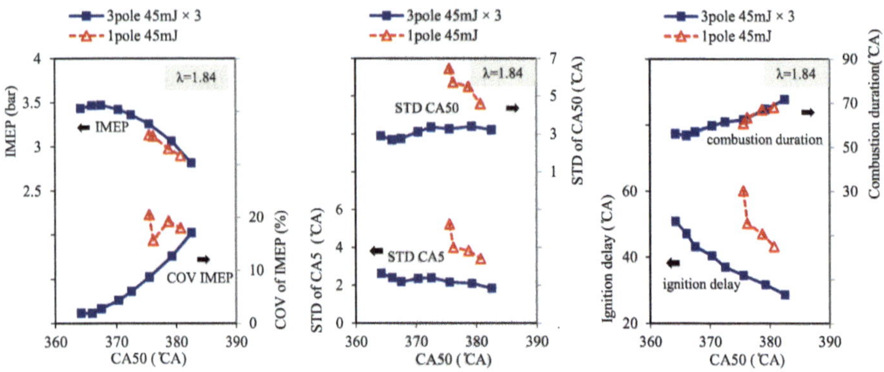

Fig. 18. Comparison of combustion parameters with one-pole and three-pole modes at λ 1.84

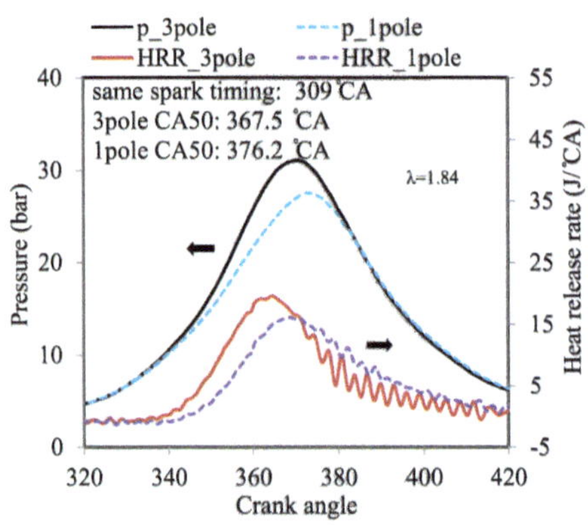

Fig. 19. Cylinder pressure and heat release rate at the maximum IMEP points for one-pole and three-pole modes

release of the three-pole ignition is significantly faster than that of the one-pole ignition. Shorter combustion durations are obtained with the three-pole ignition mode because of its ability to phase the combustion phasing in the appropriate range, which benefits fuel efficiency.

The Extension of Operable Lean Limit. By implementing ultra-lean operation, the engine efficiency is improved from stoichiometric combustion. The achievement can be assisted by enabling stable leaner operation through ignition enhancement. As afore-mentioned, the three-pole ignition can improve the combustion phasing control for the ultra-lean engine operation, thus the extension of the operable lean limit and its benefits to engine fuel efficiency are expected. Figures 20 and 21 present the data at the maximum brake torque (MBT) spark timings for each ignition strategy at the lean operable boundaries. Here, the MBT points refer to the maximum IMEP production. The stock fine-electrode spark plug, with the maximum spark energy provided, is used as baseline for comparison. With the COV of IMEP maintained below 3 %, the

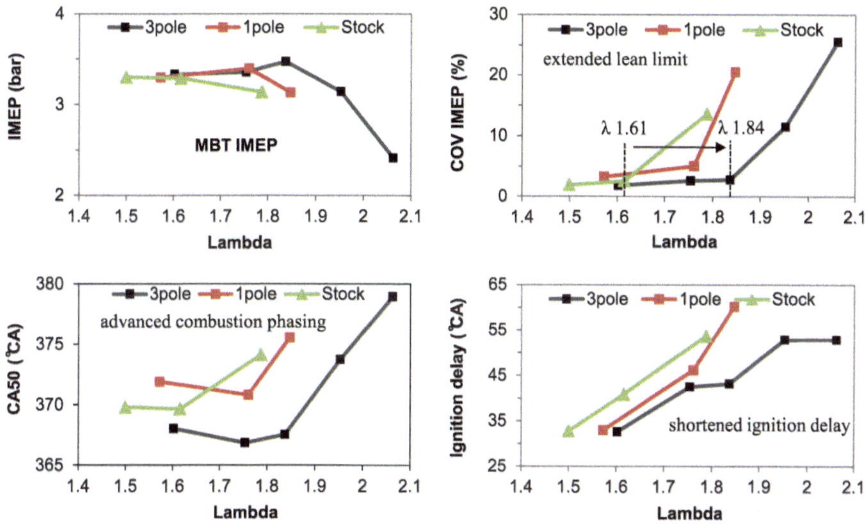

Fig. 20. Engine combustion parameters at the MBT points

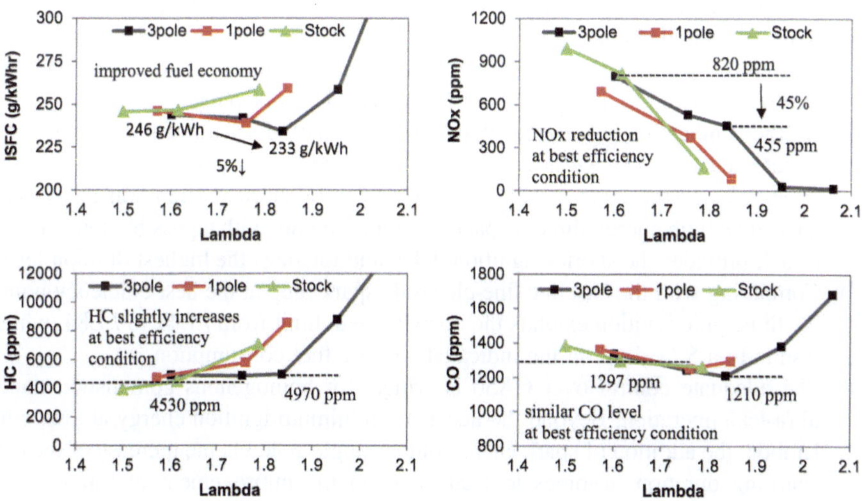

Fig. 21. Engine specific fuel combustion and emissions at the MBT points

three-pole ignition extends the lean operable limit from the baseline at λ 1.61 to λ 1.84, attributing to the adequate control over CA50. Owing to the extension of the operable lean limit, the indicated specific fuel consumption is decreased from the baseline 246 g/kWh to 233 g/kWh, which is about a 5 % drop. Correspondingly, the NOx emissions at the best efficiency points decrease substantially due to the lowered combustion temperature of the leaner mixture. For the incomplete combustion products, the HC emissions increase slightly while the CO is at a similar level. With the current ignition system and engine setup, stable operation beyond λ 2.0 is unachievable due to the deteriorated combustion efficiency.

4 Future Works

Future combustion vessel tests need to be conducted with elevated pressure and temperatures. The three-pole direct-capacitor ignition showed promising results on the combustion vessel tests, but it hasn't been tested on the engine yet. Future work will include the implementation of the direct-capacitor ignition on the engine tests. Furthermore, the tests at higher engine loads and rpm need to be conducted.

5 Conclusions

A three-pole distribution spark ignition system is introduced in this paper, and its effects on ignition improvements are preliminarily evaluated on both constant volume combustion vessels and a single-cylinder engine. Within the limited test conditions, conclusions can be drawn as below.

(1) Both the combustion vessel and engine tests indicate a clear trend that, on the basis of a similar total spark energy level, the three-pole spark distribution can shorten the ignition delay, stabilize the combustion phasing, and extend the lean limits, compared to a conventional fine-electrode spark plug.
(2) On the combustion vessel, the effectiveness of spark discharge enhancement with direct-capacitors can be augmented through simultaneously discharging of the three-pole igniter, hence resulting in significantly shortened ignition delay. The three-pole high energy direct-capacitor ignition mode, with a possible total energy of 3 J, provides the shortest ignition delay and tolerates the highest dilution level.
(3) Comparing with the baseline fine-electrode spark plug at the best efficiency point, the three-pole ignition extends the operable lean limit from λ1.61 to λ1.84, which results in a 5 % drop on the indicated specific fuel consumption.
(4) The adequate control over CA50 is critical for homogenous combustion under ultra-lean operation. Beyond the necessary minimum ignition energy at a specific lambda, the addition of spark energy onto a single-pole via the increase of the coil charging duration becomes less effective on the improvement of ignition and combustion phasing.

Definitions/Abbreviations

EGR	Exhaust Gas Recirculation
MFB	Mass Fraction Burnt (%)
CA	Crank Angle (℃A)
CA5	Crank Angle of 5 % MFB (℃A)
CA50	Crank Angle of 50 % MFB (℃A)
CA95	Crank Angle of 95 % MFB (℃A)
TDC	Top Dead Center

ATDC	After Top Dead Center
BTDC	Before Top Dead Center
CDI	Capacitive Discharge Ignition
CO_2	Carbon Dioxide
CH_4	Methane
NOx	Nitrogen Oxides
HC	Hydrocarbons
CO	Carbon Monoxide
AWG	American Wire Gauge
IGBT	Insulated Gate Bipolar Transistor
IMEP	Indicated Mean Effective Pressure (bar)
HRR	Heat Release Rate (J/℃A)
MBT	Maximum Brake Torque
ISFC	Indicated Specific Fuel Consumption (g/kWh)
COV	Coefficient of Variation (%)
STD	Standard Deviation
γ	Ratio of specific heats
λ	Excess air ratio
t_5 %	Time to 5 % MFB (ms)
t_10 %	Time to 10 % MFB (ms)
t_50 %	Time to 50 % MFB (ms)
t_90 %	Time to 90 % MFB (ms)

Acknowledgements. The research at the Clean Combustion Engine Laboratory is sponsored by NSERC, AUTO21, BiofuelNet, Canada Foundation for Innovation, Ontario Research Funds, the University of Windsor, and Ford Motor Company.

References

1. Wei, H., Zhu, T., Shu, G., Tan, L., et al.: Gasoline engine exhaust gas recirculation– a review. Appl. Energy **99**, 534–544 (2012). doi:10.1016/j.apenergy.2012.05.011
2. Gallon, E., Fontana, G., Palmaccio, R.: Effects of exhaust gas recycle in a downsized gasoline engine. Appl. Energy **105**, 99–107 (2013). doi:10.1016/j.apenergy.2012.12.046
3. Takahashi, D., Nakata, K., Yoshihara, Y., Ohta, Y., et al.: Combustion development to achieve engine thermal efficiency of 40 % for hybrid vehicles. SAE Technical Paper 2015-01-1254 (2015). doi:10.4271/2015-01-1254
4. Zheng, M., Tan, Y., Mulenga, M., Wang, M.: Thermal efficiency analyses of diesel low temperature combustion cycles. SAE Technical Paper 2007-01-4019 (2007). doi:10.4271/2007-01-4019
5. Gukelberger, R., Alger, T., Gingrich, J., Mangold, B.: Impact of operating parameters on ignition system energy consumption. SAE Technical Paper 2014-01-1233 (2014). doi:10.4271/2014-01-1233
6. Hampe, C., Kubach, H., Spicher, U., et al.: Investigations of ignition processes using high frequency ignition. SAE Tech Paper, 2013-01-1633 (2013). doi:10.4271/2013-01-1633

7. Shiraishi, T., Urushihara, T., Gundersen, M.: A trial of ignition innovation of gasoline engine by nanosecond pulsed low temperature plasma ignition. J. Phys. D Appl. Phys. **42**, 135208 (2009). doi:10.1088/00223727/42/13/135208
8. Sjöberg, M., Zeng, W., Singleton, D., Sanders, J., et al.: Combined effects of multi-pulse transient plasma ignition and intake heating on lean limits of well-mixed E85 DISI engine operation. SAE Int. J. Engines **7**(4), 1781–1801 (2014). doi:10.4271/2014-01-2615
9. Rohwein, G.J.: An efficient power-enhanced ignition system. IEEE Trans. Plasma Sci. **25**(2), 306–310 (1997). doi:10.1109/27.602504
10. Yoshida, K., Shoji, H., Tanaka, H.: Performance of newly developed plasma jet igniter. SAE Technical Paper 1999-01-3327 (1999). doi:10.4271/1999-01-3327
11. Dale, J.D., Checkel, M.D., Smy, P.R.: Application of high energy ignition systems to engines. Prog. Energy Combust. Sci. **23**(5–6), 379–398 (1997). doi:10.1016/S0360-1285(97)00011-7
12. Heise, V., Farah, P., Husted, H., Wolf, E.: High frequency ignition system for gasoline direct injection engines. SAE Technical Paper 2011-01-1223 (2011). doi:10.4271/2011-01-1223
13. Alger, T., Gingrich, J., Mangold, B., Roberts, C.: A continuous discharge ignition system for EGR limit extension in SI engines. SAE Int. J. Engines **4**(1), 677–692 (2011). doi:10.4271/2011-01-0661
14. Heywood, J.B.: Internal combustion engine fundamentals, pp. 435–437 (1988)
15. Hall, M., Matthews, R., Ezekoye, O.: Railplug ignition operating characteristics and performance: a review. SAE Technical Paper 2007-01-1832 (2007). doi:10.4271/2007-01-1832
16. Rohwein, G., Camilli, L.: Automotive ignition transfer efficiency. SAE Technical Paper 2002-01-2839 (2002). doi:10.4271/2002-01-2839
17. Suess, M., Guenthner, M., Schenk, M., et al.: Investigation of the potential of corona ignition to control gasoline homogeneous charge compression ignition combustion. P. I. Mech. Eng. D-J. Aut. **226**(2), 275–286 (2011). doi:10.1177/0954407011416905
18. Stevens, C.A., Pertl, F.A., Hoke, J.L., et al.: Comparative testing of a novel microwave ignition source, the quarter wave coaxial cavity igniter. J. Phys. D Appl. Phys. **225**(12), 1633–1640 (2011). doi:10.1177/0954407011411389
19. Yu, S., Xie, K., Yu, X., Wang, M., Zheng, M., Han, X., Tjong, J.: High energy ignition strategy for diluted mixtures via a three-pole igniter. SAE Papers 2016-01-2175 (2016)
20. Yu, S., Wang, M., Zheng, M.: Distributed electrical discharge to improve the ignition of premixed quiescent and turbulent mixtures. SAE Technical Paper 2016-01-0706 (2016). doi:10.4271/2016-01-0706
21. Xie, K., Yu, S., Zheng, M.: Investigation of multi-pole spark ignition on flame kernel development and in engine operation. In: Proceedings of the ASME 2016 Internal Combustion Fall Technical Conference, 9–12 October 2016, Greenville, SC, USA, ICEF2016-9474 (2016)

Development of Homogeneous Charged Multi-point Ignition Engine

Katsuaki Minami(✉)

Miyama Inc., 1-5-3, Inasato, Nagano-Shi, Nagano, Japan
masataka.aoki@iav.jp

Abstract. Fast combustion by devising an ignition device for simultaneous ignition from eight positions around a conventional combustion chamber, in addition to an ignition plug arranged in the center of the combustion chamber was realized. A visualized experiment on the state of combustion by mounting this into a single-cylinder 2-stroke engine was conducted. And confirmed that combustion had been stabilized to a great extent by multi-point ignition. Subsequently, the ignition device was further applied to a 4-cylinder 4-cycle engine. Optimal ignition timing was retarded by more than 10°, and torque was also increased in the wide air-fuel ratio region, compared to the case of conventional single point ignition. The lean limit was also increased by a factor of 4 in the air-fuel ratio by this fast combustion. Further, the anti-knocking property was improved, making it possible to increase the compression ratio. By the reduction in time loss and the synergetic effect of air-fuel ratio leaning and high compression ratio caused by fast combustion, thermal efficiency achieved 44.49 %.

1 Introduction

The thermal efficiency of Spark-Ignition (SI) engines originating from an Otto cycle does not come close to that of automotive diesel engines similar to a Sabathe cycle, which are supposed to be theoretically inferior. In a conventional SI engine, one ignition plug is generally arranged in the center of the combustion chamber, and the gaseous mixture present in the combustion space is burned by propagation of the flame from the ignition plug. The flame propagation time taking is the main cause as to why thermal efficiency of a theoretical Otto cycle cannot be realized. Flame propagation time results in an increase in time loss, and the occurrence of knocking, thereby narrowing the limit of air-fuel ratio leaning.

To reduce time loss and realize both an output and a thermal efficiency improvement in a homogeneous charged multi-point 4-cycle engine, we devised an ignition system for performing ignition from an additional plurality of point around the combustion space, in addition to the center of the combustion chamber. We visually confirmed the effects of this fast combustion system using a single-cylinder engine, and subsequently quantitatively analyzed the thermal efficiency improvement effect using a 4-cylinder 2-liter engine.

M. Günther and M. Sens (eds.), *Ignition Systems for Gasoline Engines*,
DOI 10.1007/978-3-319-45504-4_7

2 System Configuration of Multi-point Ignition

An ignition device is mounted between the cylinder head and a cylinder block in the form of a head gasket for ignition from the periphery of a combustion space, defined by a combustion chamber of a cylinder head and the crown surface of a piston, as shown in Fig. 1. The number of ignition gaps in the periphery is 8; these ignition gaps are coupled in series, and a secondary current is supplied from an ignition coil.

In this way, ignition from the periphery of the combustion chamber was enabled in addition to the conventional ignition from the center. This ignition device in the form of a head gasket can realize multi-point ignition without restrictions on the arrangement of intake valves, exhaust valves and the shape of the combustion chamber.

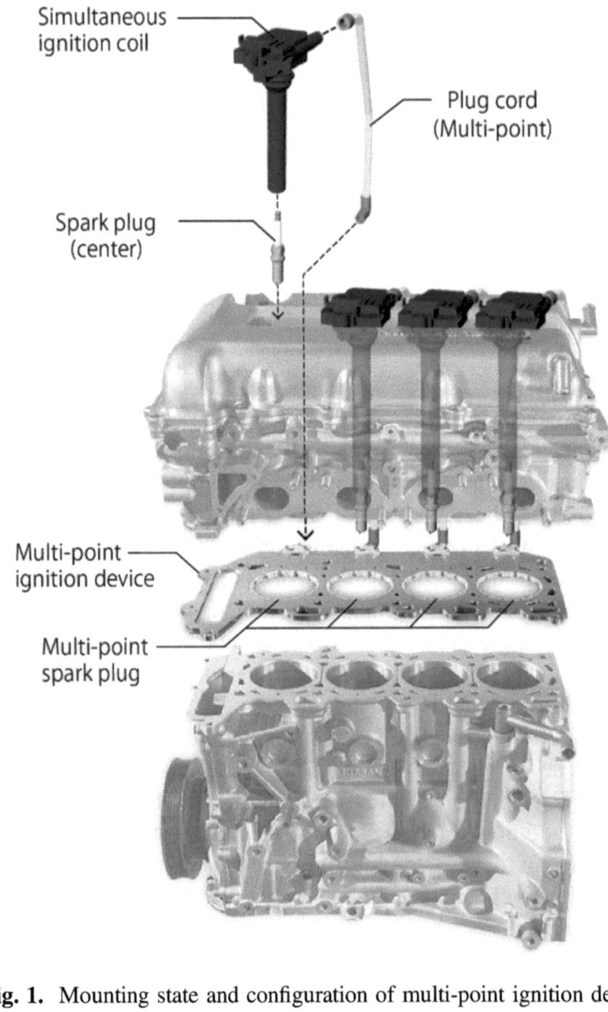

Fig. 1. Mounting state and configuration of multi-point ignition device

3 Realization of Fast Combustion by Multi-point Ignition

Prior to experimenting using a 4-stroke engine, the effect caused on fast combustion by multi-point ignition was visibly verified. A photographic comparison was made using a high-speed camera with all the conditions except the number of ignition point set equal, and a 2-stroke 250cc single-cylinder engine having a head altered to a transparent window made of reinforced glass. A comparison of results is shown below in Fig. 2.

It should be noted that, in an experimental engine, the ignition timing was set at BTDC26° more advanced than normal, assuming a 4-stroke engine, and the compression ratio was also changed to 9.0 (Table 1).

Table 1. Visualized engine specifications

Type	2stroke single cylinder	Intake system	Reed valve
Displacement	249.7cc	Scavenging	Schnurle loop scavenging
Bore × Stroke	72.8 mm × 60 mm	Air Fuel Ratio	14.7
Ignition point	8 + 1	Ignition timing	26°BTDC (fixed)
Compression ratio	9.0:1	Scavenge port open	ATDC125°
Engine speed	3500 rpm	Exhaust port open	ATDC105°

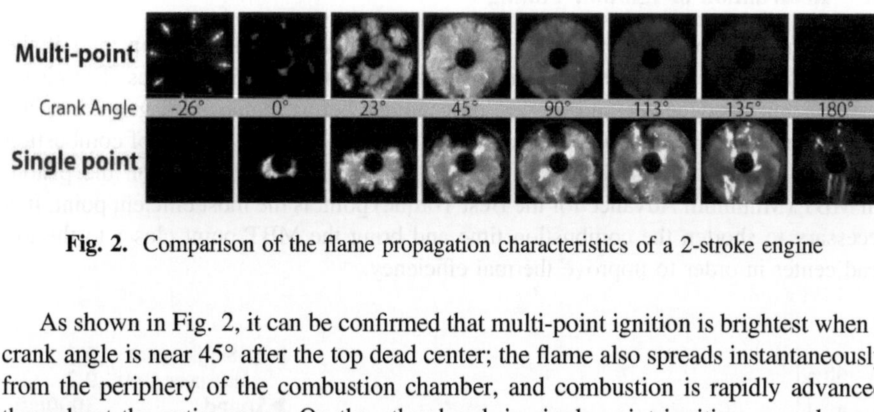

Fig. 2. Comparison of the flame propagation characteristics of a 2-stroke engine

As shown in Fig. 2, it can be confirmed that multi-point ignition is brightest when a crank angle is near 45° after the top dead center; the flame also spreads instantaneously from the periphery of the combustion chamber, and combustion is rapidly advanced throughout the entire space. On the other hand, in single point ignition, an unburned portion to where flame does not reach is shown to be dark near 45° after the top dead center, and the brightness of multi-point ignition is not seen even in a state of combustion, due to continuing combustion even near 135°. Further, even near 180°, combustion is not completed, and gas being burned, which looks bright in parts, is discharged as it is.

By multi-point ignition performing ignition not only from the center, but also from the circumference of the combustion chamber, fast combustion is realized even within a 2-stroke engine.

4 Performance Evaluation of a Multi-point Ignition Engine

The effect of an improvement in thermal efficiency when the multi-point ignition of an engine applied to a 4-stroke was analyzed, using an in-line 1998cc four-cylinder SR20DE engine produced by Nissan Motor Co., Ltd. When conducting the experiment, in order to suppress the increase of the combustion chamber volume brought about by interposing an ignition device in the form of a gasket, the bottom surface of a cylinder head for the base engine was polished. A change was made to the ignition coil for a multi-point ignition device and, in addition, an ECU was changed to a MoTeC M800 to control the ignition timing and fuel injection amount (Table 2).

Table 2. Engine specification

Type	SR20DE (Modifi)
Displacement	1998cc
Bore × Stroke	86 mm × 86 mm
Cylinder layout	Inline 4 cylinder
Compression Ratio	10 → 9.2
Ignistion point	8 + 1

4.1 Retardation of Ignition Timing

In a SI engine, combustion serves as resistance to an ascending piston, and if the ignition timing is early, a risk of inducing knocking and the like increases. The piston passes a top dead center and begins descending, before flame propagates into the combustion chamber and, thus, if the ignition timing is late, the volume of combustion space increases, and this change in pressure does not effectively act on the piston. An MBT (Minimum Advance for the Best Torque) point is the most efficient point. It is necessary to shorten the combustion time and bring the MBT point closer to the top dead center in order to improve thermal efficiency.

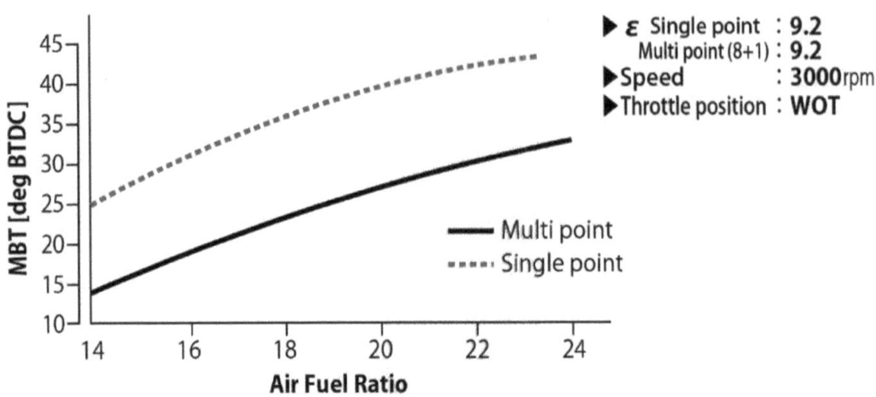

Fig. 3. Retardation effect of MBT by multi-point ignition

The results of the experiment in which the engine was operated with 3000 rpm on a dynamometer and WOT (Wide Open Throttle), and a comparison of MBT point of single point ignition and multi-point ignition are shown in Fig. 3. In the case of multi-point ignition, the MBT point is retarded by 10 to 15° in the entire air-fuel ratio region, and approaches the top dead center to a greater extent. From this, it can be confirmed that multi-point ignition causes fast combustion in the 4-stroke engine, as in the 2-stroke engine.

4.2 Realization of Lean Combustion

Fast combustion by multi-point ignition, and retardation of the required ignition timing associated with it, contributes heavily to lean combustion. The volume of combustion space of a reciprocal engine is smallest when at the top dead center. The combustion space refers to a space defined by a piston crown surface, a combustion chamber wall surface on a cylinder head side, and the upper inner surface of the cylinder. The retardation of the MBT point means a decrease in the volume at the time of ignition. The decrease in volume of combustion space leads to stable combustion during lean combustion, in which the amount of generated heat in the cylinder is reduced, due to an increase in gaseous mixture density at the time of ignition, a reduction in distance between fuel molecules, and an increase of a gaseous mixture temperature caused by adiabatic compression.

When operated at 3000 rpm, a WOT and an air-fuel ratio (A/F) 16 as shown in Fig. 4, MBT is 31° BTDC in the case of a single point ignition and 19° BTDC in the case of multi-point ignition, respectively. The crank angle and the volume of combustion space are shown as in Fig. 4, and there is a difference 1.36-fold in volume of combustion space at each MBT point, between multi-point ignition and single point ignition.

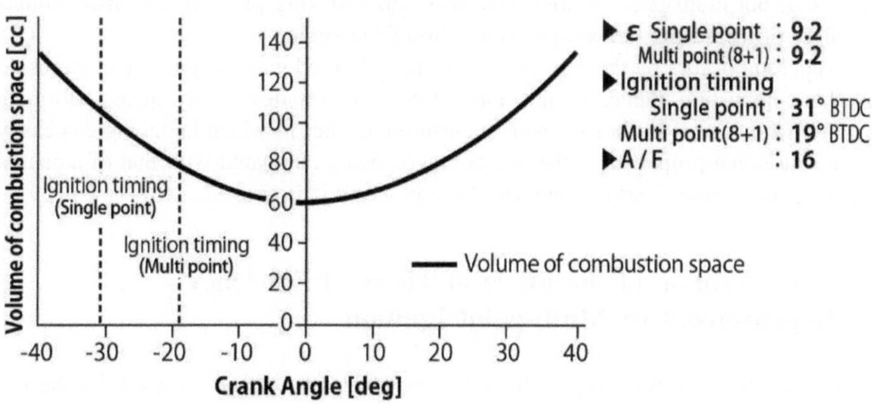

Fig. 4. Ignition timing of volume of combustion space reduce by multi-point ignition

A comparison of the brake torque and lean limit at each air-fuel ratio of a single point ignition and multi-point ignition under the same conditions as in the aforementioned experiment is shown in Fig. 5.

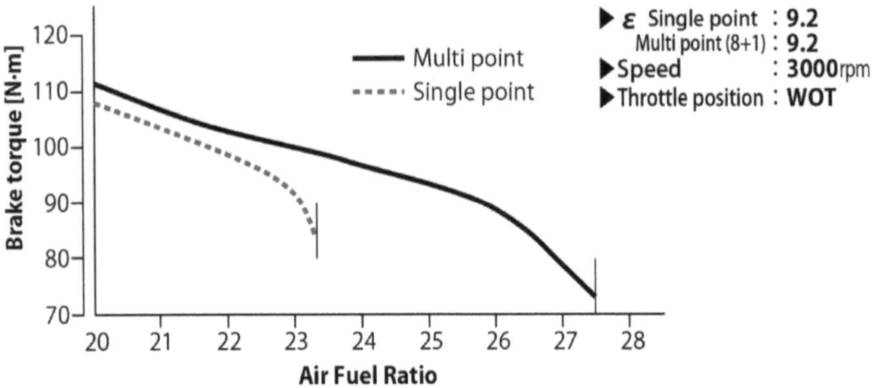

Fig. 5. Increase of torque by multi-point ignition and expansion of lean limit

The air-fuel Ratio (A/F) when a brake torque of 100 N·m is generated is A/F21.5 in single point ignition, whereas it is A/F 23 in multi-point ignition. As such, it can be understood that the air-fuel ratio is leaner when the same brake torque is generated. Further, even when the torque is increased within the whole air-fuel ratio region to a value less-lean than this, a lean limit is also increased to A/F27 as compared to A/F23 in single point ignition, and it can be confirmed that multi-point ignition has superiority in realizing lean combustion.

4.3 Realization of High Compression Ratio by a Multi-point Ignition

A higher compression ratio of an engine increases the risk of the occurrence of knocking, but multi-point ignition enhances anti-knocking property and also exhibits an effect of increasing the compression ratio of the engine.

Knocking occurs by the self-ignition of the unburned gaseous mixture compressed by the combustion flame. In this multi-point ignition device, by also performing ignition from a wall surface of the combustion chamber in which knocking can easily occur, the flame propagation distance is halved when compared with that of a central single point ignition, and the anti-knocking property is improved.

5 Consideration of the Effect of Thermal Efficiency Improvement by Multi-point Ignition

Theoretical thermal efficiency of the Otto cycle with perfect gas is given by the following formula. An increase of the compression ratio and specific heat ratio are also indispensable for the improvement of thermal efficiency in SI engines originating from the Otto cycle.

$$\eta_{th} = 1 - \frac{1}{\varepsilon^{\kappa-1}}$$

η_{th}: theoretical thermal efficiency [%]
ε: compression ratio
κ: specific heat ratio of gas

Fast combustion by multi-point ignition is thought to reduce time loss, and largely contribute to increases of the compression ratio and specific heat ratio by leaning of the engine, and realizing high thermal efficiency by a synergetic effect of these.

Further, the SR20DE engine was changed to have the optimal shape for multi-point ignition, and an experiment was conducted in order to verify the thermal efficiency improving effect brought about by multi-point ignition (Table 3). In the engine, the upper surface of the cylinder head and the bottom surface of the cylinder block were grinded, a piston having a convex-shaped crown surface shaped was incorporated, and the compression ratio was changed from 9.2 to 12.9. Operation conditions of the engine were 3000 rpm with a WOT in both cases.

Table 3. Engine specifications

Type	SR20DE (Modifi)
Displacement	1998cc
Bore × Stroke	86 mm × 86 mm
Cylinder layout	Inline 4 cylinder
Compression Ratio	9.2 → 12.9
Ignistion point	8 + 1

A comparison of gas pressure in cylinder waveforms of multi-point ignition engine with a high compression ratio and a single point ignition at A/F20 is shown in Fig. 6. A maximum gas pressure in cylinder (Pmax) was 4.15 MPa in single point ignition, but increased to 7.74 MPa in multi-point ignition, and Pmax including a compression ratio increasing effect was improved by 86 %.

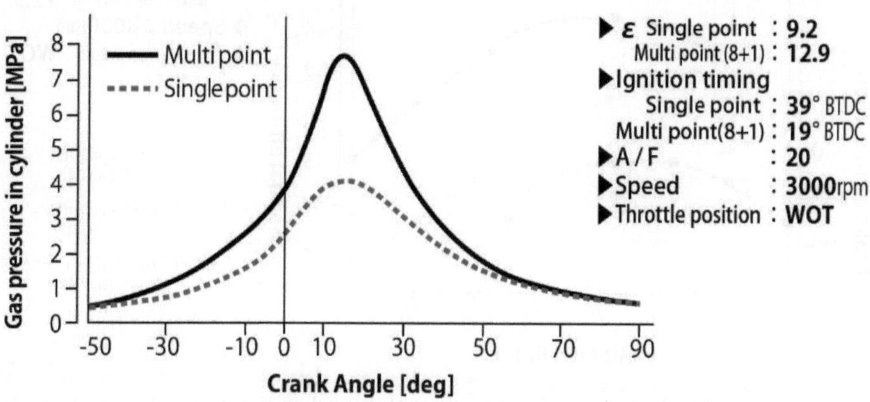

Fig. 6. Increase of Pmax by multi-point ignition

Also in a PV diagram comparison under the same conditions shown in Fig. 7, it can be confirmed that the area enclosed by the multi-point ignition curve surface has increased. Including the compression ratio increase effects in this experiment, the shown average effective pressure was improved by 28 %.

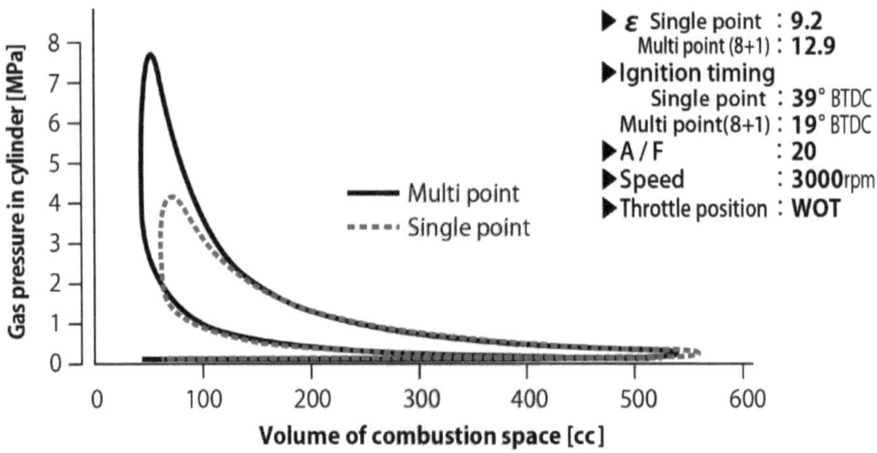

Fig. 7. Increase of effective gas pressure in the cylinder by multi-point ignition

The output and combustion effect of a multi-point ignition engine is shown in Fig. 8, in which the compression ratio has been increased from 9.2 to 12.9. In an experiment measuring the brake torque and thermal efficiency at each air-fuel ratio under the same conditions as in the aforementioned experiment, a high brake torque was also generated at a leaner air-fuel ratio, and a thermal efficiency of 44.49 % was also achieved in multi-point ignition, compared to single point ignition. Furthermore, the lean limit was increased by the higher compression ratio.

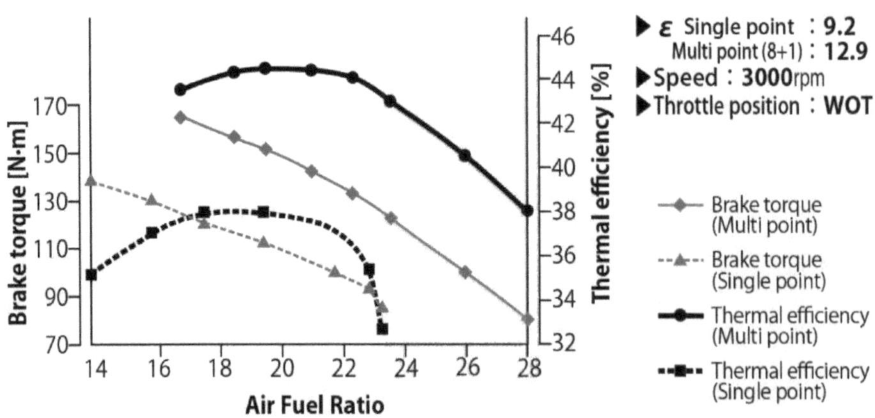

Fig. 8. Increase of brake torque and improvement of thermal efficiency at each air-fuel ratio by number of ignition point

6 Conclusion

We obtained the following conclusions from this study.

1. Fast combustion was obtained by multi-point ignition and the MBT point approached the top dead center.
2. This increased the torque at the same air-fuel ratio and improved the thermal efficiency.
3. Furthermore, the lean limit was largely increased.
4. The anti-knocking property was improved and the compression ratio was able to be increased.
5. By a synergetic effect of these, the lean limit was increased and thermal efficiency was further improved, achieving 44.49 %.

7 Closing Remark

A multi-point ignition engine has a high effect in improving the thermal efficiency of a SI engine without including a direct fuel injection, or the changing shapes of the combustion chamber and the pistons associated with this. On the one hand, since the high gas pressure in cylinder by multi-point ignition also results in an increase in the possibility of machine loss, it is thought to be possible to realize higher thermal efficiency, for example by designing an engine optimal for multi-point ignition, so that the ignition effect can be maximally exhibited.

Reference

1. Minami, K.: Development of Homogeneous Charged Multi-point Ignition Engine, Japan Land Engine Manufacturers Association, No. 516

Development of an Ignition Coil Integrated System to Monitor the Spark Plugs Wear

Filipa Fernandes(✉), Marco Lönarz, and Peter Weyand

Delphi Powertrain Systems, Bascharage, Luxembourg
filipa.fernandes@delphi.com

Abstract. The most recent spark plug development comprehends an increasing usage of rare precious metal alloys like Iridium and are aiming to a lifetime extension of the device, as well as the improvement of the ignition efficiency. However, they are also contributing to the higher cost of the latest spark plugs.

These were the drivers for the present work, in which Delphi is proposing the monitoring of the spark plug wear in order to enable the vehicle to detect a predetermined critical level of degradation. This wear compromises not only the combustion but also the surrounding components. The real-time monitoring of the plugs eliminates the supplier requirement to replace the plug after - usually conservative - predetermined exchange cycles. The achieved extended lifetime leads to a reduced waste of precious metals, and is answering the environmental challenge as well.

This work is based on ignition coils that are using on-board electronics which include a microcontroller. The programmable controller adds new functions and more independence to the ignition itself through the flexible software solution. The paper explains the method for measuring the breakdown voltage by making use of the microcontroller software and peripheral hardware which control both the acquisition of that data and the transmission of consolidated information to the ECU. The result can then be displayed as service information in the car. Usage of the ignition coil integrated system allows to overcome the difficulty in directly measuring high voltage. This goal is achieved coil-internally and there is no need for an external device to measure this high voltage, which would be hard to place around the engine and definitely a more expensive option.

Keywords: Spark plug · Wear · Ignition coil · Microcontroller · Software · Punch-through voltage · ECU

1 Introduction

Horizon 2020 is a long-term programme, set by European Union, which aims to a customer-driven technology and innovation leadership by funding R&D in public and private sectors as well as developing European policies accordingly, [1]. In the wake of this comprehensive program, and, more importantly, of European Union's Regulation (EC) No 443/2009, there has been a cumulative effort to achieve the average consumption target of 95 g CO_2/km set for the newly registered fleet, phasing in from 2020, [2, 3]. Automotive industry is already coming up with innovative solutions at the engine level such as downsizing of Otto-engines, Lean Burn Processes (usage of mixtures

M. Günther and M. Sens (eds.), *Ignition Systems for Gasoline Engines*,
DOI 10.1007/978-3-319-45504-4_8

inside the cylinder with a high air/fuel ratio), Stratified Combustion and High Dilution Exhaust Gas Recirculation (EGR), [3, 4]. With these combustion processes, however, a diversity of drawbacks related to boundary conditions comes along. For that same reason, Horizon 2020 has been a driving force likewise for development of injection and ignition systems which can compensate for some of those shortcomings, [4].

The effects of spark ignition regarding combustion and engine performance and vice-versa have been a subject of many researches, [5]. Nevertheless, it is known that the latest developments in the automotive industry, especially in response to emission limits and cost reduction, have imposed unfavourable conditions to the components, such as overheating, higher pressures, downsizing and poorer air/fuel mixtures which must be overcome through design changes and the usage of new materials, [4, 6].

The ignition coil integrated system is an example of such evolution. Leaner air/fuel mixtures, exhaust emission controls and suchlike goals require stronger and/or multiple sparks and longer burning times, [3, 4, 7]. This, in turn, leads to severe wear of spark plug electrodes, besides the inevitable damage these materials undergo due to extreme pressures and temperatures generated during the normal ignition process inside the combustion chamber, [5, 6]. Thereby, one can observe an increasing trend to use precious metals, namely iridium, platinum and tungsten, which significantly increase manufacturing costs, [5].

Even so, with the usage of these materials, it is inevitable to observe electrodes gap growth, roughness of the electrodes surface, changes in the geometry of the electrodes and coatings formed by the products of combustion. Figure 1 shows the coexistence of all the aforementioned phenomena, by comparing a brand new spark plug with one that had been used for 47800 km in a car.

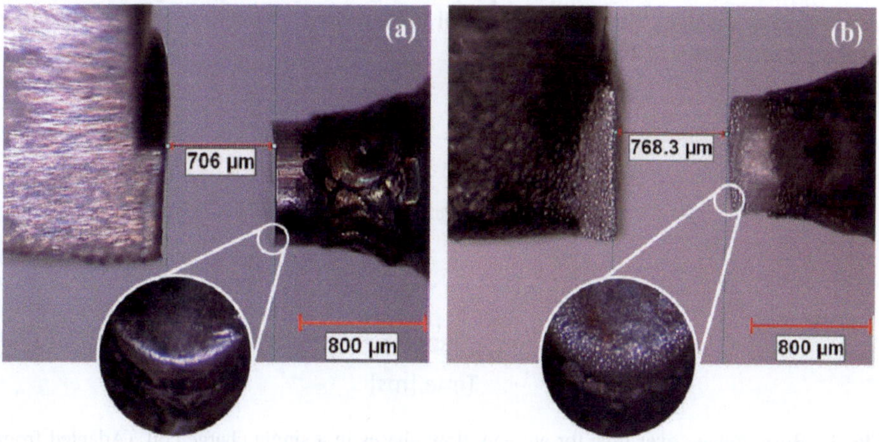

Fig. 1. Comparison between (a) new spark plug electrodes and (b) worn spark plug electrodes after 47800 km.

2 Ignition Process

Rudolf Maly proposed a division of the spark generation into four phases in 1984, [8]. These were named pre-breakdown phase, breakdown phase, arc discharge and ultimately glow discharge, [8, 9]. Pre-breakdown phase extends for a few microseconds and is characterized by an increasing electric field between the spark plug electrodes. During breakdown phase (in nanoseconds range), a self-created space charge is concentrated in a narrow channel between the electrodes wherein a partial volume has higher conductivity allowing for a higher current flux to be carried out, [8]. The temperatures consequently increase in the range of 50,000 K to 60,000 K, [9]. This step of the ignition event is the most relevant for the combustion process, [5]. Following the Breakdown Phase, high current arc discharge arises, lasting for a few microseconds, [5, 8]. During the referred interval, when comparing to breakdown phase, there is a drop of plasma temperature (5,000 K to 6,000 K), [9]. For this arc to be sustained, electrons must be emitted from the cathode surface, specifically from pools of molten material named hot cathode spots, [8]. These are the outcome of high current densities at the electrodes. The drawback of this mandatory event is the significant damage of the electrode surface, therefore this discharge phase is the most critical regarding the wear of the spark plug electrodes, [5, 8, 9]. It is possible to define a transition period wherein one can observe a frequent commutation from arc to glow discharges and vice-versa. Arc discharge arises whenever a strong electric field and/or high pressure occur in the electrodes gap therefrom establishing high current densities in the cathode, [9]. The major distinction between arc and glow discharges is the cold cathode of the last, since feedback electrons are generated by ion impact, thereby it is desired a rapid transition to low current glow discharge in order to minimize the electrode wear (essentially caused by sputtering throughout burning time), [8, 9]. Usually, glow discharge extends for several milliseconds, [8]. Arc discharge and glow discharge are shown in Fig. 2.

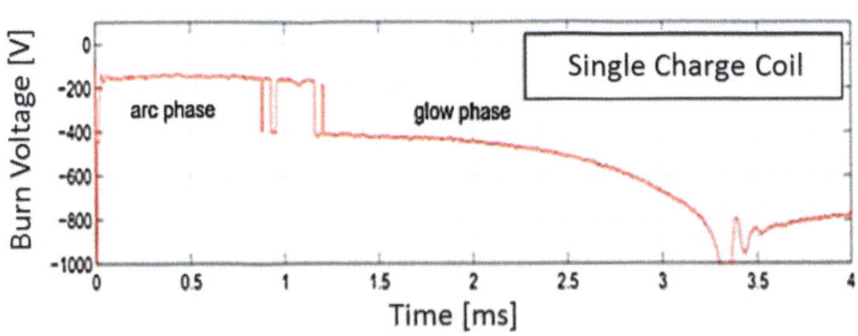

Fig. 2. Burn voltage over time for arc and glow phases in a single charge coil. (Adapted from Source: Advanced Ignition Systems: Technical Possibilities and Limitations, p. 4)

Spark plug tips are optimized to work in the range between 400 °C and 850 °C. Below this range, the temperature is not enough to evaporate the products of combustion (mainly carbon and oil) that are deposited on the electrodes surface during consecutive cycles. The consequence of this fouling condition is misfire. Above 850 °C, erosion becomes more relevant and the air/fuel mixture is more prone to pre-ignition. In both situations the electrodes are presenting a non-uniform wear and therefore, gap growth is also heterogeneous. For a full load condition and high rpm in a gasoline internal combustion engine, center electrode is generally hotter than ground electrode (the discrepancy between both temperatures depends on load and frequency of rotation, rpm) which leads to a greater erosion rate on the center electrode, [10].

3 Breakdown Voltage Determinant Factors

The maximum voltage value obtained between the spark plug electrodes – hereinafter referred to as breakdown voltage, V_B – depends on the following parameters (not listed by order of relevance), [5]:

- Air/fuel ratio
- Fuel type
- Geometry of the electrodes
- Pressure within the spark plug gap
- Width of spark plug gap
- Temperature of the electrodes
- Fouling of electrodes and insulator tip
- Internal impedance of the spark plug
- Voltage increase rate

For example, the required breakdown voltage is higher for lean mixtures, since the greater the air fraction present in the mixture, the lower is the electrical conductivity thereof. For the same reason, fuels that demand high air/fuel ratio call for higher breakdown voltages (e.g. compressed natural gas, CNG, compared to gasoline), [5].

The electrodes shape play an important role in determining V_B as downsizing of both the centre and the ground electrodes allows for a higher strength of the electrical field therein, thus leading to a decrease of the required breakdown voltage. However, this energy concentration on the electrode tip must not reach such level that is generating a critical temperature on the electrode tip. In such circumstances, the electrodes erosion builds up and the spark plug lifetime is shortened, [5].

High pressures (result of load conditions in the car, which varies with driving conditions), on the other hand, entail as well higher voltages, because it means that a greater density of molecules is present in the spark plug gap, *i.e.* the mean free path decreases, thus the electrons must have a greater acceleration in order to ionize those same molecules, [5, 7]. Greater acceleration is equal to a higher electrical field strength. Otherwise, a conductive streamer cannot be established in the spark gap and the arc does not develop. A wider spark plug gap implies as well a greater breakdown voltage, since an electrical voltage potential difference is required that maintains the field strength for streamer generation over a larger distance. Lean burn or diluted mixtures are presenting comparably less gasoline molecules per air- or inert gas molecules. In these two cases - besides the field strength requirement responsible for the initial ionization - the energy introduction into the air fuel mixture has to be sufficient to start and maintaining an exothermic reaction. The ignition energy provided to the mixture has to be high enough to allow the flame kernel to develop and being sustained up to a suitable temperature. The spark plug gap, however, is growing over time due to erosion that takes place during discharge. Simultaneously, the deposits on the insulator tip or a small voltage build-up rate reduce the voltage impressed in the electrodes gap, demanding a higher V_B. Figure 3 is enlightening about such phenomenon, showing the results of preliminary tests which aimed to distinguish a new plug from a worn one through V_B (an extra capacitor was added to the setup used in the laboratory, representing load capacitance, C_{Load}).

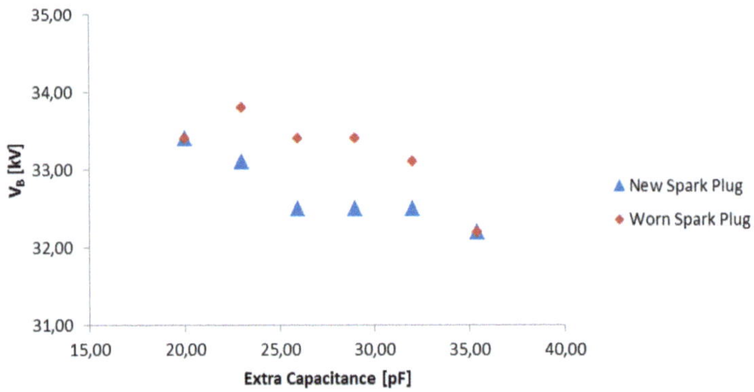

Fig. 3. Breakdown voltage as a function of capacitance for a new and a worn spark plug.

Of course, the supply voltage of the ignition coil has a limit. If gap growth brings about a breakdown voltage above this limit, misfire events can arise, potentially causing damage to the catalytic converter and discomfort to the driver, [5].

During driving, speed and load conditions vary continuously. It is known that both of them are affecting the breakdown voltage. However, the quantification of this influence is yet to be determined in this work.

3.1 Paschen Law

It was in 1889 that Friedrich Paschen established the relation between breakdown voltage, pressure and spark plug gap, [11]. The Paschen Law can be translated into the equation below.

$$V_B = \mathrm{B}\frac{pd}{ln(Apd) - ln(ln(1+\gamma^{-1}))} \quad (1)$$

where V_B is the breakdown voltage, p and d represent the pressure and width of the spark plug gap, respectively, γ is the secondary emission coefficient (or second Townsend coefficient) and A *and* B are constants that depend on the gas between the electrodes, [12, 13]. It is possible to ascertain that the required breakdown voltage is lower for a smaller gap. Nonetheless, reducing the spark plug gap might bring on a decrease in the plug performance when dealing with mixtures with high dilution or high air/fuel ratios, unless this limitation is considered in the electrodes design, because there is a lower amount of fuel molecules in the electrodes gap [5]. Equation 1 is translated into the Paschen Curve (Fig. 4).

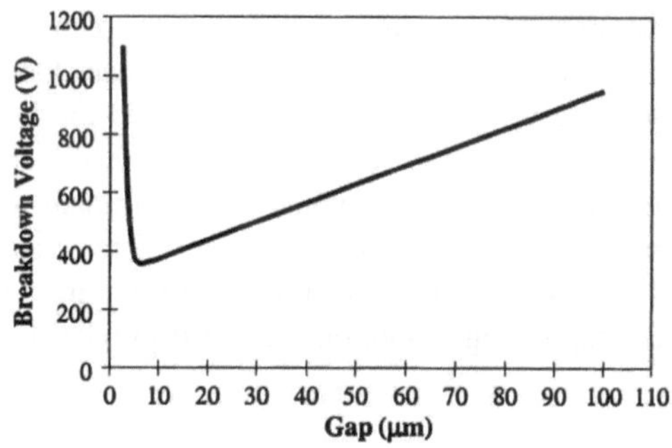

Fig. 4. Paschen curve for air: gaseous breakdown voltage as a function of the distance between two conductors, at 1 atm. (Adapted from source: Electrical breakdown and ESD phenomena for devices with nanometer-to-micron gaps, p. 88)

Paschen curve depends on the gas in which the breakdown occurs and on the electrodes material, although the latter has not much influence, [13]. Paschen Curve enables the study of electrodes gap through V_B measurements. [14] Therefrom, it is theoretically possible to deduct wear conditions of spark plug electrodes. The difficulty of this analysis mainly consists in isolating other wear sources that might affect the breakdown voltage since, for example, when comparing two spark plugs with equal electrodes gap, the voltage is usually higher for the fouled plug, [5].

4 Microcontroller-Based Ignition Control

Digital electronics have been doubtlessly playing a major role in instrumentation and control for automotive technology, in which the engine system is one of the most significant control applications. Low cost and real-time adaptability are the leading advantages of utilizing a microprocessor-based engine control system, [8]. An engine control unit (ECU) of this sort opens doors to the implementation of new coil-internal features by the usage of a microcontroller implying minimal or even any change to the peripheral circuitry but software improvements instead.

Currently, Delphi's ignition coils have been making use of microcontrollers for a wide range of functions, such as transmitting data regarding ignition parameters (e.g. charging current, spark duration, spark current or even information related to Multi-Charge mode), [15]. Furthermore, this control unit carries out diagnosis and measurement data acquisition which can be transmitted back to the ECU or be used for further operations inside the ignition coil.

4.1 Indirect Measurement of Breakdown Voltage

First and foremost, it was a purpose of the present work to develop a method to measure V_B through the microcontroller contained in the ignition coil. The link between this data and spark plug wear is still to be quantified, although preliminary tests indicate such correlation (Fig. 3). The goal was to use the signals already collected by the microcontroller in order to minimize or eliminate the need for using additional electrical components. Within the scope of this aim, two solutions were brought up to discussion concerning the indirect measurement of breakdown voltage. Option A consisted of looking at collector voltage, $V_{C\,-}$, while Option B focused on the voltage across the capacitor that integrates the filter of the voltage across the secondary current sensor – a shunt resistor providing a voltage signal which is proportional to secondary current (hereinafter this voltage signal will be represented as V_{Is} for reasons of practicality). Figure 5 demonstrates the main functions of a microcontroller in an ignition PCB circuit as well as the nodes corresponding to Options A and B previously explained.

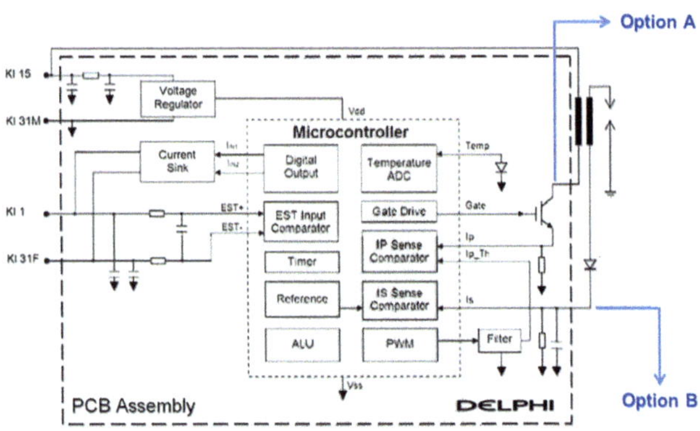

Fig. 5. Ignition block diagram focusing on microcontroller-based PCB circuit and nodes where one can measure $V_{C\,-}$ (Option A) and V_{Is} (Option B).

In order to measure $V_{C\,-}$ and V_{Is}, it was used the setup shown in Fig. 6.

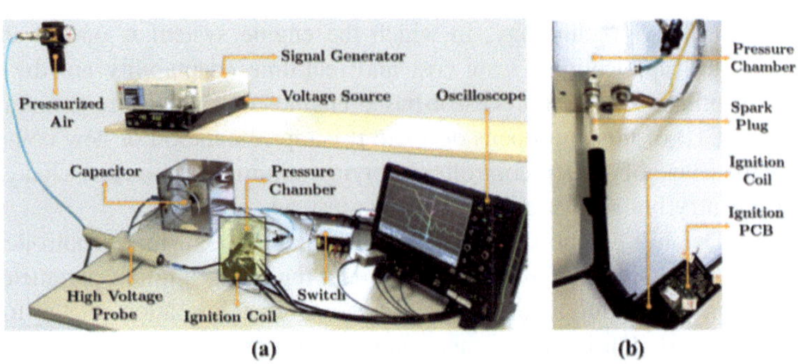

Fig. 6. Laboratory setup for direct measurement of V_B: (a) complete assembly and (b) detail of the highlighted area of the image on the left.

The resulting signals are simultaneously presented in Fig. 7. IGBT (Insulated Gate Bipolar Transistor) gate signal was collected solely as a reference, since its step exhibits the instant when primary winding of the ignition transformer is disconnected from ground, thus transferring the energy to the secondary winding to initiate the discharge.

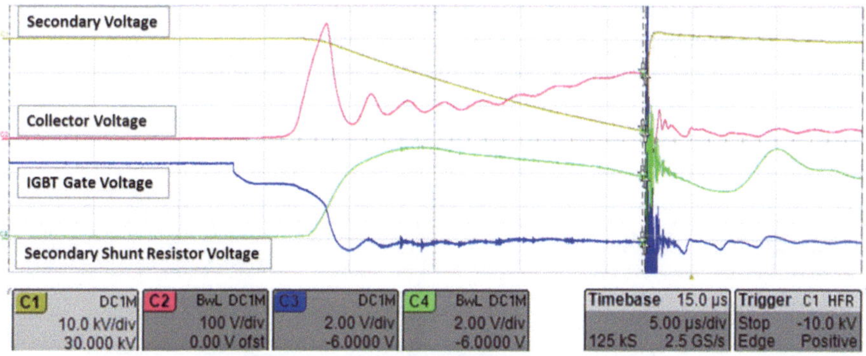

Fig. 7. Secondary voltage (yellow), V_{C-} (red), IGBT gate voltage (blue) and V_{Is} (green) during breakdown event, for V_B = 27 kV and C_{Load} = 30 pF.

For higher breakdown voltages, the secondary voltage rise time is longer and, for that reason, all the signal lengths are essentially extended, have not been detected any change in the curves course (assuming load capacitance is kept constant).

Table 1 allows for a decision between Options A and B based on a practical point of view concerning not only the simplicity and reliability of the process but also the associated costs to the respective implementation.

Table 1. Comparison of the implementation requirements for Options A and B.

	Option A	Option B
Measured variable	Collector voltage	Secondary current
Implementation	• Sample and hold circuit • D/A converter	• Software modification • I_S sensing circuit (if not already present)
Implementation cost	Development and additional components costs	Development costs
Measured variable depend on	• Pressure at the spark gap • Load capacitance	• Pressure at the spark gap • Load capacitance
Breakdown timing detection	Available	Not available

Since the present work was developed based on a MultiCharge system, which comprise already a shunt resistor to measure I_S and respective filter, for Option B alone no hardware modifications need to be undertaken. Therefore, Option B represents the lower cost option. It is the preferred embodiment, as long as reliability specifications are fulfilled. Although V_{Is} does not allow for the exact breakdown timing identification, this does not represent an obstacle for the implementation.

Within this scope, further tests were performed to determine the interrelationship between load capacitance and breakdown voltage and certain coil parameters, namely secondary current, IGBT gate voltage and filtered EST ("electronic spark timing") voltage. Figures 8, 9 and 10 are demonstrating specifically how the following timings vary with the change in V_B and C_{Load} ($C_{Load} = 35$ pF ∨ $C_{Load} = 15$ pF since those were the limits of the extra capacitor used in the laboratory tests setup):

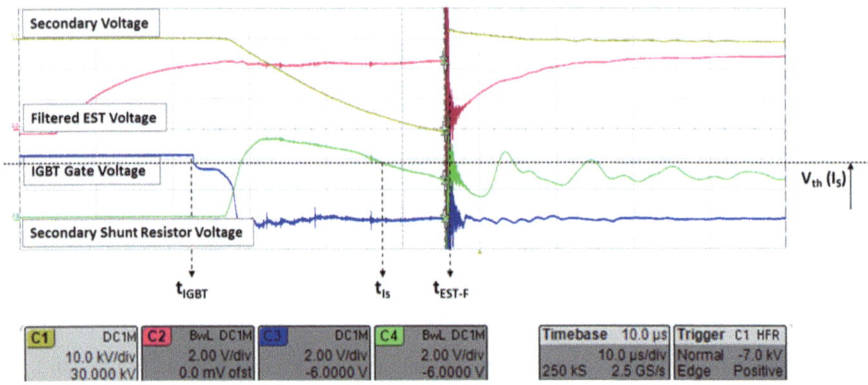

Fig. 8. Secondary voltage (yellow), filtered EST voltage (red), IGBT gate voltage (blue) and V_{Is} (green) during breakdown event, for $V_B = 31$ kV and $C_{Load} = 35$ pF.

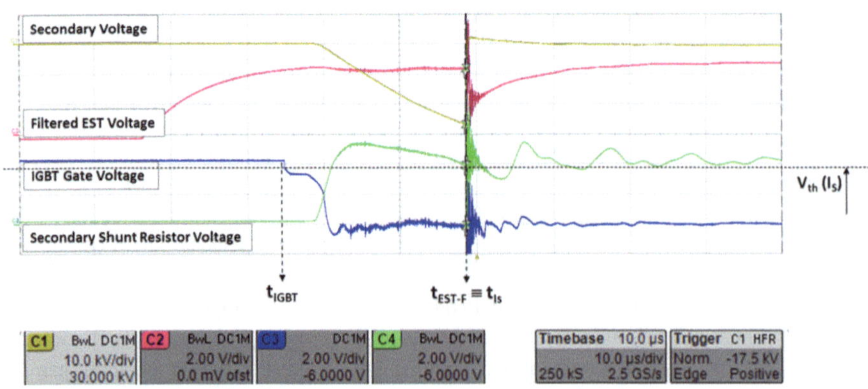

Fig. 9. Secondary voltage (yellow), filtered EST voltage (red), IGBT gate voltage (blue) and V_{Is} (green) during breakdown event, for $V_B = 27$ kV and $C_{Load} = 35$ pF.

- t_B: always positioned at the breakdown instant, this signal can serve as a reference
- t_{IGBT}: start trigger for all the timing measurements since it is internally generated

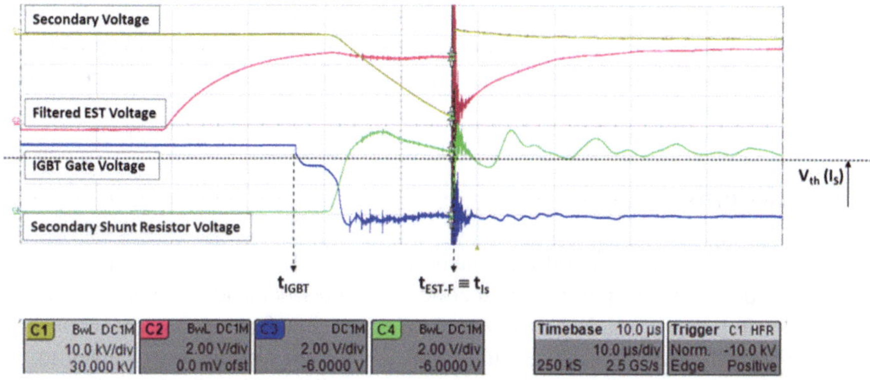

Fig. 10. Secondary voltage (yellow), filtered EST voltage (red), IGBT gate voltage (blue) and V_{Is} (green) during breakdown event, for V_B = 27 kV and C_{Load} = 15 pF.

- t_{Is}: moment when I_S drops below a certain threshold, $V_{th}(I_S)$; the threshold is predefined according to the breakdown voltage value that one considers critical (for greater critical breakdown voltages, a lower threshold must be selected).

It can be observed that the time intervals t_B - t_{IGBT} and t_{Is} - t_{IGBT} increase with C_{Load} and V_B, although the latter has a much greater influence on such parameter, as it was expected. It is also shown that the amplitude of the filtered EST voltage peak increases with V_B and decreases with C_{Load}. This signal was expected to be a constant line in this discharge phase. However, the electrical components that filter the EST signal before it is fed into the microcontroller create a kind of "antenna" that increases the amplitude of the noise originated in the high voltage end of the secondary winding. As a result, filtered EST signal serves as a detector circuit for the exact breakdown timing. It should be noticed that the existence of this filtered EST voltage peak was confirmed not only in laboratory conditions but also in real driving conditions.

Whenever the microcontroller is performing the measurements and thereafter providing the relevant timings that allow to interpret the ignition cycle, it is considered $t_{IGBT} = 0$ (moment when the microcontroller starts the measurement), therefore t_B and t_{Is} represent the interval since the IGBT turns to off-state until a specific event takes place. t_{out} represents the maximum timing for which the microcontroller looks for the predetermined conditions. To make sense of the results, three main conditions should be established regarding a primary condition $t_{Is}, t_B \in [0; t_{out}]$.

1. $t_{Is} > t_B V_{Is} > V_{th}(I_S) \forall t \in [0; t_{out}] \Rightarrow V_B < V_{B,crit}$
2. $t_{Is} = t_B \Rightarrow V_B = V_{B,crit}$
3. $t_{Is} < t_B \Rightarrow V_B > V_{B,crit}$

Once the microcontroller continuously transmits to the ECU one of the previous conditions per each cycle, the ECU should be able to, under statistical data processing, determine whether the spark plug is undertaking critical breakdown voltage conditions (i.e. the third condition is obtained in hundreds of almost consecutive measurements). It should be noticed that a single measurement does not allow for any reliable conclusion, particularly regarding the unfavorable set of conditions that the ignition system undergo in the car (lack of clean environment, inconsistent speed, load and fuel quality, etc). At the moment, tests are still being carried out to quantify the influence of load and speed in these results.

5 Conclusions

The development of ignition coil integrated system has the potential to respond to the most recent demands of emissions reduction while cost demands do not slow down. The monitoring of the spark plugs wear is a promising feature that can contribute thereto, especially if it does not entail significant design and/or cost changes to the current devices thanks to the usage of the microcontroller. Although electrodes wear is an inherent process to the ignition system, tracking it through breakdown voltage is a preventive measure that aims the protection of the surrounding components.

A monitoring of such parameter without the need of additional components is a large step, since it creates a customer-driven solution which is easier to implement. It was shown in the present work that a secondary current sensing circuit may also serve as a breakdown voltage sensor, presenting reliable results for different values of breakdown voltage and load capacitance. Still, the most challenging hurdle consists in isolating the gap enlargement from other events that also contribute to an increase of the breakdown voltage, such as engine load and speed, spark plug fouling and pressure and temperature in the electrodes gap. Further tests in the car aim to overcome this, since the real driving conditions are not possible to be reproduced in the laboratory. Once reliable results are obtained, a correlation between these and the spark plug wear must be quantified.

References

1. Executive Summary of the Impact Assessement Accompanying the Communication from the Commission "Horizon 2020 - The Framework Programme for Research and Innovation. Brussels (2011)
2. Setting Emission Performance Standards for New Passenger Cars as part of the Community's Integrated Approach to Reduce CO2 Emissions from Light-duty Vehicles, Official Journal of the European Union, European Parliament and Council of the European Union (2009)
3. Geringer, B.: Demands on gasoline engines with focus on the stringent fuel economy and emission limits until 2020. In: Expert Verlag GmbH (Ed.), Adv. Ignition Syst. Gasol. Engines, Die Deutsche Bibliothek, Renningen, pp. 1–17 (2013)
4. Weyand, P., Lorenz, F., Schilling, S., Systems, D.P.: Adaptive continuous spark ignition as enabler for high dilution EGR operation. In: 2nd Conference on Advanced Ignition Systems, Berlin (2014)
5. Javan, S., Hosseini, S.V., Alaviyoun, S.S., Ommi, F.: Effect of electrode erosion on the required ignition voltage of spark plug in CNG spark ignition engine. J. Engine Res. **26**, 31–39 (2013)

6. Yoshinaga, T., Igashira, T., Mori, K., Kawai, H., Morino, S.: Ignition System for Internal Combustion Engines, US 4,350,137 (1982)
7. Lorenz, F., Weyand, P., Jacque, E., Schilling, S.: Advanced Ignition Systems: Technical Possibilities and Limitations. In: SIA Conference on Spark Ignition Engine Future, Strasbourg (2013)
8. Maly, R.: Spark ignition: its physics and effect on the internal combustion process. In: Hilliard, J.C., Springer, G.S. (Eds.) Fuel Econ. Road Veh. Powered by Spark Ignition Engines. Springer Science+Business Media, New York (1984). doi:10.1007/978-1-4899-2277-9
9. Lorenz, F.: Untersuchung und Entwicklung von MultiCharge-Zündsystemen für direkteinspritzende Ottomotoren, Fachhochschule Aachen (2009)
10. Javan, S., Hosseini, S.V., Alaviyoun, S.S.: An experimental investigation of spark plug temperature in bi-fuel engine and its effect on electrode Erosion.pdf. Int. J. Automot. Eng. 2 (2012). http://www.iust.ac.ir/ijae/browse.php?a_code=A-10-98-2&slc_lang=en&sid=1
11. Paschen, F.: Ueber die zum Funkenübergang in Luft, Wasserstoff und Kohlensäure bei verschiedenen Drucken erforderliche Potentialdifferenz. Ann. Phys. **273**, 69–96 (1889). doi:10.1002/andp.18892730505
12. Löffler, F., Siewert, C.: Homogeneous coatings inside cylinders. Surf. Coatings Technol. **177–178**, 355–359 (2004). doi:10.1016/j.surfcoat.2003.09.026
13. Burm, K.T.A.L.: Calculation of the townsend discharge coefficients and the paschen curve coefficients. Contrib. Plasma Phys. **47**, 177–182 (2007). doi:10.1002/ctpp.200710025
14. Wallash, A.J., Levit, L.: Electrical breakdown and ESD phenomena for devices with nanometer-to-micron gaps. Proc. SPIE. **4980**, 87–96 (2003)
15. Eisen, S.-M., Bolz, S., Schmauss, H. Reuther, A., Goetzenberger, M.: Ignition Device for an Internal Combustion Engine and Method for Operating an Ignition Device for an Ignition Device for an Internal Combustion Engine, US 2013/0263835 A1 (2013)

Components

Fatigue Life Simulation and Analysis of an Ignition Coil

Ravish Masti[1(✉)], Pedro Carreira[2], Chandrashekhara SM[1], and Ravi Kumar[1]

[1] Technical Center India, Delphi Automotive Systems, Bangalore, India
{Ravish.Masti,Chandrashekhar.SM, Ravi.Kumar}@delphi.com
[2] Technical Centre Luxembourg, Delphi Automotive Systems, Luxembourg, Luxembourg
Pedro.Carreira@delphi.com

Abstract. An ignition coil is subjected to a very dynamic loading environment during its operational service. Some of these loads expose the ignition coil to mechanical fatigue. In the process of designing a reliable coil, it is beneficial to make an accurate durability assessment of the coil design even before the prototype is built. A case study is presented here that explores into the thermal fatigue and durability assessment of the primary wire of a coil by using CAE simulation. The simulation presents the effects of the coil assembly loads and operating service loads upon the fatigue and durability of the wire. The simulation prediction is validated by the lab test. A validated simulation would help in quickly identifying the critical design parameters and exploring the design space to develop a reliable and durable ignition coil.

Keywords: Ignition coil · Wire · Fatigue · Durability · Thermal · Simulation

1 Introduction

An ignition coil works on the principle of electromagnetic induction and its primary function is to deliver a high voltage to initiate the breakdown of the spark plug gap and provide sufficient energy to efficiently sustain the combustion process across the entire range of the engine operating conditions. With the aid of the engine controller and engine-conditioning sensors, the coil works in a closed-loop control system and assists to meet the performance goals and regulatory emission norms [1].

An ignition coil, installed in a vehicle, during its service life is subjected to vibration, temperature cycling, thermal shock, magnetic hysteresis, environmental/climatic factors, etc. Prior to its exposure to the in-service loads, the coil parts also undergo mechanical stresses/strains during the assembly of the coil. Therefore, the design and development of a reliable ignition coil should account for the loading effects during the assembly and in-service conditions.

This paper focusses on the computer simulation and prediction of the mechanical stress/strain response and fatigue durability of the primary wire of the coil when the wire is exposed to the loading conditions occurring during the coil assembly and the

M. Günther and M. Sens (eds.), *Ignition Systems for Gasoline Engines*,
DOI 10.1007/978-3-319-45504-4_9

thermal cycling conditions during the service life of the coil. Establishment of a validated simulation procedure is beneficial in ascertaining the analytical confidence level, improving accuracy of predictive design methods, meeting the customer quality requirements, reducing design cycle time and prototype building plus validation costs.

2 Ignition Coil

The ignition coil that is the subject of study in this paper consists of the primary and secondary wire, epoxy filling, casing, magnetic core laminations, coil spool, plastic over-mold, electrical connector (IDC), etc. Figure 1 provides an exploded view of the coil assembly.

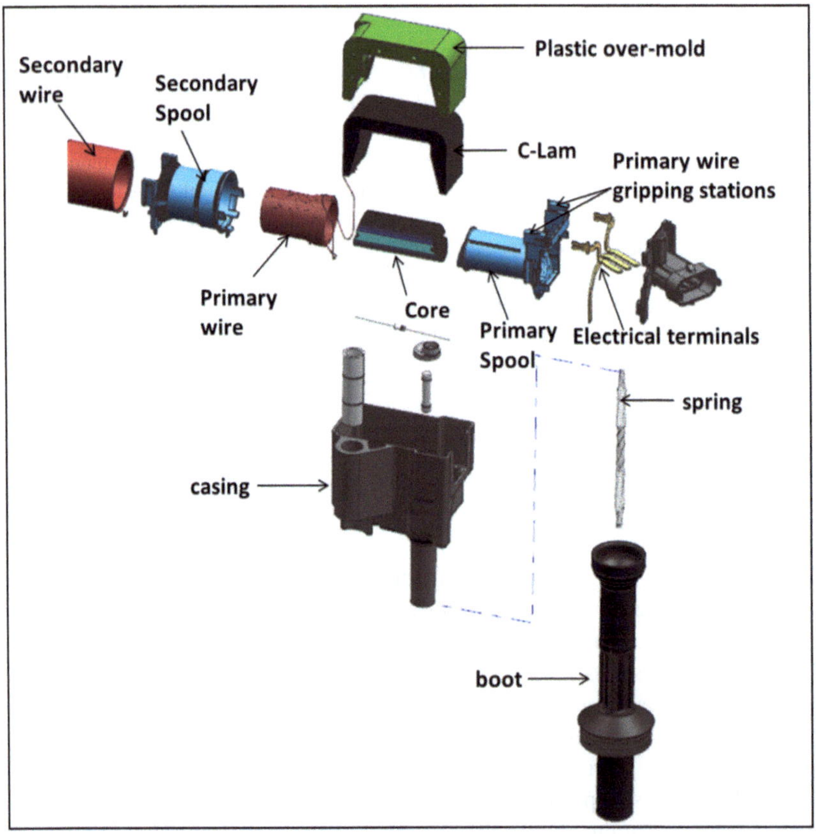

Fig. 1. Ignition coil assembly–an exploded view

3 Thermal Fatigue and Durability of Primary Wire

In an ignition coil, there are surface contact interactions among the adjoining parts. By virtue of the different co-efficient of thermal expansion of the adjoining parts, thermal stresses/strains are induced in the parts and they undergo variation in response to the varying temperature conditions. Fluctuating thermal stresses/strains cause material fatigue of the coil parts.

The thermal strain (ε) in a material is related to the coefficient of thermal expansion (α) and the temperature change (ΔT) by the following relation:

$$\varepsilon = \alpha * \Delta T. \tag{1}$$

The primary wire interacts with the spool, epoxy and other neighboring parts. The wire is subjected to thermal fatigue due to the exposure to temperature cycling during the in-service conditions of the coil. The primary wire is also mechanically pre-stressed/strained during the assembly of the coil. These stresses raise the mean stress level of the fatigue loading cycle and reduce the fatigue life of the coil. Excessive fatigue loading of the wire leads to crack formation in the wire and deterioration of the coil performance. It could even finally lead to breakage of the wire and complete breakdown of the ignition coil function.

4 Assembly Loads and In-Service Thermal Operating Conditions

An ignition coil is required to be designed to operate dependably and reliably under all operating conditions and throughout the entire engine speed range to help the OEMs meet the highest engine performance, exhaust emissions, and fuel economy objectives.

During its service life the coil is exposed to harsh vibration, mechanical shock, thermal cycling, temperature shock, electrical voltage/field reversals and transients, environmental/climatic factors such as humidity, corrosive conditions, etc., RFI (Radio Frequency Interferences) and EMI (ElectroMagnetic Interferences), and other conditions [1, 2].

During the coil development program, the Design Validation (DV) and Product Validation (PV) stages employ customer-compliant and in-house product standard test procedures to qualify the ignition coil for environmental, durability and capability tests under the mechanical, electrical, chemical, environmental, etc. compatibility requirements.

As the focus in this paper is to predict the thermal fatigue and durability of the primary wire, those assembly operations causing sufficient pre-loading stresses/strains on the wire and the thermal loadings during the in-service coil operation are taken into account.

Figure 2 shows a CAD model assembly of few of the coil parts. There are 2 stations on the primary spool that grip the ends of the primary wire.

During assembly, the free ends of the primary winding on the spool are pulled out and wound around the gripping stations. In order to ensure that the wire conforms

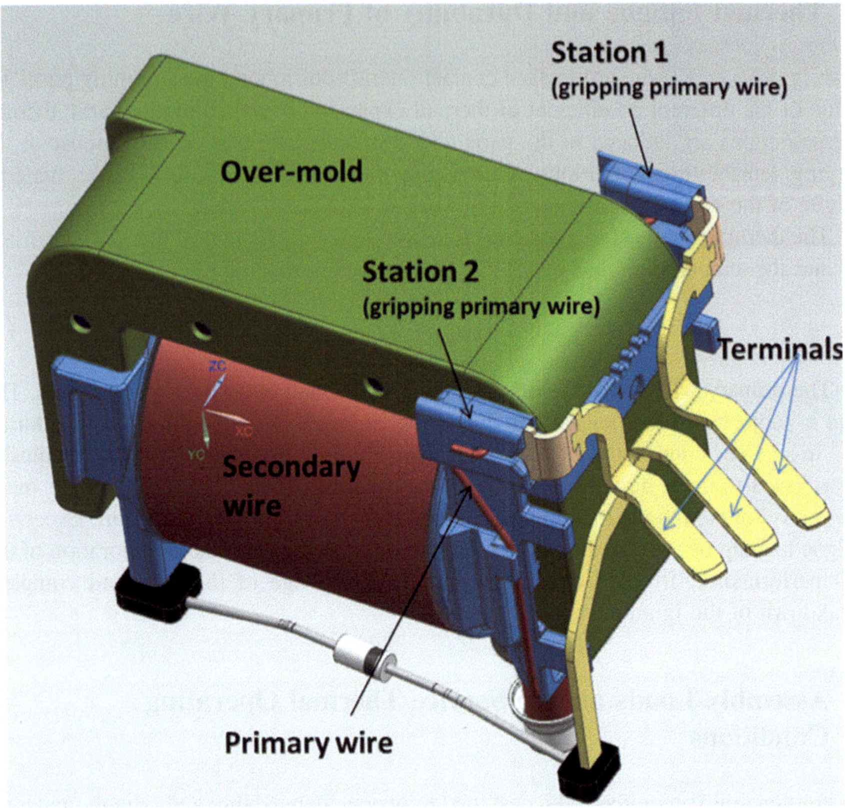

Fig. 2. Ignition coil–few of the parts in assembly

properly to the shape of the stations (where it comes in contact), a sufficient amount of pretension in the wire is maintained during the winding process. Then the electrical connector (IDC) is introduced to establish a reliable electrical connection between the wire and the terminals connected to the external power source. The parts are presoaked through a temperature profile and the epoxy is potted at a suitable temperature.

The coil under the hood, during its service life, is exposed to the varying degrees of temperature conditions due to fluctuating climatic conditions, varying operating conditions of the engine and engine-mounted components, internal heating, etc. To facilitate lab testing in short time during product development stages, the real-life temperature conditions are appropriately transformed into an accelerated temperature cycling profile.

5 Computer Simulation Modelling and Analysis for Predicting Thermal Fatigue Durability

A simulation-driven-design activity employs analytical software tools for modelling and analysis of the design even before the product development goes through a DV stage. Delphi employs analytical design procedures right from the concept stage of the ignition coil development.

The details presented in the sections below describe the steps for evaluating the thermal fatigue life of the primary wire that was part of a base design of a coil conceptualized during the initial stages of the coil development. Establishment of a validated simulation procedure is useful in ascertaining the analytical design confidence level, preparing for any improvement requirement of prediction accuracy and improving the robust design of the coil. Therefore the intention was to predict the thermal fatigue life (cycles) of the coil for crack initiation and validate the simulation results by the lab tests that were conducted to know the 'Test to failure' cycles of the DV prototype samples of the coil.

The simulation activities in this study comprised three main stages:

1. Primary wire assembly
2. In-service thermal cycling of coil assembly
3. Thermal fatigue life evaluation of the primary wire

The simulations were conducted using finite element (FE) modelling and analysis approaches as implemented in the commercial FE solver products ABAQUS CAE and FeSafe from Dassault Systèmes. The first two stages as outlined above were conducted using Abaqus CAE and the fatigue life evaluation was conducted using FeSafe. A portion of the model building activity was also done using Hypermesh software from Altair HyperWorks.

5.1 Simulating Primary Wire Assembly

Figure 3 shows a meshed model of one of the 2 stations (for gripping wire) on the primary spool and the primary wire. As explained in Sect. 4, the primary wire is initially maintained at a suitable tension and then wound around the station. The electrical connector (IDC) is then pushed into the station to lock with the wire (to functionally establish an electrical connection between the wire and the external electrical terminals). The mechanics of these operations involve non-linear contact interactions. In addition to boundary non-linearity, the simulation takes into account geometric and material (elasto-plastic behavior) nonlinearities of the mating parts. The FE analysis used is a non-linear static analysis with finite-sliding contact interaction [3] among the mating parts. The finite-sliding formulation, models the separation and sliding of finite amplitude and any arbitrary rotation of the contact surfaces. In order to accurately capture the contact interactions, the surface-to-surface contact discretization is employed. Frictional effects are included in the model as part of the contact property definition. Figure 4 shows the initial and final snap-shots of the FE-based assembly simulation of the primary wire installation onto the station.

Also shown is a good correlation of the simulated primary wire deformation with that of a dissected prototype sample after installation.

5.2 Simulating in-Service Thermal Cycling of Coil Assembly

Figure 5 shows a part of the CAD model (without epoxy for clarity) and the meshed model (with epoxy) used for the thermal cycling analysis. The model symmetry was taken into advantage to keep the model size smaller and reduce the computational time. The deformed model configuration from the previous analysis (Sect. 5.1) was carried forward into the model building step of this simulation stage and assembled with the other relevant parts. The filler spaces among the array of the parts that correspond to the epoxy volume are extracted using Hypermesh and reassembled with the rest of parts in Abaqus CAE. The Contact Tie conditions were imposed between the parts at their contact surfaces. The static thermal stress analysis approach [3] was employed to simulate the thermal deformation and corresponding stresses/strains in the parts. The thermal cycling profile approved by the coil development program was employed for the temperature loadings. It was sufficient to simulate only for a few cycles until a reasonable stable stress/strain response of the primary wire was observed. Figure 6 shows a chain of strain responses of the primary wire, when the wire was subjected to one cycle of the thermal profile loading.

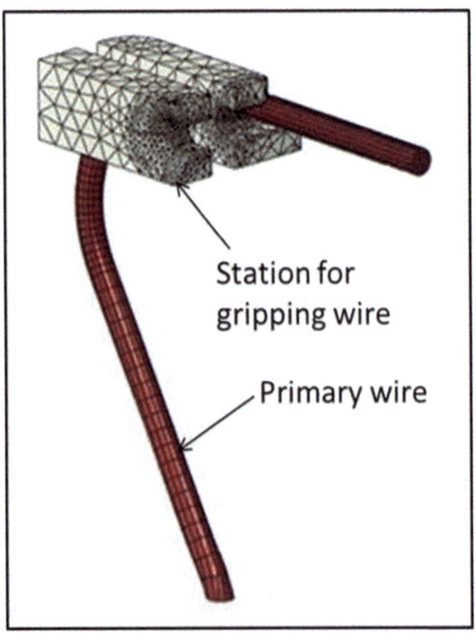

Fig. 3. Meshed parts of the primary wire and spool station

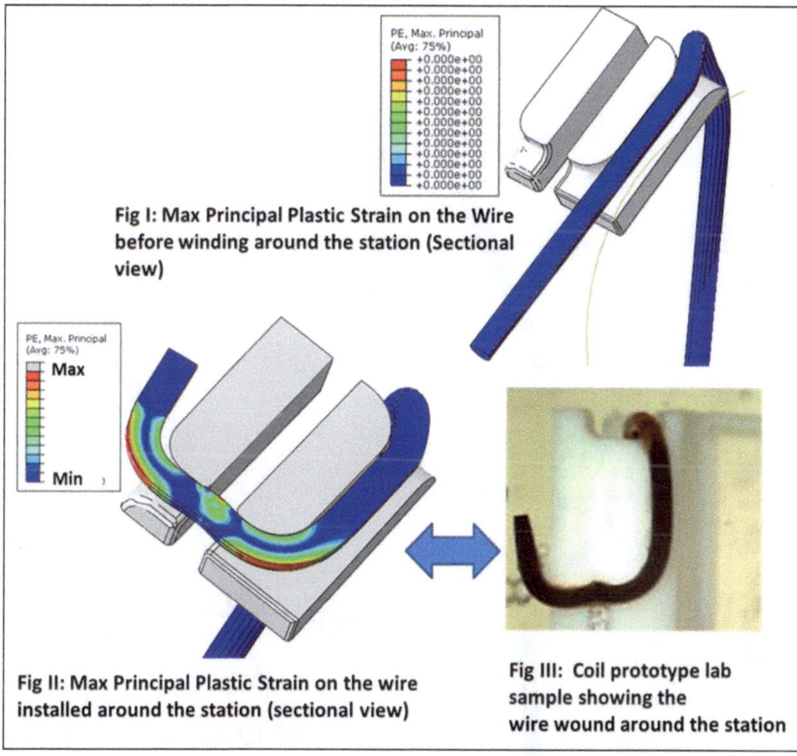

Fig. 4. Primary wire installation around a spool station

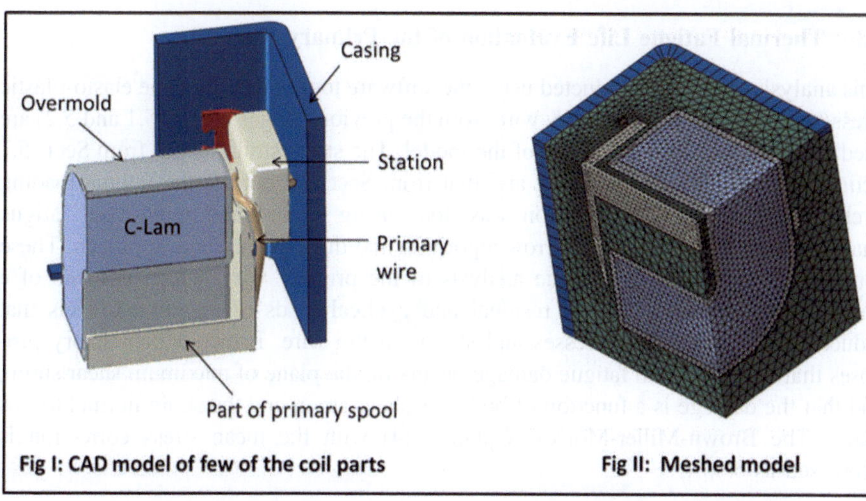

Fig. 5. Thermal cycling: few CAD parts and meshed model

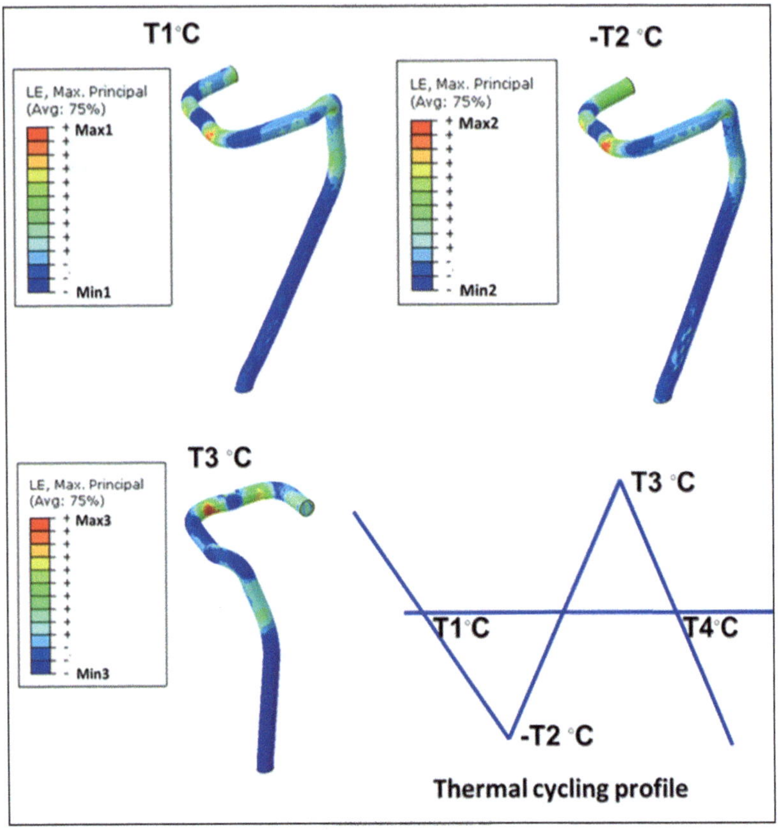

Fig. 6. Thermal cycling: strain contours at different temperature points

5.3 Thermal Fatigue Life Evaluation of the Primary Wire

This analysis stage was conducted using the software tool FeSafe [4]. The elasto-plastic stress/strain results of the primary wire from the previous stages (Sects. 5.1 and 5.2) are used in the fatigue load definition of the model. The stress/strain output from Sect. 5.1 defines the residual stress loading and that from Sect. 5.2 defines the fatigue loading cycle. The fatigue life calculation was done using strain-based multi-axial fatigue analysis using Brown-Miller-Marrow algorithm and the Critical Plane approach. These settings are preferable for fatigue analysis of the primary wire, which is made of a ductile metal and subjected to residual and cyclical loads of magnitude-levels that induce multi-axial plastic stresses and strains in the wire. Brown-Miller theory proposes that the maximum fatigue damage occurs on the plane of maximum shear strain, and that the damage is a function of both this shear strain and the strain normal to this plane. The Brown-Miller-Morrow equation [4] with the mean stress correction is expressed as below:

$$\frac{\Delta\gamma_{max}}{2} + \frac{\Delta\varepsilon_n}{2} = C_1 \frac{\left(\sigma_f' - \sigma_{n,m}\right)}{E} \left(2N_f\right)^b + C_2 \in_f' \left(2N_f\right)^c. \quad (2)$$

Where, $\frac{\Delta\gamma_{max}}{2}$ is the shear strain amplitude, $\frac{\Delta\varepsilon_n}{2}$ is the normal strain amplitude, $C_1 = 1.65$ and $C_2 = 1.75$ are the constant values widely used, σ_f' is the fatigue strength coefficient and $\in_f'$ is the fatigue ductility coefficient of the material, $2N_f$ is the number of cyclic reversals, b and c are the fatigue strength and ductility exponents of the material respectively. $\sigma_{n,m}$ is the mean normal stress on the plane and E is the modulus of elasticity of the material.

At every FE node of the wire model, the Critical Plane approach resolves these strains onto a number of planes, and calculates the damage on each plane. This helps in identifying the region on the wire with the least fatigue life.

Figure 7 shows the fatigue life contour on the wire and shows that the region of least fatigue life i.e. the region of earliest crack initiation is on the inner bend towards the spool side. Also shown is the location of crack initiation likely to start on the right inner bend towards the free side of the wire.

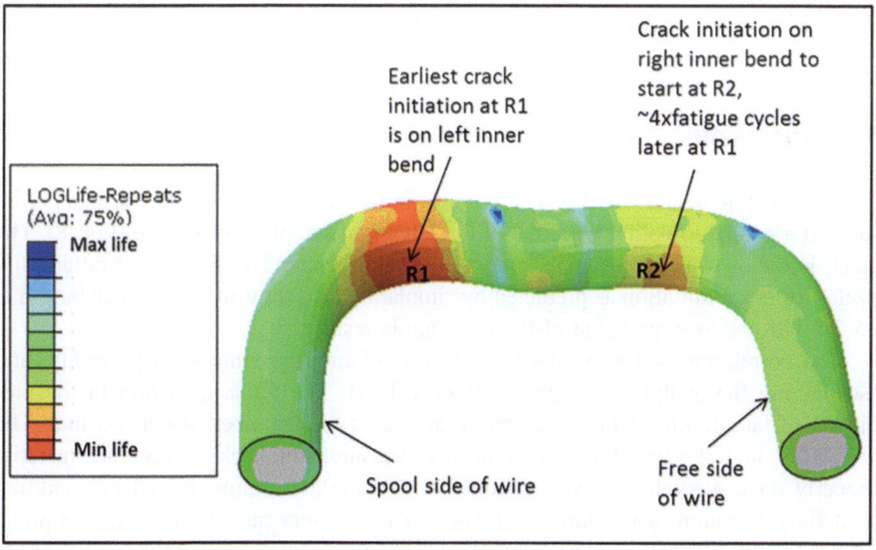

Fig. 7. Fatigue life contour on the primary wire

6 Lab Test Results and Comparison with Simulation Results

To validate the simulation results for fatigue durability of the wire, coil prototypes were built and were subjected to intentional 'Test to Failure' thermal cycles in the laboratory. The thermal cycles were continued up to crack formation. Figure 8 shows the crack location regions on a sample, as seen using an electron microscope. Cracks were found

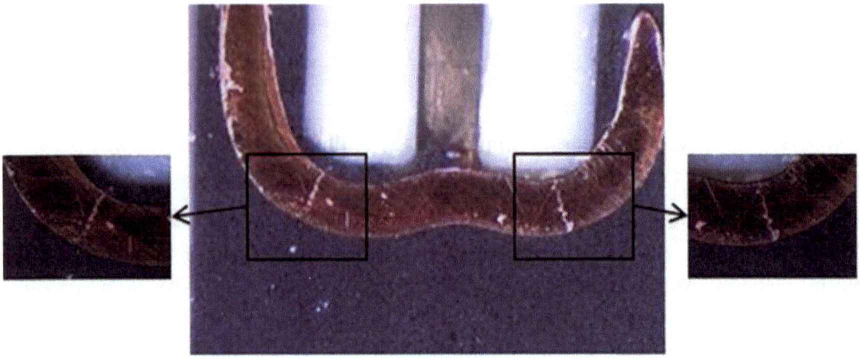

Fig. 8. Lab test results for crack locations on a tested coil sample

on the inner side of the bends at two locations and happen to correlate well with locations R1 and R2 found by simulation as in Fig. 7. All the tested samples showed repeatedly the same location for the earliest crack initiation. However the fatigue life cycles from the test measurements varied between the samples and hence an average of these values was considered. The number of fatigue life cycles for crack initiation as predicted by simulation matched 85 % with the average value of the 'Test to Failure' cycles.

7 Conclusions

The simulation results in Fig. 7 for crack initiation location on the primary wire showed a good correlation with that of the lab tested samples of the wire (Fig. 8). The crack initiation was on the inner side of the wire bends. The number of fatigue life cycles for crack initiation as predicted by simulation also showed a good matching (i.e. 85 %) with the average value of the test sample results.

The correlation of the results helped in validating the simulation procedure and ascertaining the analytical design confidence level. The 15 % deviation in the prediction of fatigue life of the wire can be attributed to a number of reasons like; The variability in test sample construction and measurement, lack of accurate material property data, prediction accuracy limitations of the algorithms, boundary condition modelling deviations and a number of others. These factors can be corrected to improve further the prediction accuracy.

The simulation approach presented in this paper can be employed to make a reliable prediction of the thermal fatigue and durability of the ignition coil parts to know upfront the validation and field failure behavior of the coil, in the early part of the design cycle itself. Using this approach with the design iterations based on the framework of Six-Sigma and DOE methodologies will help in identifying critical design parameters and arriving at a robust design of the product. All these steps will help in meeting the customer quality requirements, reducing product development time and prototype building plus validation costs.

Acknowledgments. The authors acknowledge with thanks to Etienne Jacqué and Peter Weyand for their approvals, inputs and support for the entire part of this product development study and the publication of this paper. The authors thank Ravi Salagame for his constant support and encouragement. All the above mentioned sponsors are from Delphi Corporation.

References

1. Delphi Automotive Systems: Delphi Ignition Systems Application Manual (2005)
2. Aye, A., Jacqué, E., Lorenz, F., Weyand, P., Schilling, S.: Simulation as integral part in the product development process of ignition coils. In: 1st IAV Conference: Advanced Ignition Systems for Gasoline Engines (2012)
3. Dassault Systémes: Simulia Abaqus ver 2016 Documentation (2016)
4. Dassault Systémes: Simulia FeSafe ver 2016 Documentation (2016)

Visualization of Ignition Processes

Calorimetry and Atomic Oxygen Laser-Induced Fluorescence of Pulsed Nanosecond Discharges at Above-Atmospheric Pressures

Benjamin Wolk([✉]) and Isaac Ekoto

Combustion Research Facility, Sandia National Laboratories,
Livermore, CA 94551, USA
bmwolk@sandia.gov

Abstract. The conversion efficiency of secondary electrical energy into thermal energy was measured in air using an optically accessible spark calorimeter for high-voltage (28 kV peak) pulsed nanosecond discharges with secondary streamer breakdown (SSB) and similar low-temperature plasmas (LTP) without. Initial pressures were varied between 1 and 5 bar absolute, with the anode/cathode gap distances likewise varied between 1 and 5 mm. Secondary electrical energy was measured using an in-line attenuator, with the thermal energy determined from pressure-rise calorimetry measurements. The SSB probability at each initial pressure and gap distance was also recorded. The calorimetry measurements confirm that, similar to inductive spark discharges, SSB discharges promote ignition by increasing the local gas temperature. LTP discharges, on the other hand, had very little local gas heating, with electrical-to-thermal conversion efficiencies of ~ 1 %. Instead, the LTP was found to generate substantial O-atom populations — measured using two-photon laser-induced fluorescence near the anode where electric field strengths were strongest — that persisted for 100's of microseconds after the discharge. The influence of 10 repetitive pulses spaced 100 μs apart was also evaluated for a fixed 5 mm electrode gap distance, with the conditional SSB probability for each pulse evaluated using an available photodiode, with the SSB probability found to have increased for each successive pulse. The influence of chemical and thermal preconditioning by the preceding LTP pulse was evaluated, with the increase in SSB occurrence attributed predominantly to mild gas heating that decreased number densities between the electrodes and hence the gas resistance for the subsequent pulse.

Keywords: Spark calorimetry · Pulsed nanosecond discharge · Low-Temperature plasma · Atomic oxygen Two-Photon laser induced fluorescence

1 Introduction

Plasmas — defined here as globally neutral gaseous mixtures of positively- and negatively-charged particles that interact by electric forces in a collective manner — have long been used to control ignition and combustion processes. Conventional inductive coil sparks that feature thermal plasmas with electron energies in equilibrium

M. Günther and M. Sens (eds.), *Ignition Systems for Gasoline Engines*,
DOI 10.1007/978-3-319-45504-4_10

with the bulk gas energy have fallen out of favor for next-generation automotive engines due to excessive electrode erosion rates from high-temperature arc discharges and poor tolerance to dilution by air or exhaust-gas recirculation (EGR) [1]. Conversely, ignition induced by low-temperature plasma (LTP), where an imbalance exists between ion and electron temperatures, leads to enhanced reaction kinetics as a result of elevated electron energies without the associated translational gas temperature increases that cause electrode erosion. Automotive low-temperature plasma-assisted ignition (PAI) systems extend dilution tolerance limits through the formation of active radicals that shorten ignition delays [2] and rapid heating via electron energy transfer processes [3–5]. Ignition kernel expansion speeds may also be influenced by hydrodynamic instabilities and ionic winds [6, 7]. LTP generated atomic oxygen (O-atom) in particular has been identified as an important radical species that accelerates hydrogen abstraction from the fuel, which in turn increases overall mixture reactivity [8]. For automotive engine combustion, these attributes reduce ignition delay times, suppress knock inducing end-gas auto-ignition, alter fuel reactivity and diffusivity characteristics, and improve noxious emission oxidation [9].

LTP are generally classified by the reduced electric field strength, or the electric field strength (E) normalized by the gas number density (N). Broadly speaking, PAI systems sorted by increasing E/N include: microwaves [3, 10–12], high-frequency resonant discharges [7, 13–16], and pulsed nanosecond discharges (PND) [2, 4, 17–20]. Since LTP related ionization and dissociation rates increase with higher E/N due to limited collisional electron energy transfer effects [21], several recent PAI experiments have focused on PND systems. These igniters typically have E/N greater than 100 Townsend (Td) and electron temperatures above 10 electron volts (eV), while pulse energies remain modest (10 s of mJ) due to the truncated discharge [22–25]. Note that secondary streamer breakdown (SSB), where the inter-electrode gas becomes highly electrically conductive, can occur if the gas resistance is sufficiently low (i.e., from low pressures, high temperatures, short electrode gap distances, or gases with low ionization potential) [26, 27]. As a result, a SSB discharge more closely resembles a thermal plasma spark with 2 key differences: (1) the electrical-to-thermal energy transfer is generally more efficient [1, 28], and (2) electrode wear greatly increases due to the elevated discharge current.

While electrode erosion issues can be mitigated through the use of dielectric electrode insulators that repel the discharge current [29, 30], the electric field strength becomes compromised with an associated decrease in radical formation rates. Instead it may be preferable to generate LTP without the use of dielectrics, with SSB instead inhibited through the selection of appropriate electrode geometry and pulse discharge characteristics. It is important to note that at engine-relevant pressures, LTP streamer propagation deviates from the well-established self-similar laws (i.e., increased streamer branching, thinner leader structures, and slower than expected streamer velocities) with unknown impact on discharge kinetics [31–33]. Potential causes for the scaling deviation at high-pressure include a change in electron recombination physics [31], less effective streamer cooling by thermal conductivity [34], and photoionization by excited-state nitrogen relaxation [32].

Previous studies have investigated the use of multi-pulse operation, where close-coupled PND with pulse dwell times of ~100 μs are used to continuously augment radical production while ideally avoiding SSB [22, 23]. However, direct discharge imaging during stable engine operation revealed that SSB almost always occurred at some point during the pulse burst; an example image of the luminous SSB event from a 10-pulse burst from Sjöberg et al. [23] is shown in Fig. 1. Moreover, once SSB occurs during the burst, each successive discharge in the pulse-burst exhibits SSB biased towards the same cathode site as the first SSB event. The results suggest some sort of ambient gas preconditioning between the electrodes has occurred. Whether the preconditioning mechanism is chemical or thermal remains an open question.

The current paper describes experiments performed in a custom-built, optically accessible spark calorimeter used to evaluate high-voltage PND. For all experiments,

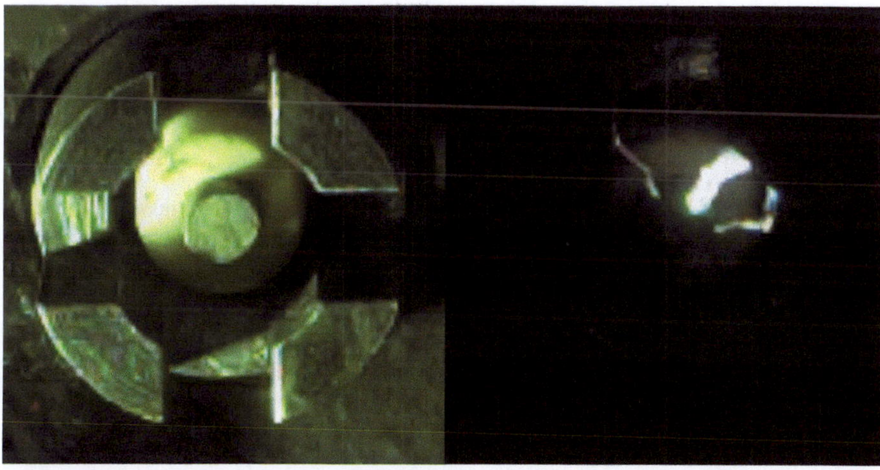

Fig. 1. Image of a 4-prong PND plug in a single-cylinder research engine along with imaging for 10-pulse PND bursts with a 100 μs dwell between successive pulses. Intense luminosity between the central anode and one of four separated cathode sites is indicative of SSB.

the test gas was ultra-air with a point-to-point electrode configuration selected so that the local E/N, and hence active radical formation, was maximized near the anode and cathode. Pressure-rise calorimetry was used to measure the electrical-to-thermal energy conversion efficiency for single PND discharges as a function of ambient pressure and electrode gap distance. Calorimetry results were further used to determine the SSB probability for each pressure and gap distance; the electric-to-thermal efficiency was conditionally averaged based on whether the PND remained an LTP or transitioned to a SSB discharge. The SSB probability as a function of initial pressure was further determined for a 10-pulse burst with a fixed 5.0 mm electrode gap distance. Post-discharge *in situ* O-atom populations were also directly measured via two-photon absorption laser induced fluorescence (TALIF) [35, 36] for select LTP-only conditions. Finally, the calorimetry results are used to evaluate potential thermal and chemical mechanisms believed to be responsible for SSB.

2 Experiment Description

All experiments were performed in a custom-built, optically accessible, spark calorimeter illustrated in Fig. 2. The calorimeter body is constructed of 316 stainless steel, with a cylindrical 28 cm^3 internal chamber volume ($\mathcal{V}_{cal}$). An inlet port is used to fill ultra-air into the chamber up to the desired pressure (up to 10 bar absolute), with an embedded type-K thermal couple used to measure the initial gas temperature prior to ignition. The calorimeter body also features viewing access of the discharge event through an optical port on the front side wall along with input and output laser access ports that facilitate acquisition of laser-based spectroscopy measurements.

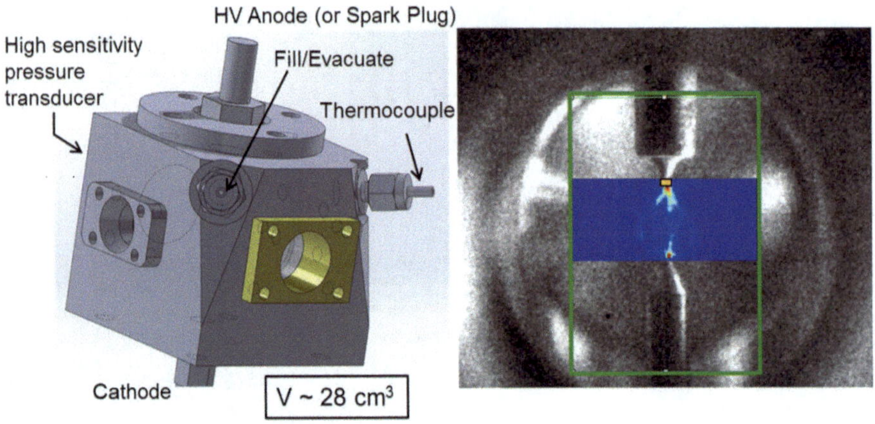

Fig. 2. Illustration of the optically accessible spark calorimeter (left) and an image of the spark gap as visualized through the optical viewing port (right). An image of filtered natural luminosity from the discharge has been superimposed between the electrodes. The TALIF interrogation region is indicated by the orange rectangle near the anode.

2.1 Pressure-Rise Calorimetry

To generate the PND within the spark calorimeter, a Transient Plasma Systems Inc. SSPG-101-HF high-voltage (28 kV peak) pulse generator with a 12 ns full width at half max (FWHM) pulse with and a 5 ns rise time was used. An inline attenuator was used to monitor pulse voltage and current for each discharge event. The high-voltage anode used a modified NGK DP7EA-9 size M12 non-resistive sparkplug that was centrally positioned on the calorimeter top. The anode tip was machined to a rounded point so that local electric field strengths were maximized, while maintaining relatively repeatable discharge characteristics. The plug J-hook and the top 1 cm of material from the outer body ground was removed, with the cathode replaced by a sharpened 3.18 mm diameter steel rod installed from the calorimeter base and secured in place by a Swagelok fitting. Multiple rod lengths were fabricated so that the inter-electrode distance between the anode and cathode could be varied between 1.0 and 5.0 mm. Key calorimeter specifications and operating conditions are summarized in Table 1.

Table 1. Calorimeter specifications and operating conditions.

Calorimeter internal volume [cm^3]	28.0
Purge and fill cycle time [s]	90
Initial temperature [K]	294 ± 5
Initial pressures [bar]	1.08 – 5.0
Inter-electrode distance [mm]	1.0, 2.1, 2.9, 4.0, 5.0

Initial chamber pressures were varied between 1 and 5 bar absolute with initial temperatures fixed at ~294 K. For these conditions, the chamber gas number densities are equivalent to in-cylinder engine conditions with pressures ~3 times greater due to compressive heating. At each gap distance and initial pressure, 30 calorimetry runs were performed. To ensure that residual discharge products did not persist between runs and possibly influence the subsequent spark, chamber contents after each run were purged using fill and vacuum solenoid valves connected to the ultra-air supply and an available vacuum turbo-pump, respectively. Chamber fill pressure was accurately controlled using a Proportion-Air pressure control valve. Differential chamber pressures were measured using a PCB 106B52 high-sensitivity pressure transducer (725 μV/Pa) for initial pressures up to 4 bar and a PCB 106B51 high-sensitivity pressure transducer (145 μV/Pa) for higher initial pressures. Each transducer was secured to the calorimeter using the same acetal mounting adapter for electrical isolation with the signal amplified by a Kistler 5134B coupler. All data were recorded using a LeCroy HDO 6054 500 MHz high-definition oscilloscope. Recorded pressures were filtered by a 400 Hz low-pass filter during post-processing. The SSB probability at each initial pressure and gap distance was determined from the resultant pressure record for the single-pulse experiments. The influence of repetitive discharges (up to 10) spaced 100 μs apart were also evaluated for a fixed 5.0 mm electrode gap at pressures up to 2.5 bar. The conditional breakdown probability for each discharge event in the pulse-burst was determined from light emission recorded by a Thorlabs DET210 high-speed photodiode placed an available window port.

Electrical-to-thermal conversion efficiency — defined here as the thermal energy deposited into the gas (E_{therm}) normalized by the secondary electrical energy delivered (E_{elec}) to the igniter — was determined for each discharge. Secondary electrical energy was obtained by direct integration of the instantaneous electrical power (W), calculated from the product of the inline attenuator voltage (V) and current (I) measurements over the length of each pulse (t_p):

$$E_{elec} = \int_0^{t_p} W(t)dt = \int_0^{t_p} V(t)I(t)dt \tag{1}$$

Gas thermal energy deposition was determined from the volumetric integration of the change in internal energy (δu):

$$E_{therm} = \int \delta u d\mathcal{V} = \int c_v \Delta T d\mathcal{V} = \left(\frac{c_v}{R}\right) \mathcal{V}_{cal} \Delta P = \frac{\mathcal{V}_{cal} \Delta P}{\gamma - 1} \quad (2)$$

where R, c_v, and γ respectively are the gas constant, volumetric heat capacity, and specific heat ratio for air. The pressure-rise (ΔP) was the highest recorded differential pressure measurement from a calorimetry experiment. Due to the small pressure-rises encountered in the experiments, a constant $\gamma = 1.4$ was assumed.

2.2 O-Atom Two-Photon Laser Induced Fluorescence

O-atom TALIF measurements within the spark calorimeter were enabled by 16 mm diameter entrance and outlet laser windows with a 12.9 mm clear aperture that were installed on opposing calorimeter sidewall ports. A larger viewing widow with a 20 mm clear aperture was placed on the sidewall orthogonal to the laser windows. Calorimeter window ports included a 45° beveled seat and copper gasket that mated with a complementary bevel on the front window surface to minimize mechanical stresses that could induce window fracture when the calorimeter was pressurized. A metal retainer was used to secure the windows into the calorimeter via 4 screw holes that mated with tapped holes around the window ports; another copper gasket was placed between the retainer and the window back surface to again minimize mechanical stresses. Two Kalrez O-rings were placed into machined O-ring groves during window installation to eliminate spurious leakage when the calorimeter was pressurized.

O-atom TALIF was excited by a deep-UV laser beam with a 12 ns FWHM pulse width that was generated by an optical parametric oscillator (Spectra-Physics MOPO SL) pumped by an injection-seeded Nd-YAG laser (Spectra-Physics Pro-270); a schematic of the setup is provided in Fig. 3. An integrated frequency doubler tuned to 225.7 nm generated the deep UV probe beam. A single plano-convex spherical lens ($f = 100$ mm) was used to focus the input laser beam just below the spark anode, with the beam diameter ~500 μm at the interrogation region. Laser energy delivered into the spark calorimeter was 1.2–2.5 mJ per pulse, and was monitored using a back reflection off of the spherical lens by a Coherent J-25 MB-LE shot resolved laser energy meter. For each operating condition, 50 TALIF images were recorded through the viewing port window via a Princeton instruments PI-MAX 4: 1024 EM intensified camera with a 1024 × 1024 pixel imaging array. The camera intensifier gate time set to 300 ns. To prevent intensifier damage, only consistent LTP conditions (i.e., SSB probability equals zero) were imaged. A single narrow band filter centered at 850 nm (50 nm FWHM) was used to isolate the $3p^3P \rightarrow 3s^3S$ transition located near 844.87 nm. The filter did allow stimulated emission of electronically excited atomic nitrogen and oxygen produced by the discharge to pass through (see Fig. 2 for a sample image). Accordingly, the earliest O-atom TALIF measurements were recorded 15 μs after the discharge so that all discharge induced luminosity could subside. The maximum camera gain and 2×2 super-pixel binning were used to amplify the signal-to-noise ratio. Flat field and intensifier nonlinear response corrections were implemented using an integrating sphere and arc lamp [37]. To make the results semi-quantitative, TALIF images were corrected for electronic bias, background illumination, and the square of the laser irradiance (to account for the 2-photon absorption process). True quantitative number densities require a more accurate identification of the laser wavelength, which will be accomplished in future TALIF experiments by direct measure of the fundamental wavelength before the frequency doubler by an integrated wavemeter.

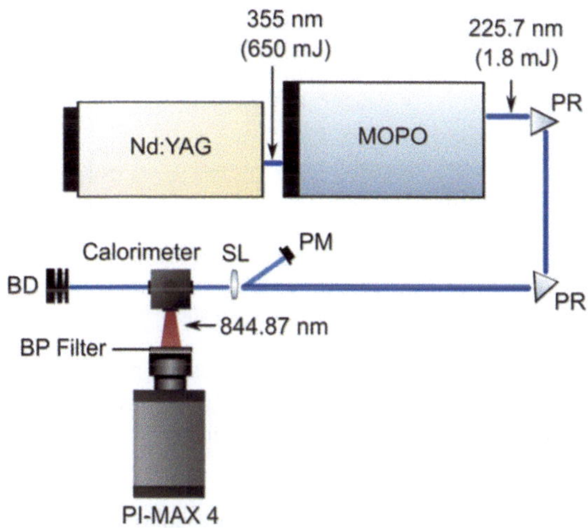

Fig. 3. Diagram of the O-atom TALIF imaging setup into the calorimeter.

High laser irradiances required for sufficient TALIF signal meant the laser entrance and outlet windows could be susceptible to surface etching and/or drilling, which could limit usable window lifespans and lead to costly downtime while replacements are installed. Several design features were accordingly adopted to extend the window service life and simplify the replacement process. A laser shutter was installed to limit laser strikes on the windows to only those needed for the measurements. Standard UV grade quartz (Corning 7980) windows were used because of their superior tolerance to laser damage. Front and back window surfaces were polished with ¼ wave surfaces and had a specified wedge angle of less than 30 s. Finally, care was taken to limit the peak irradiance on all window surfaces below 20 MW/cm^2.

3 Results and Discussion

3.1 Pressure-Rise Calorimetry

Ensemble averaged pressure-rise time histories normalized by the peak recorded pressure-rise are plotted for 60 ms after the discharge in Fig. 4 for a fixed 5.0 mm electrode gap with initial pressures between 1.08 and 1.52 bar. Pressure-rise traces were conditionally averaged depending on whether they were from LTP (evidenced by peak pressure-rise values below 12 Pa), or from PND that underwent SSB (indicated

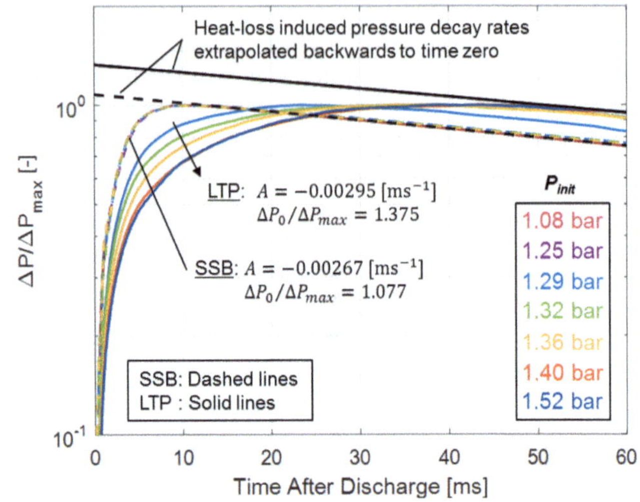

Fig. 4. SSB (dashed) and LTP (solid) post-discharge pressure rise time histories normalized by the peak recorded pressure for a 5.0 mm gap in air for initial pressures between 1.08 and 1.52 bar. Pressure decay rates are also plotted for each discharge.

by peak pressure-rise values that were 100 s of pascals). Heat-loss induced pressure decay rates (A) for both the LTP and SSB discharges were extrapolated to time zero using a logarithmic function (Eq. 3) fit to the respective normalized pressure-rise time histories between 40 and 60 ms after the discharge:

$$\log\left(\frac{\Delta P}{\Delta P_{max}}\right) = A \cdot time(ms) + \log\left(\frac{\Delta P_0}{\Delta P_{max}}\right) \tag{3}$$

Best fit values for A and the normalized regression pressure at time zero ($\Delta P_0/\Delta P_{max}$) for both the LTP and SSB discharges are included in Fig. 4. Decay rates for the LTP and SSB discharges are in good agreement with each other, with similar results observed from the pressure-rise time-histories for the other gap distances.

Normalized SSB discharge pressure-rise time histories collapsed to a common curve, with a single steep pressure-rise that lasted ~8 ms and followed by the heat-loss induced pressure decay. For the LTP discharges, the pressure-rise displayed two distinct stages: (1) a steep initial pressure rise similar to what was observed for the SSB discharges, followed by (2) a shallower, extended pressure rise that lasted another ~20–30 ms. The first stage is believed to result from local gas heating by the discharge and a possible increase in number density due to O_2/N_2 dissociation, while the second stage is believed to be from exothermic recombination of dissociated radicals or electronic state relaxation back to the ground state. The constant pressure decay rates observed 40 ms beyond the discharge suggest that de-excitation of long-lived electronic states such as $O_2(a1)$ was largely complete for all conditions. Unlike the SSB discharges, normalized LTP discharge pressure-rise curves for initial pressures between 1.29 and 1.4 bar did not collapse to a common curve. For these conditions, as the initial pressure increased there was an associated decrease in the proportion of first-stage gas heating. For the two highest evaluated initial pressures (1.4 and 1.52 bar), the traces once again were self-similar, with roughly equal energy deposition into the gas from stages 1 and 2. These results suggest the existence of a LTP-to-SSB transition regime, which will be explored in further detail below.

Single-Pulse SSB Occurrence Probability. The SSB occurrence probability was determined for a range of electrode gap distances and initial pressures up to 5.0 bar for a single-pulse, with the results plotted in Fig. 5. At all gap distances above a threshold initial pressure there was a relatively steep decline in SSB occurrence probability, with the fall-off sharper for larger electrode gap distances. For the 1.0 mm gap, 100 % SSB occurrence probabilities were observed for initial pressures below 3 bar, with at least some marginal chance for SSB occurrence at all initial pressures explored in the current study. When the gap distances were increased, the initial pressures where SSB was observed dropped. With the largest 5.0 mm gap, no SSB was observed for initial pressures above 1.4 bar. Note that for the 5.0 mm gap, the 3 initial pressures in the transition regime where inconsistent SSB were observed are the same LTP-to-SSB transition conditions identified from the pressure-rise time histories in Fig. 4. These results suggest that complete SSB did not occur for before the discharge was terminated.

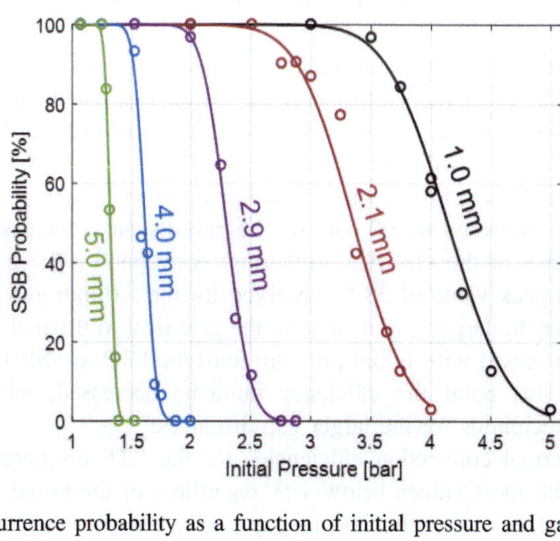

Fig. 5. SSB occurrence probability as a function of initial pressure and gap size for a single 28 kV discharge in air.

Electrical-to-Thermal Energy Conversion Efficiency. Calculated pulse electric energy (Eq. 1) for the SSB and LTP discharges from the in-line attenuator discharge voltage and current measurements are plotted in Fig. 6a. LTP discharge electrical energy — regardless of the initial pressure, electrode gap distance, or whether it overlapped with the LTP-to-SSB transition regime — was relatively consistent at around 15 mJ. For SSB discharges in the LTP-to-SSB transition regime (highlighted in the dashed rectangle in Fig. 6a), the discharge electrical energy more than doubled to

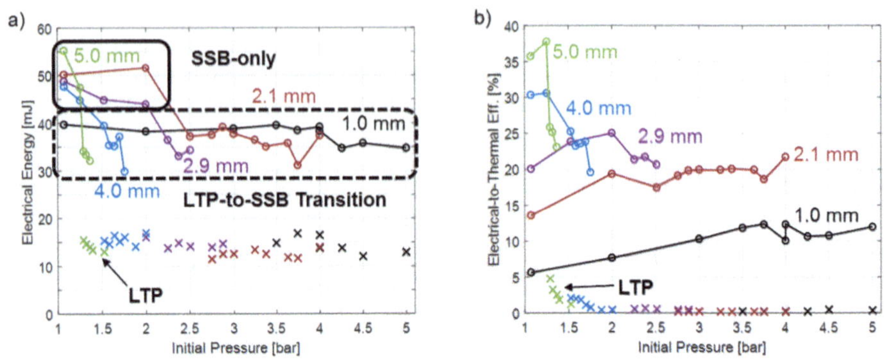

Fig. 6. Plots of (a) electrical energy and (b) electrical-to-thermal energy conversion efficiency for SSB discharges (O) and for LTP discharges without SSB (x) as a function of gap size and initial pressure.

around 38 mJ as the significant decrease in inter-electrode gas resistance led to a large jump in current flow between the anode and cathode. For SSB discharges outside of the LTP-to-SSB transition regime, the delivered discharge electrical energy further jumped to around 48 mJ; the exception was for the 1.0 mm gap where the discharge electrical energy remained around 40 mJ.

The electric-to-thermal conversion efficiency was calculated for all LTP and SSB conditions using the discharge electrical energy from Eq. 1 and the thermal energy deposited in the gas from Eq. 2, with all results plotted in Fig. 6b. When the initial pressure was 1.08 bar, the SSB discharge electric-to-thermal conversion efficiency for the 1.0 mm gap was just over 5 %, with this value monotonically increasing to 12 % as the initial pressure increased to 5.0 bar. As the gap distances increased, there was a proportional increase in the electrical-to-thermal conversion efficiency for the SSB discharges, with a peak value of 38 % observed for the 5.0 mm gap with a 1.25 bar initial pressure. For the larger gap distances, the electrical-to-thermal conversion efficiency likewise increased with initial pressure until the LTP-to-SSB transition region was reached. At this point the efficiency numbers decreased, with the decrease becoming more precipitous for the larger gap distances.

Electrical-to-thermal conversion efficiencies for the LTP discharges were almost uniformly low, with most values below 1 % regardless of the initial pressure or gap distance. The exception was for the 5.0 mm gap (and to a lesser extent the 4.0 mm gap), where in the LTP-to-SSB transition regime the LTP electrical-to-thermal conversion efficiencies increased to ~5 % for the 1.29 bar initial pressure condition. These results confirm that LTP discharges at this condition flirt with a transition to SSB.

Multi-Pulse SSB Occurrence Probability. The SSB occurrence probability for a 10 pulse burst with a 100 μs dwell between each pulse was evaluated for a fixed 5.0 mm gap at a range of initial pressures up to 2.5 bar. The first pulse where SSB occurred was identified by whether or not luminous emission was observed by a photodiode positioned at the laser exit window. Note that after the first observance of SSB, SSB was observed in all successive discharge events in the pulse train. Accordingly, only the conditional probability for each pulse is plotted in Fig. 7, where discharges preceded by SSB were excluded from the statistics. Note further that the SSB occurrence probability as a function of pressure for the first pulse was nearly identical to the SSB occurrence probability in Fig. 5 for the 5.0 mm gap obtained from the pressure-rise calorimetry measurements. The good agreement suggests that (1) the first discharge in the 10-pulse burst was well matched to the single-pulse results, (2) the experiment setup is highly repeatable, and (3) the photodiode is capable of reproducing similar SSB probability results as the calorimetry.

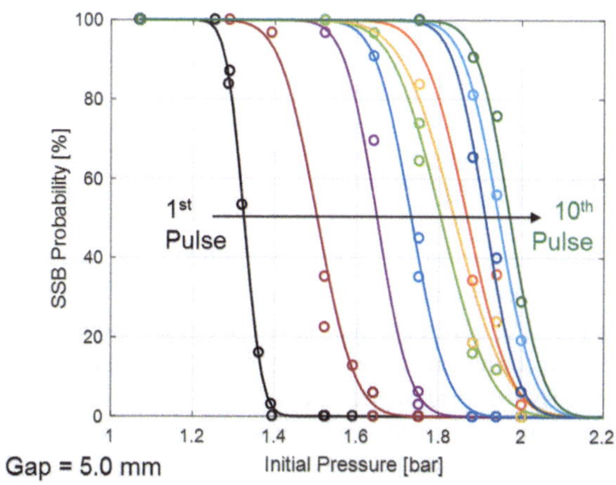

Fig. 7. SSB probability for each pulse in a 10-pulse burst with a 100 μs dwell between successive discharges as a function of initial pressure and with a fixed 5.0 mm electrode gap distance. SSB probability was zero for all pulse numbers at 2.5 bar initial pressure.

If subsequent discharges in the 10-pulse burst had no dependence on gas preconditioning from the previous pulse, then the conditional averaging process should produce identical SSB probability curves as a function of initial pressure. However, from Fig. 7 this is not the case with each successive pulse resulting in a rightward shift of the SSB probability curve towards higher initial pressures. The highest initial pressure where SSB was observed shifted from 1.25 bar for the first pulse to around 2.1 bar for the tenth pulse. The magnitude of the shift lessened with each successive pulse, with data from the 10th pulse nearing an asymptote. The results demonstrate that with close-coupled multi-pulse discharges, SSB can occur at conditions where there otherwise would have been no chance for SSB occurrence due to some form of thermal or chemical preconditioning from previous pulses.

To more clearly illustrate the discharge characteristics for multi-pulse operation in the LTP-to-SSB transition regime, a scatter plot of the delivered pulse electrical energy for the first 5 pulses of each test when the initial pressure was 1.6 bar is provided in Fig. 8a. Respective regions where the discharges were clearly LTP (i.e., less than 20 mJ delivered electric energy) or SSB (i.e., greater than 35 mJ delivered electric energy) are labeled. Pulses in between the dashed lines are believed to be from the LTP-to-SSB transition discharges.

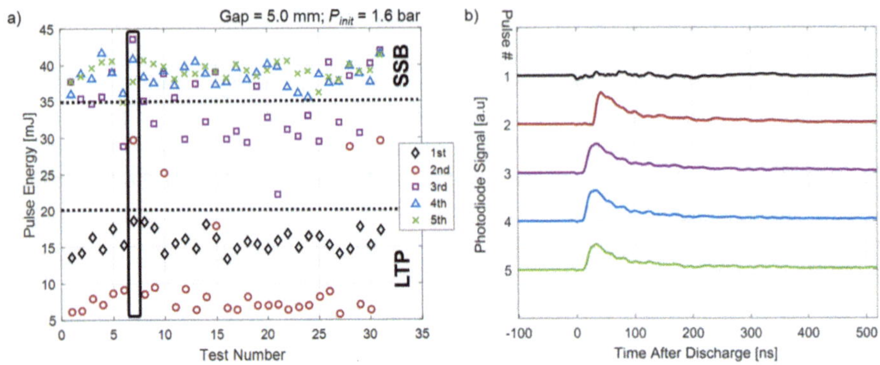

Fig. 8. (a) Discharge electrical energy for the first 5 pulses in the 10-pulse burst; LTP and SSB regions are identified by the dashed lines. (b) The photodiode signal time history for each pulse in Test Number 7, highlighted in Fig. 8a.

For the test sequence in Fig. 8a, each first pulse in the burst was LTP with all delivered pulse electric energies around 15 mJ. In most instances the second pulse likewise remained LTP, although the pulse energy roughly halved to around 8 mJ. There were 4 recorded instances where the second pulse transitioned into the LTP-to-SSB regime. By the third pulse, all discharges had delivered pulse energies that were characteristic of the LTP-to-SSB transition regime or were full SSB discharges. For all subsequent pulses, the discharges were firmly in the SSB regime.

In Fig. 8b, the photodiode time-history for each pulse from Test 7 in the sequence (highlighted in Fig. 8a) is shown for each discharge event. Based on the delivered pulse energies for this test, the first pulse was LTP, with the second pulse in the LTP-to-SSB transition region and all subsequent pulses in the SSB regime. As expected, there was very little luminosity from the LTP of the first pulse. For each subsequent pulse, there was a rapid rise in photodiode signal shortly after the discharge indicative of SSB; the peak recorded photodiode signal is relatively consistent for each discharge. However, intense luminosity was not detected by the photodiode until ~30 ns after the discharge for the second pulse, which roughly doubled the ~15 ns dwell time observed for all subsequent SSB pulses. The delayed luminosity for the second pulse indicates a delay in the onset of SSB, which is characteristic of LTP-to-SSB discharges.

In Fig. 9, discharge power, current, and voltage traces for Pulse 2 (LTP-to-SSB discharge) and Pulse 3 (SSB discharge) from Test Number 7 in Fig. 8a are overlaid with the corresponding photodiode signal from Fig. 8b. For both pulses, the voltage initially experienced a rapid rise (~5 ns) shortly after the discharge as the primary streamer head advanced toward the cathode. Free electrons generated by the ionization processes within the streamer flooded back toward the anode, leading to a large increase in discharge current (~125 amps) and a rapid rise in voltage potential (~20 kV). Once the primary streamer connected to the cathode, there was a rapid drop in voltage potential as a negative charge "return wave" flooded the primary streamer channel with electrons from the cathode. The discharge current likewise experienced a rapid decline. For both pulses, most discharge energy was delivered during this primary streamer process.

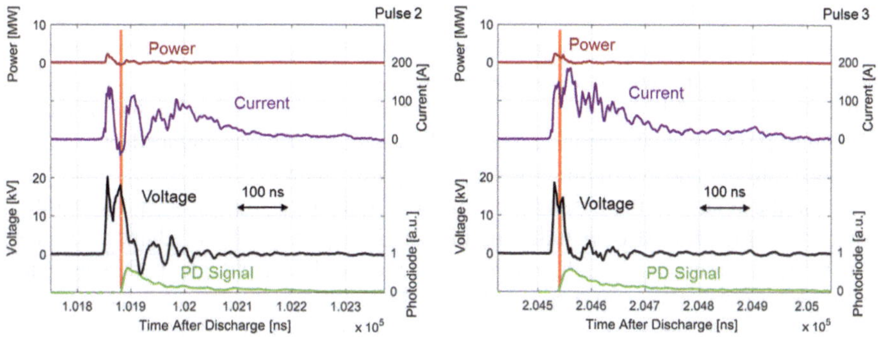

Fig. 9. Power, current, voltage, and photodiode time-history traces for Pulses 2 and 3 from Test Number 7 in Fig. 8a.

Concurrent with the rapid primary streamer advance and attachment to the cathode was the slower secondary streamer propagation also from the anode to the cathode, with the propagation speeds dictated by the local electric field strength. A second voltage spike that resulted from secondary streamer ionization processes is observed in the respective voltage time histories, with peak values roughly 20–30 ns after the discharge. For both Pulse 2 and Pulse 3, the peak voltage from the secondary streamer induced voltage spike closely corresponded with the appearance of photodiode luminosity indicative of SSB (highlighted by the orange vertical line in both plots).

The central difference between the two discharge events is the apparent slower secondary streamer propagation speeds for Pulse 2, which meant the primary streamer current was effectively zero by the time the secondary streamer connected with the cathode. While the current subsequently increased sharply due to the decrease in gas resistance between the electrodes, there was a commensurate drop in voltage potential as the supplied electrical pulse had already terminated. The net result was that there was very little delivered electrical energy after the SSB event. Pulse 3 on the other hand, exhibited a secondary streamer voltage spike with a sharper rise rate that is indicative of faster secondary streamer propagation velocities. As a result, the secondary streamer was able to connect to the cathode while the overall gap current remained relatively high. The combination of elevated voltage and current values that coincided with the SSB discharge led to an associated increase in electrical energy delivered to the gap.

3.2 LTP Two-Photon Laser Induced Fluorescence

O-atom TALIF was performed in the optically accessible calorimeter for the 5.0 mm gap with an initial pressure of 1.7 bar. The condition was selected because there was no observance of SSB for either the first or second pulse from the pressure-rise calorimetry experiments. As mentioned in the experiment description section, the laser beam focus was positioned as close to the anode as possible; it was obvious when the beam clipped the anode as luminous bits of metal fluorescence was observed in the TALIF images. The region below the anode was selected because this is where the largest electric field strengths are located ($E/N \approx 536$), and because it was the only location where consistent O-atom TALIF was observed. Signal was recorded for 50 consecutive discharges, with purge and fill cycles performed between each discharge as described earlier. O-atom TALIF images were recorded at discrete times after the discharge, beginning at 15 μs post-discharge (selected to avoid discharge induced emission as described earlier) and continuing until 200 μs post-discharge.

A sample image of the ensemble averaged TALIF measurement near the anode is provided in Fig. 10. Also included in Fig. 10 is a plot of the irradiance corrected TALIF signal intensity as a function of time after the discharge; both single and double-pulsed TALIF signal data are shown. Each data point represents the ensemble average of the 2 super pixels with the highest recorded signal from each frame. Although the data are not quantitative O-atom number densities, they are nonetheless useful to compare relative concentrations as a function of time after the discharge and how a single or double pulse impacts O-atom formation rates.

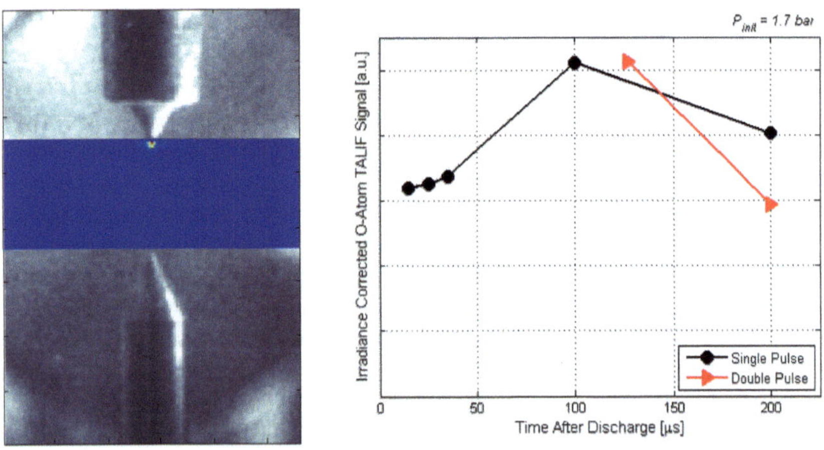

Fig. 10. Point (left) image of O-atom TALIF. Plotted on the right are the irradiance corrected TALIF signal data as a function of time after the discharge for single (black) and double (red) pulse operation.

O-atom TALIF signal increased roughly 60 % between 15 and 100 μs after the discharge, with a decline thereafter. These results are consistent with the measured O-atom number densities observed by Pendleton et al. [35] for a similar discharge in ambient air, although they observed a sharp decline in O-atom concentrations by 100 μs after the discharge, with no observable concentrations by 200 μs post-discharge. These results are somewhat surprising since recombination rates for the current setup are expected to be faster with the higher initial pressure. However, it is important to note that the pulse characteristics for the Pendleton et al. study were somewhat different, with a shallower rise-time along with lower peak voltages and currents. It is possible that larger O-atom concentrations at elevated electronic states were created with the current setup, with ground state O-atom populations being continuously fed by the collisional de-excitation of these species. Additional measurements are needed to verify this result.

Double pulse data were also acquired, with the results plotted in Fig. 10 referenced to the first of the two pulses. It is interesting that the O-atom populations at 127 and 200 μs after the first pulse are nearly identical to the single-pulse values acquired at the same time after the pulse. Possible reasons for this observation include: (1) the second pulse induced some fluid mechanic process that displaced some of ground state O-atom populations generated by the first pulse, (2) ground state O-atom populations were completely saturated due to a lack of available diatomic oxygen, or (3) products created by the second pulse consumed much of the ground state O-atom generated by the preceding pulse. It is also important to note that the first discharge event resulted in some nominal gas heating between the electrodes. As a result, the total number density at the anode during and after the second discharge would likely be lower than for the first condition. Gas heating by the preceding discharge event process will be evaluated in the next section.

3.3 Discussion

It remains unclear why subsequent discharges from multi-pulse operation led to higher SSB probabilities relative to the previous discharge. There are two likely mechanisms that led to some sort of gas preconditioning between the electrodes by the preceding pulse that make the subsequent pulse more sensitive to breakdown. For the first mechanisms, we postulate that there is a chemical effect where the preceding LTP pulse discharge leads to the formation of new species (e.g., O-atom, atomic nitrogen, nitrogen oxide, ozone) or excited states that persist well after the original discharge and effectively lower the overall gas ionization potential (IP) for the next discharge. For the second mechanism, we consider the effect of modest gas heating by the preceding LTP discharge, which would reduce the local number density and hence the effective resistance of the gas between the electrodes.

To evaluate the chemical mechanism, the IP of commonly generated LTP species from discharges in air are considered, with the values tabulated in Table 2. Note that the only electronic excited state considered was $O_2(a1)$, which has decay lifetimes on the order of milliseconds in ambient air; all other possible electronic excited states have lifetimes that are no longer than a few nanoseconds. From Table 2, it is notable while N_2 has the highest IP relative to the other species, O_2 is among the lowest with ground state O-atom and ozone (O_3) having IP higher than O_2. Moreover, while the dissociation of N_2 into atomic nitrogen (N) would decrease the IP on a molar basis, the increase in number density would decrease the local reduced electric field. Only substantial production of nitrogen oxide (NO) or electronically excited $O_2(a1)$ would appear to reduce the gap IP in an appreciable way without leading to an associated reduction in E/N. Future experiments will seek to measure NO concentrations between the gap via laser-induced fluorescence. It should be noted that chemical modeling simulations using the GRI 3.0 NOx [38] chemistry mechanism indicate that most dissociated N would recombine as N_2 rather than form NO. Moreover, NO equilibrium formation rates are $\sim$1 ms, which is much longer than the 100 μs dwell between

Table 2. Ionization potentials for oxygen, nitrogen, and other LTP generated species.

Species	IP [eV]
O_2	12.1
N_2	15.6
NO	9.3
O_3	12.5
O	13.6
N	14.5
O_2(a1)	11.1

successive pulses used for the experiments here. It is thus concluded that while chemical effects are important, they are likely not a major driver for the increased incidence of SSB.

The thermal mechanism was evaluated by making the assumption that all thermal energy measured from the LTP-discharge calorimetry experiments was initially deposited into a cylindrical region that spanned the electrode gap. A uniform temperature increase (ΔT) in the volume was then estimated as the thermal energy normalized by the total heat capacity of the volume following Eq. 4.

$$\Delta T = E_{Therm} / \left(\rho c_p \mathcal{V}_{spark}\right) \tag{4}$$

Here, ρ and c_p are the density and heat capacity at constant pressure for air, while $\mathcal{V}_{spark}$ is the estimated initial cylinder volume of the spark channel. The spark channel diameter was estimated to be 625 μm based on the natural luminosity images acquired during the discharge event.

The initial temperature profile in the radial direction was then imposed as a boundary equation for a one-dimensional heat equation (in cylindrical coordinates), with the surrounding fluid set to the initial temperature prior to the discharge. To eliminate a discontinuity in the boundary condition between the cylinder and the surrounding ambient fluid, the initial profile was specified as an error function. The evolution of the temperature profile was then computed at discrete time intervals using the Matlab Partial Differential Equations tool box. An example profile of the of the initial temperature and the corresponding profile 100 μs later is provided in Fig. 11 for the calorimeter test with a 5.0 mm gap and a 1.4 bar initial pressure. For this condition, about 240 μJ of thermal energy was measured from the single-pulse calorimetry experiments, which from Eq. 4 led to a modest temperature increase in the initial spark volume ($\Delta T = 93\,\text{K}$). By 100 μs after the first discharge the centerline temperature was essentially unchanged, which meant the centerline number density had effectively decreased ~24 % between the first and second pulses.

From Fig. 5, the SSB occurrence probability was nearly zero when the initial pressure was 1.4 bar. By the subsequent pulse, the 24 % decrease in number density due to the increased temperature would have resulted in an effective decrease in initial pressure for the second pulse to 1.07 bar, which for the single-pulse data would have resulted in 100 % SSB occurrence probability. This SSB probability in fact agrees very

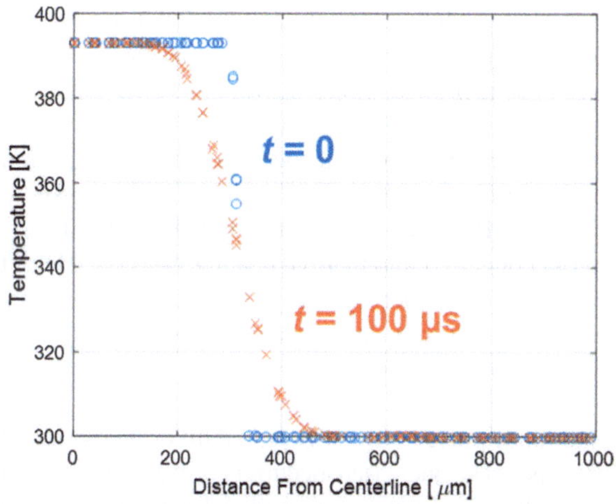

Fig. 11. Calculated temperature profiles at 0 and 100 μs after the discharge for a single-pulse calorimetry experiment with a 1.4 bar initial pressure and a 5.0 mm electrode gap. The initial profile temperature was set from Eq. 4 assuming that a 625 μm diameter cylinder extended between the electrodes and $E_{therm} = 240$ μJ from the calorimetry experiments.

well with the second pulse data for the 1.4 bar condition. Similar effects were observed for the other conditions. Based on these results, the proposed thermal mechanism appears to be predominantly responsible for the gas preconditioning between successive close-coupled PND pulses.

4 Conclusions

The conversion efficiency of delivered electrical energy into thermal energy in air was measured in a spark calorimeter modified for optical viewing and laser access for high-voltage (28 kV peak) PND (5 ns rise time, 12 ns FWHM) that underwent SSB or remained LTP. Initial pressures were varied between 1 and 5 bar, with the gap distances between a sharpened anode and cathode varied between 1 and 5 mm. Secondary electrical energy was measured using an in-line attenuator, while the thermal energy deposited into the gas was determined from the pressure-rise calorimetry measurements. The probability for single-pulse SSB occurrence was evaluated using the calorimetry measurements. The SSB occurrence probability for a 10 pulse burst with a 100 μs dwell between each pulse was also evaluated for a fixed 5.0 mm gap at a range of initial pressures up to 2 bar. Here, SSB was identified by luminous emission recorded by a photodiode positioned at an available laser exit window. Finally, for LTP-only discharges with either single or double-pulse operation, in situ O-atom TALIF was performed near the anode where electric field strengths were strongest. Major findings are as follows:

- The SSB occurrence probability for single-pulse PND decreased with larger gap distances and lower initial pressures. For all gap distances evaluated, LTP-to-SSB transition regions were identified where intermittent SSB would occur.
- Electrical energies for single-pulse LTP were ~ 15 mJ with less than 1 % of the electrical energy converted into thermal energy, except for the 5 mm gap where the thermal energy conversion increased to 5 % in the LTP-to-SSB region. Electrical energy in the LTP-to-SSB regime more than doubled (~ 38 mJ), with up to 25 % of the electrical energy converted into thermal heating. SSB outside of the LTP-to-SSB transition region had even higher delivered electrical energies (45 mJ+) and conversion efficiencies (30 %+) at the larger gap distances. Differences in SSB behavior were attributed to slower secondary streamer propagation rates for discharges in the LTP-to-SSB regime.
- For close-coupled multi-pulse PND, SSB probabilities were found to increase with each successive pulse. The increased occurrence was attributed mostly a thermal preconditioning mechanism from the preceding LTP pulse, modest heating of the gas between the electrodes lowered the number density and hence the electrical resistance for gases within the electrode gap.
- Significant O-atom populations were measured via TALIF just below the anode beginning ~ 15 μs after a LTP discharge for an initial pressure of 1.7 bar; the results agreed well with previous measurements from a similar setup, but at atmospheric pressure. Double-pulse operation was found to have negligible influence on the post-discharge O-atom populations near the anode. Additional planar O-atom TALIF imaging is needed to determine if pulse-induced advection displaced O-atom generated by the previous pulse.

The results suggest that to maximize active radical formation via LTP when PND are used, dwell times between successive pulses should be long enough to allow for the inter-electrode gas temperatures to relax back to ambient, or that successive pulse energies should be lowered to lessen the possibility of SSB. Moreover strong tumble flows could enable rapid and continuous refreshment of cooler ambient gases provided that the electrode design does not obstruct flows between the anode and cathode.

Acknowledgements. This work was performed at the Combustion Research Facility, Sandia National Laboratories, Livermore, CA. Financial support was provided by the U.S. Department of Energy, Office of Vehicle Technologies. Sandia National Laboratories is a multi-program laboratory managed and operated by Sandia Corporation, a wholly owned subsidiary of Lockheed Martin Corporation, for the U.S. Department of Energy's National Nuclear Security Administration under contract DE-AC04-94AL85000. The authors gratefully acknowledge the hardware support and fruitful technical discussions from Daniel Singleton and Jason Sanders of Transient Plasma Systems Inc. We further acknowledge technical engine data provided by Magnus Sjöberg of Sandia National Laboratories along with James Sevik, Riccardo Scarcelli, and Thomas Wallner of Argonne National Laboratory.

References

1. Dale, J.D., Checkel, M.D., Smy, P.R.: Application of high energy ignition systems to engines. Progr. Energy Combust. **23**(5–6), 379–398 (1997). doi:10.1016/S0360-1285(97)00011-7
2. Sun, W.T., Uddi, M., Won, S.H., Ombrello, T., Carter, C., Ju, Y.G.: Kinetic effects of non-equilibrium plasma-assisted methane oxidation on diffusion flame extinction limits. Combust. Flame **159**(1), 221–229 (2012). doi:10.1016/j.combustflame.2011.07.008
3. Wolk, B., DeFilippo, A., Chen, J.Y., Dibble, R., Nishiyama, A., Ikeda, Y.: Enhancement of flame development by microwave-assisted spark ignition in constant volume combustion chamber. Combust. Flame **160**(7), 1225–1234 (2013). doi:10.1016/j.combustflame.2013.02.004
4. Mintusov, E., Serdyuchenko, A., Choi, I., Lempert, W.R., Adamovich, I.V.: Mechanism of plasma assisted oxidation and ignition of ethylene-air flows by a repetitively pulsed nanosecond discharge. Proc. Combust. Inst. **32**, 3181–3188 (2009). doi:10.1016/j.proci.2008.05.064
5. Kosarev, I.N., Aleksandrov, N.L., Kindysheva, S.V., Starikovskaia, S.M., Starikovskii, A. Y.: Kinetics of ignition of saturated hydrocarbons by nonequilibrium plasma: CH_4-containing mixtures. Combust. Flame **154**(3), 569–586 (2008). doi:10.1016/j.combustflame.2008.03.007
6. Jaggers, H.C., von Engel, A.: The effect of electric fields on burning velocity of various flames. Combust. Flame **16**(3), 275–285 (1971). doi:10.1016/S0010-2180(71)80098-6
7. Leonov, S.B., Yarantsev, D.A.: Plasma-induced ignition and plasma-assisted combustion in high-speed flow. Plasma Sources Sci. Technol. **16**(1), 132–138 (2007). doi:10.1088/0963-0252/16/1/018
8. Ju, Y.G., Lefkowitz, J.K., Reuter, C.B., Won, S.H., Yang, X.L., Yang, S., et al.: Plasma assisted low temperature combustion. Plasma Chem. Plasma Process. **36**(1), 85–105 (2016). doi:10.1007/s11090-015-9657-2
9. Ju, Y.G., Sun, W.T.: Plasma assisted combustion: progress, challenges, and opportunities. Combust. Flame **162**(3), 529–532 (2015). doi:10.1016/j.combustflame.2015.01.017
10. Nishiyama, A., Ikeda, Y.: Improvement of lean limit and fuel consumption using microwave plasma ignition technology. SAE International 2012-01-1139 (2012). doi:10.4271/2012-01-1139
11. Wang, Q., Zhang, G., Liu, Y., Hou, L., Liu, C.: Microwave plasma ignition process of methane and air mixture under high pressure. In: 2013 19th IEEE Pulsed Power Conference (PPC), 16–21 June 2013, pp. 1–3 (2013)
12. Michael, J.B., Chng, T.L., Miles, R.B.: Sustained propagation of ultra-lean methane/air flames with pulsed microwave energy deposition. Combust. Flame **160**(4), 796–807 (2013). doi:10.1016/j.combustflame.2012.12.006
13. Chintala, N., Bao, A.N., Lou, G.F., Adamovich, I.V.: Measurements of combustion efficiency in nonequilibrium RF plasma-ignited flows. Combust. Flame **144**(4), 744–756 (2006). doi:10.1016/j.combustflame.2005.08.040
14. Pertl, F.A., Smith, J.E.: High-level modeling of an RF pulsed quarter wave coaxial resonator with potential use as an SI engine ignition source. SAE International 2008-01-0089 (2008). doi:10.4271/2008-01-0089
15. Hampe, C., Kubach, H., Spicher, U., Rixecker, G., Bohne, S.: Investigations of ignition processes using high frequency ignition. SAE International 2013-01-1633 (2013). doi:10.4271/2013-01-1633

16. Mariani, A., Foucher, F., Moreau, B.: The effects of a radio frequency ignition system on the efficiency and the exhaust emissions of a spark-ignition engine. SAE International 2013-24-0053 (2013). doi:10.4271/2013-24-0053
17. Bozhenkov, S.A., Starikovskaia, S.M., Starikovskii, A.Y.: Nanosecond gas discharge ignition of H_2- and CH_4- containing mixtures. Combust. Flame **133**(1–2), 133–146 (2003). doi:10.1016/S0010-2180(02)00564-3
18. Kosarev, I.N., Aleksandrov, N.L., Kindysheva, S.V., Starikovskaia, S.M., Starikovskii, A. Y.: Kinetic mechanism of plasma-assisted ignition of hydrocarbons. J. Phys. D Appl. Phys. **41**(3) (2008). doi:10.1088/0022-3727/41/3/032002
19. Adamovich, I.V., Choi, I., Jiang, N., Kim, J.H., Keshav, S., Lempert, W.R., et al.: Plasma assisted ignition and high-speed flow control: non-thermal and thermal effects. Plasma Sources Sci. Technol. **18**(3) (2009). doi:10.1088/0963-0252/18/3/034018
20. Won, S.H., Jiang, B., Dievart, P., Sohn, C.H., Ju, Y.G.: Self-sustaining n-heptane cool diffusion flames activated by ozone. Proc. Combust. Inst. **35**, 881–888 (2015). doi:10.1016/j.proci.2014.05.021
21. Starikovskiy, A., Aleksandrov, N.: Plasma-assisted ignition and combustion. Progr. Energy Combust. **39**(1), 61–110 (2013). doi:10.1016/j.pecs.2012.05.003
22. Sevik, J., Wallner, T., Pamminger, M., Scarcelli, R., Singleton, D., Sanders, J.: Extending lean and EGR-Dilute operating limits of a modern GDI engine using a low-energy transient plasma ignition system. In: Proceedings of the Asme Internal Combustion Engine Division Fall Technical Conference 2015, vol. 1 (2016)
23. Sjöberg, M., Zeng, W., Singleton, D., Sanders, J.M., Gundersen, M.A.: Combined effects of multi-pulse transient plasma ignition and intake heating on lean limits of well-mixed E85 DISI engine operation. SAE Int. J. Engines **7**(4), 1781–1801 (2014). doi:10.4271/2014-01-2615
24. Breden, D., Raja, L.L., Idicheria, C.A., Najt, P.M., Mahadevan, S.: A numerical study of high-pressure non-equilibrium streamers for combustion ignition application. J. Appl. Phys. **114**(8) (2013). doi:10.1063/1.4818319
25. Shiraishi, T., Urushihara, T., Gundersen, M.: A trial of ignition innovation of gasoline engine by nanosecond pulsed low temperature plasma ignition. J. Phys. D Appl. Phys. **42** (13) (2009). doi:10.1088/0022-3727/42/13/135208
26. Bastien, F., Marode, E.: Breakdown simulation of electronegative gases in non-uniform field. J. Phys. D Appl. Phys. **18**(3), 377–393 (1985). doi:10.1088/0022-3727/18/3/007
27. Ono, R., Oda, T.: Formation and structure of primary and secondary streamers in positive pulsed corona discharge - effect of oxygen concentration and applied voltage. J. Phys. D Appl. Phys. **36**(16), 1952–1958 (2003). doi:10.1088/0022-3727/36/16/306
28. Teets, R.E., Sell, J.A.: Calorimetry of ignition sparks. SAE International (1988). doi:10.4271/880204
29. Gibalov, V.I., Pietsch, G.J.: The development of dielectric barrier discharges in gas gaps and on surfaces. J. Phys. D Appl. Phys. **33**(20), 2618–2636 (2000). doi:10.1088/0022-3727/33/20/315
30. Shiraishi, T., Urushihara, T.: Fundamental analysis of combustion initiation characteristics of low temperature plasma ignition for internal combustion gasoline engine. SAE International 2011-01-0660 (2011). doi:10.4271/2011-01-0660
31. Babaeva, N.Y., Naidis, G.V.: On streamer dynamics in dense media. J Electrostat. **53**(2), 123–133 (2001). doi:10.1016/S0304-3886(01)00135-8
32. Liu, N., Pasko, V.P.: Effects of photoionization on similarity properties of streamers at various pressures in air. J. Phys. D Appl. Phys. **39**(2), 327–334 (2006). doi:10.1088/0022-3727/39/2/013

33. Lin, Y.H., Singleton, D., Sanders, J., Kuthi, A., Gundersen, M.A.: Pressure effects on transient plasma discharge in air. In: 2013 19th IEEE Pulsed Power Conference (PPC), 16–21 June 2013, pp. 1–5 (2013)
34. Tardiveau, P., Marode, E., Agneray, A., Cheaib, M.: Pressure effects on the development of an electric discharge in non-uniform fields. J. Phys. D Appl. Phys. **34**(11), 1690–1696 (2001). doi:10.1088/0022-3727/34/11/321
35. Pendleton, S.J., Bowman, S., Carter, C., Gundersen, M.A., Lempert, W.: The production and evolution of atomic oxygen in the afterglow of streamer discharge in atmospheric pressure fuel/air mixtures. J. Phys. D Appl. Phys. **46**(30) (2013). doi:10.1088/0022-3727/46/30/305202
36. Zhang, S., van Gessel, A.F.H., van Grootel, S.C., Bruggeman, P.J.: The effect of collisional quenching of the O 3p P-3(J) state on the determination of the spatial distribution of the atomic oxygen density in an APPJ operating in ambient air by TALIF. Plasma Sources Sci. Technol. **23**(2) (2014). doi:10.1088/0963-0252/23/2/025012
37. Williams, T.C., Shaddix, C.R.: Simultaneous correction of flat field and nonlinearity response of intensified charge-coupled devices. Rev. Sci. Instrum. **78**(12) (2007). doi:10.1063/1.2821616
38. GRI-Mech 3.0, The Gas Research Institute. http://www.me.berkeley.edu/gri_mech/

Comparing Visualization of Inflammation at Transient Load Steps Comparing Ignition Systems

Olaf Toedter[1(✉)], Alexander Heinz[1], Christian Disch[1], Thomas Koch[1], and Stefan Seefeldt[2]

[1] Institut für Kolbenmaschinen, Karlsruher Institut für Technologie, Karlsruhe, Germany
olaf.toedter@kit.edu
[2] LaVision GmbH, Anna-Vandenhoeck-Ring 19, 37081 Goettingen, Germany

Abstract. The comparison of spark ignition systems and corona ignition systems is performed by changing a spark plug to a corona igniter in an existing engine configuration on an Engine-in-the-Loop test bed. In this investigation visual inspection of the inflammation is done comparing visible image with spectral resolved two-dimensional information investigating the for flame initiation relevant OH^*, CH^* and C_2^* chemiluminescence. A general comparison between optical and thermodynamic analysis is performed. The observations are then used to evaluate transient cycle-to-cycle comparisons of the two ignition systems.

1 Introduction

The use of alternative ignition systems in new consumption and emission optimized combustion systems often yields to a chicken egg problem regarding industrial applications.

Alternative ignition systems often lack a detailed understanding about the features and the limits of operating conditions. Therefore it is often not possible to define realistic requirements for alternative ignition systems and to optimize their behavior in a whole engine context for a wide range of operating points.

A well-established method to investigate flame initiation is the use of emission spectroscopy of chemiluminescence which can be combined with classic in cylinder pressure indication and emission measurements in order to investigate the combustion process. Two-dimensional information can be used to characterize flame development. Since transient operation conditions will become more important with real driving emissions as a focus of automotive development, all measurements have to be synchronized between different measurement systems down to the cycle-to-cycle level [1]. Thermodynamic and optical analysis using visual light have been part of investigating ignition by corona in different publications regarding standard and advanced combustion processes [2–4].

M. Günther and M. Sens (eds.), *Ignition Systems for Gasoline Engines*,
DOI 10.1007/978-3-319-45504-4_11

2 Experimental Setup

2.1 Requirements for Engine Testing in Transient Operation

In contrast to steady-state engine operation, the transient operation requires a measurement setup which allows to consistently investigate the engine related sub-processes: gas exchange, combustion process and the emissions formation/reduction [5]. For this reason, the used test bed setup should be able to analyze the processes with crank angle resolution for hundreds of consecutive combustion cycles. In order to evaluate the flame initiation process as function of the ignition system under these transient conditions an optical visualization is preferred. Additionally, it is important that a synchronization of the measurement equipment in the millisecond range is ensured including the camera setup. Two optical accesses allow parallel observation of the flame kernel development at two different spectral ranges in parallel and with synchronized setup.

2.2 Engine in the Loop as Test Bed with Synchronized Measurements

Following the idea to transfer maneuvers of real world driving cycle into a laboratory environment, a customized engine in the loop (EiL) setup is applied. The EiL test bed ensures reliable investigations with highest reproducibility and repeatability. The multi-cylinder engine used, is characterized by the technical data shown in Table 1.

In addition to the optical systems which are described below the EiL combines several measurement systems which are synchronizable in order to obtain cycle resolved values. This includes gaseous exhaust emissions for cylinder 6 and total engine out emissions after turbocharger for CO, CO_2, NO and HC as well as a particle size spectrometer [5].

The synchronization between the camera systems and the pressure indication is critical to obtain the needed information for this investigation. The cylinder head includes two optical access points at cylinder 6 for endoscopic optical investigation as

Table 1. Technical data of the multi-cylinder test engine

Engine data		
Type		Gasoline
Cylinder		Six
Displacement (total)	V_H	2979 cc
Max. engine power	P_{max}	225 kW @ 5800 rpm
Max. engine torque	M_{dmax}	400 Nm @ 1200 rpm
Compression ratio	ε	10.2 : 1
Valve timing		Variable
Charging system		Turbocharger
Fuel		RON98
Fuel injection nozzle		Bosch HDEV 5.2
Combustion mode		Homogeneous
Engine Control		Bosch MEVD 17

well as pressure indication in all cylinders. The pressure signals are used to evaluate start and end of combustion as well as heat release. The resolution is 0.5° CA (degree crank angle).

Figure 1 shows an overview of the engine on the EiL test bed as well as all measurement systems. The configuration shows the corona setup as well as the most important systems for this investigation highlighted in blue.

The optical systems as well as indication are triggered via the EiL automation system. Furthermore the recording span of the camera systems is logged to the indication system. The ignition is recorded via current clamp as well as the resulting delayed trigger signal for the CCU. These signals are also used to check the correct alignment of image and pressure data since the ignition timing changes during transient load steps.

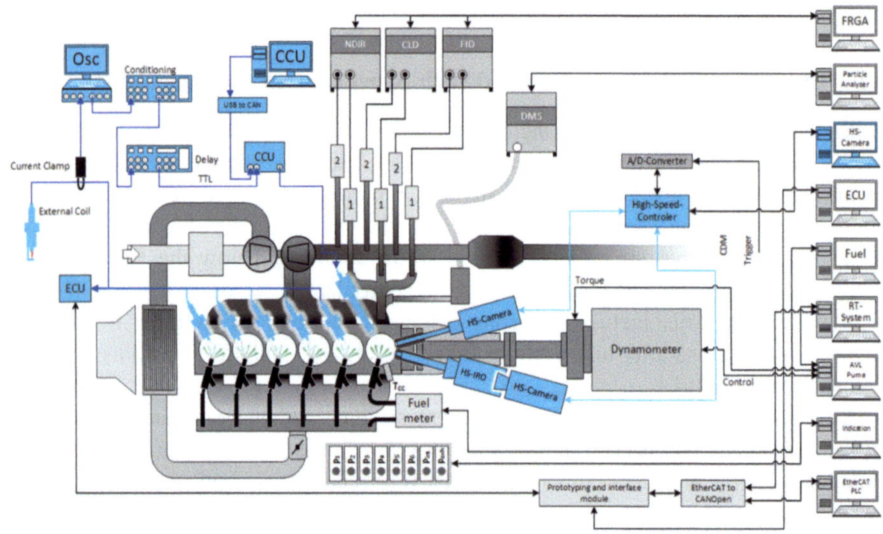

Fig. 1. Overview of the arrangement of measurement systems at the Engine-in-the-Loop test bed [5]

Different setups for the optical systems were used. A CMOS high-speed camera (LaVision, HSS6) was used in combination with an intensifier and three different optical bandpass filters to investigate chemiluminescence for different types of species. In case of UV investigation for OH this system was additionally paired with a second high speed camera to obtain visual data simultaneously at the second window. For all measurements hypersampling was used at a rate of 0.5° CA for the investigations in visible light and 1° CA for UV light range. This ensures the synchronization between optical data and indication. Since the investigation of UV needed different optical properties in the optical path two different endoscopes were used.

3 Investigated Transient Processes

Switching from coasting mode into load demand is an emission critical maneuver for most engines. It is well known that the local lambda variation at the ignition system varies most in the first cycles after switching from coasting to an acceleration phase. These types of cycles are critical regarding mixture formation and therefore critical with respect of the ignition system tolerance and functionality.

In order to evaluate these transient cycles including their transient behavior and on the other side to achieve high reproducibility the engine is run on the EiL test bed with conditioned oil and water temperatures at 85° C.

Each experiment is performed using load step test cycle, consisting of two phases as shown in Fig. 2.

- Conditioning phase
- Tip-in switching phase of eleven tip-in

In conditioning phase the engine is run at 1500 rpm and constant accelerator pedal position of 20 % to achieve reproducible temperature condition at engine wall and piston. The tip-in switching phase is characterized by switching between 0 % pedal position and 25 % pedal position for each 5 s both at a constant engine speed of 1500 rpm.

The tip-in in of the experiment is a positive load step beginning in motored operation followed by the demand of a higher engine load with the shortest possible delay (0,001 s). This is an emission critical maneuver due to the previous cooled down conditions of the combustion chamber surfaces during motored operation.

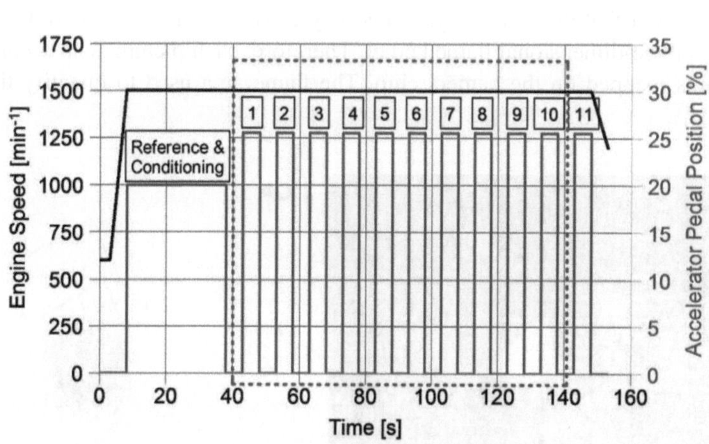

Fig. 2. Exemplary load step test cycle with eleven tip-in cycles

The engine has to react to each tip-in by switching injection mass and adapting air intake. Both actions will affect the local and global lambda values and therefore ignitability of the mixture. The lambda value has a strong effect on ignition (burning)

delay and flame speed. Therefore, we expect this transient cycle to be an authentic test cycle comparing ignition systems.

Each tip-in phase is dominated by a very transient mixture formation combined with adapted valve timing in the first cycles. The investigation will therefore firstly focus in overall comparison of mean values starting with the fifteenth fired cycle in order to investigate the used evaluation methods. At this point the variable valve timings are already adapted by the ECU to fit the load requirements to the engine.

Starting from the fifteenth fired cycle each individual cycle is investigated by its pressure and its optical results. The high reproducibility of these ramps is the fundamental base for a detailed analysis. For the first cycles exemplary results starting with the third fired cycle will be discussed with the spread of results.

4 Camera Measurements of Chemiluminescence as Tool for Flame Kernel Investigation

Figure 3 (left) shows the corona ignition in the combustion chamber without injected fuel as a color picture taken in visible spectral range, nitrogen emission is dominant. It shows the viewable part of the combustion chamber with the corona igniter in the center and intake valves from the left. This is also the viewing angle for all investigations using a filter and intensifier setup. Figure 3 (right) shows the flame kernel using the UV setup for a spark plug ignition. Since a different endoscope is used there is a slight difference in the observed part of the combustion chamber. This leads to a higher resolution but smaller observable area.

Observing the visible flame kernel development by a camera system we have to take into account that the camera system is only capable to acquire a two-dimensional picture of a three-dimensional flame kernel. Therefore, each picture is an accumulation of all beams mapped on the camera chip. The flame area used to quantify the flame

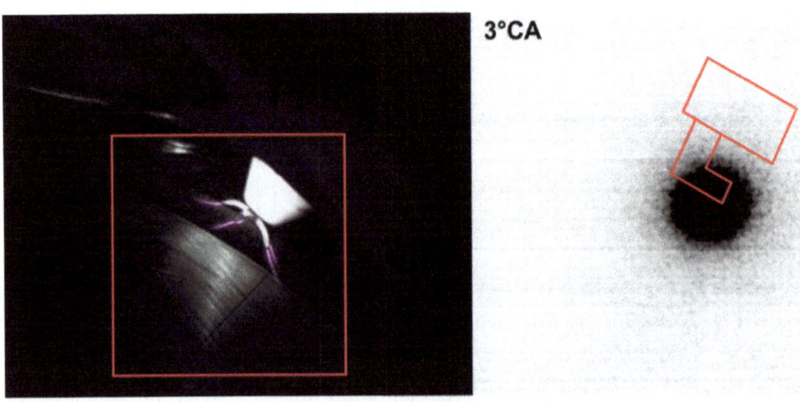

Fig. 3. Exemplary pictures: visible spectrum corona w/o fuel (left) with approximated field of view of the UV endoscope flame kernel optical filtered at 307 nm ± 5 nm with approximated spark plug position (right)

progression is therefore not the surface area of the flame, but the two-dimensional projection of the flame in direction of the camera chip.

There are several possible approaches to evaluate the camera data. By image processing a flame contour can be evaluated from the camera pictures. In a second step the flame area is calculated of this flame contour. In this case the parameter is the threshold between low and high intensity where a high threshold reduces noisy effects but decreases the maximum slew rate of the area observed. A parameter sensitivity study yields a threshold of 30 counts for the OH* images and 7 counts for the CH* and C_2* images while peak intensities vary between 1000 and 2000 counts of the maximum of 4096 counts at the chosen intensifier configuration. The obtained cycle resolved areas are normalized in order to account for the different imaging properties of the optical paths with different filters and endoscopes. In order to better compare the inflammation processes of the ignition systems all cycle resolved data including the indicated pressures and related values will be given with regard to crank angle degree after ignition timing. Figure 4 shows an example of the described processing steps.

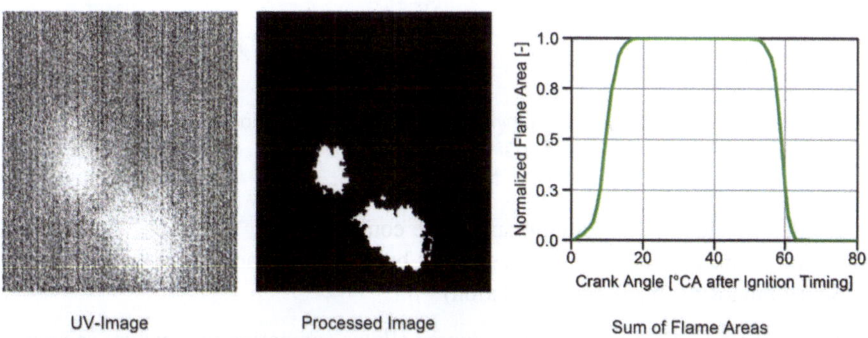

Fig. 4. Image processing steps to evaluate flame area with exemplary data of corona

A different possibility to extract information by this camera images is to visualize the speed of the flame area propagation with colors. After image processing the flame contour is plotted for each cycle in a diagram showing the projected chip area. The advantage of these pictures is clearly pointing out the direction of the flame area propagation. On the other side the graph tends to be overloaded if visualizing too much cycles. An exemplary evaluation of the OH* signal in a spark plug configuration is shown in Fig. 5. The calculated flame boundaries for several pictures and the resulting propagation rate are shown with 2° CA between consecutive pictures.

The rate is much higher in this early phase towards the outlet valves which can be expected because of higher temperatures in this part of the combustion chamber. The detected areas for show.

Each camera picture represents a light accumulation for a time period defined by the gate of the camera system or the multi-channel plate image intensifier. The light corresponds to several in-cylinder sources [6, 7]:

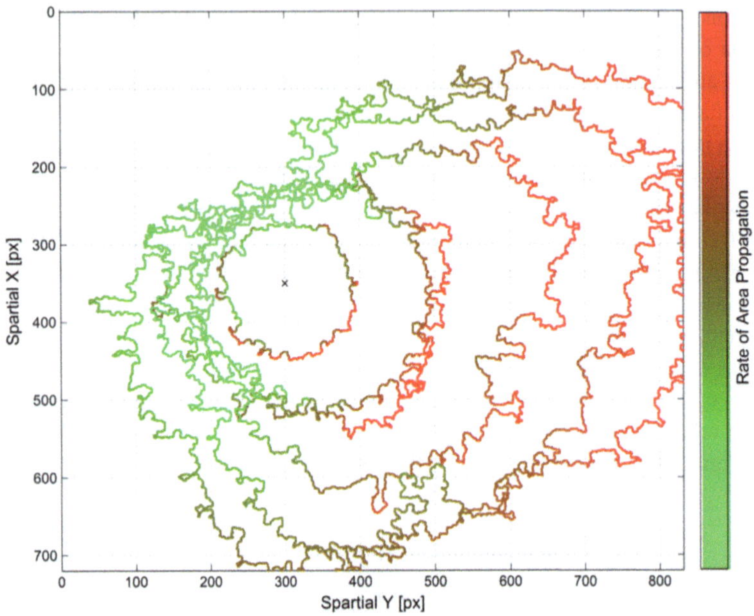

Fig. 5. Flame kernel contour plot showing rate of area propagation for consecutive pictures using the spark plug and OH signal at 2°CA resolution

- Excited species by temperature and flame conditions
- Chemiluminescence of the species exited by chemical reactions
- Soot luminosity (black-body radiation)

Using chemiluminescence as light source allows comparing two picture sequences both based on the same timing conditions when detecting light with two camera systems in parallel. The evaluation of these signals is based on a level sensitive decision regarding light intensity observed. Using the same parameters for the post processing functions as described above allows a fair comparison of the conditions and the ignition systems. Signal evaluation software has to take care of camera noise especially using an image intensifier not to misinterpret the measurement fluctuation as increased values. Additionally the observed room angle is a relevant input to compare the intensities of two parallel pictures.

The goal of the investigation is the evaluation of the flame initiation. Therefore, we decided to investigate the following species:

- OH^* 308,5 nm, filtered at 307 ± 5 nm, 60 % transparency
- CH^* 431,4 nm, filtered at 430 ± 5 nm, 85 % transparency
- $C2^*$ 473 nm, filtered at 471 ± 5 nm, 85 % transparency

Figure 6 shows an exemplary emission spectrum with the used optical filters marked in grey. The peaks of the chemiluminescence of the species lay well in the filter ranges. All three species are used in literature as relevant species indicating the flame front propagation [9], as well as air-fuel ratio and heat release [10].

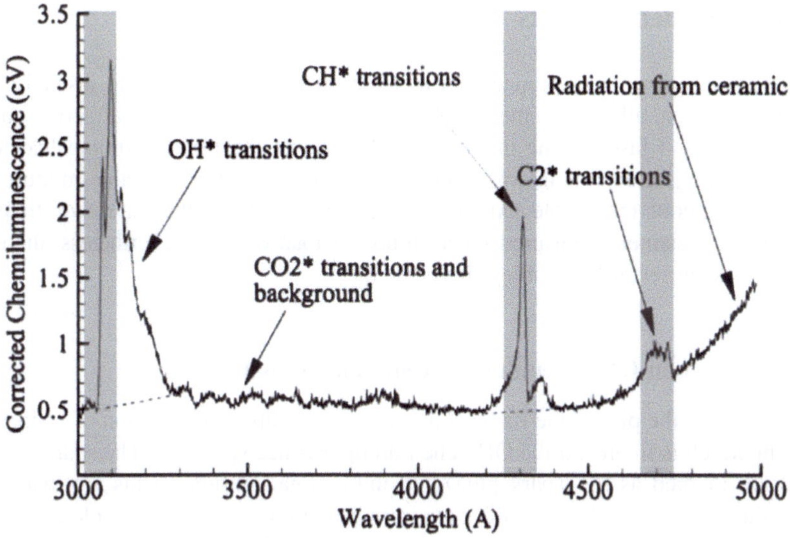

Fig. 6. Flame spectra of hydrocarbon flame caused by chemiluminescence after [8] with used filter wavelength marked in grey

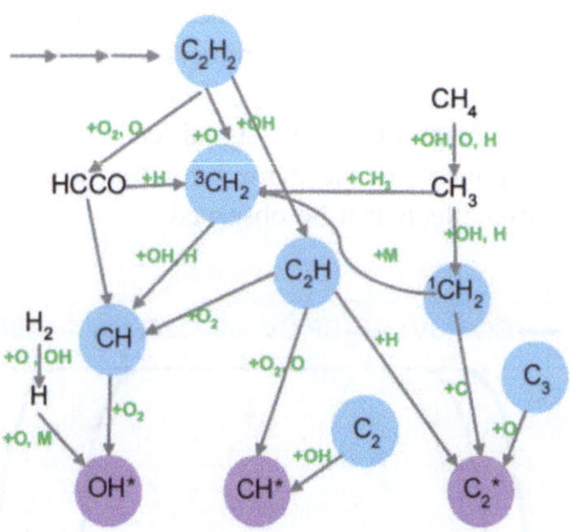

Fig. 7. Reaction scheme of chemiluminescence species in acetylene chemistry [8]

Other investigations are using these species as base for determining as described in Fig. 7 reaction kinetics especially regarding flame propagation [8]. Therefore these chemiluminescence signals are clear indicators to compare ignition systems.

5 Measurement Results

In order to compare flame propagation as result of the ignition system instead of the combustion timing all measurements were performed with a fixed mass fraction burned (MFB50). Due to faster flame initiation of the corona the ignition timing had to be delayed. The injection timing was fixed in order to achieve comparable mixture formation throughout the whole experiment. However, due to the nature of transient operation, this cannot be ensured for all transient load cycles. Nevertheless, there are some significant results describing the behavior.

5.1 Comparison OH*Chemiluminescence and Visual Light

Figure 8 shows the plot of the flame contour area in a direct comparison of visible and UV light which is filtered at the OH* chemiluminescence spectrum. The spark ignition results are marked as SP (spark plug) and the corona ignition results are marked as COR values. The index Vis declares data from camera images of the visible light range (≈ 400–750 nm) and the index UV stands for the OH* chemiluminescence signal (filtered at 307 ± 5 nm). All contour areas are referenced to the maximum area in order to compare their behavior with no respect to individual signal strengths defined by the optical path and the camera system.

The results are averaged over 26 cycles of the load step test cycle as described in Fig. 2. The high reproducibility of the measurement yields to very similar results for each cycle.

It is obvious, that the average of the OH-signal leads the visible signal as the OH radicals are initiating components in the flame initiation process. Comparing Corona and spark plug ignition two effects can be observed.

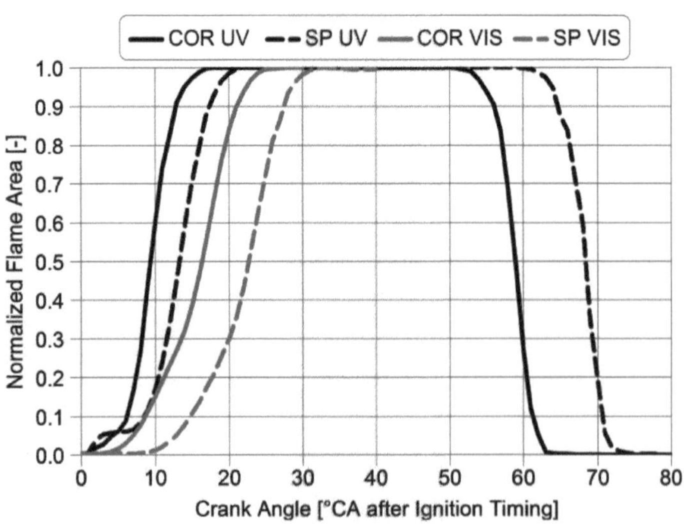

Fig. 8. Contour area plot comparing corona and spark plug ignition

Both OH-signals start comparable but the spark ignited mixture shows a delay of about 5° CA after the first signal occurs which may show the cold flame part for this time slot. Corona ignited mixtures do not show this cold flame part ramping up the OH density directly to full extension. When the thesis of a non-existing cold flame region with corona discharge is reliable there is a potential reason for the faster ignition with this alternative system.

The visible signals do not show this specific behavior neither at spark-based or corona ignition.

In general, we could observe a shorter burning duration based on the OH* signals for the corona ignited mixture compared to the spark ignited mixture. There are lots of potential root causes for this measurement relating to the combustion timing and its influence of the flame speed, local mixture formation etc. When adapting a comparison to similar ignition timing and resulting in a shorter ignition delay there is a shift in combustion timing unpreventable. Directly correlating the optical methods with thermodynamic analysis, such as mass fraction burned or heat release rate, will be shown in further analysis below. However a clear tendency can be observed with the optical signal alone.

5.2 CH* and C_2* Results

The CH* and C2* images show a comparable behavior to the OH and Vis images. As in the previous experiment the corona ignition is clearly faster with regard to start of combustion. Figure 9 shows the results for the 430 nm and 470 nm filters.

As in the previous experiment the corona ignition is clearly faster with regard to start of combustion. There is a good correlation between the calculated flame area propagation and the heat release rate for both experiments. For CH* the flame

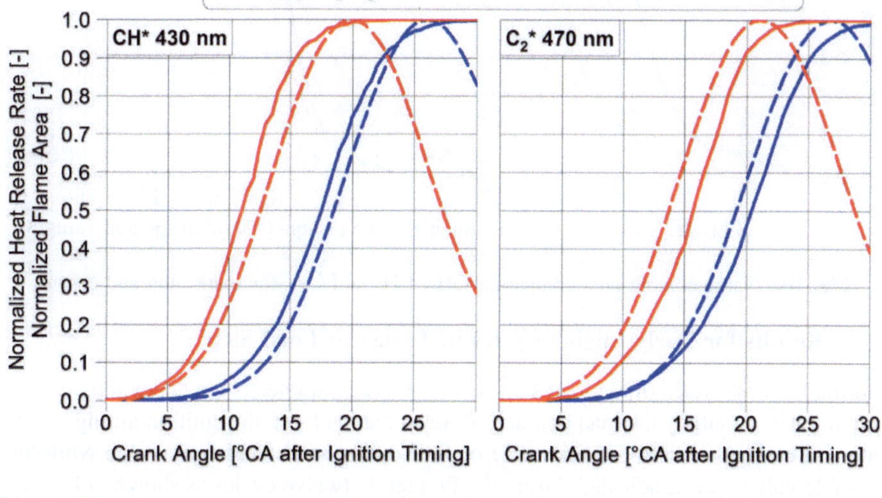

Fig. 9. Flame contour area spectral filtered to CH* and C_2* comparing spark plug and corona

propagation is slightly ahead of the heat release curve. This has to be expected as CH* occurs earlier in the reaction scheme and the exothermic reactions which lead to CO_2 are major contributors to the overall heat release of the reaction. This correlation is observed in both ignition systems and true up to the point of maximum heat release. At this point the optical signal cannot give further insight since it is optimized to show early combustion phases. In order to investigate at higher overall intensities different configurations for the calculations with regard to threshold and intensifier would be needed.

The results for the C_2* 470 nm filter also show a correlation between heat release and optical signal. In contrast to the 430 nm results the heat release precedes the optical flame detection. C_2* should not be detected much later than CH*. One possible explanation can be a weak C_2* chemiluminescence. The optical signal would then be determined strongly by chemiluminescence from CO_2*. Further investigations with other filters (e.g. with 450 nm) are needed in order to quantify the CO_2 influence.

Figure 10 shows the findings of the different species wavelength in an overview. The differences between the two ignition systems can be observed in all three but are most prominent in the OH* signal as explained above. The differences between the OH* and CH* signals are also more prominent with the spark plug ignition. This is another indicator for larger cold flame regions using a spark plug.

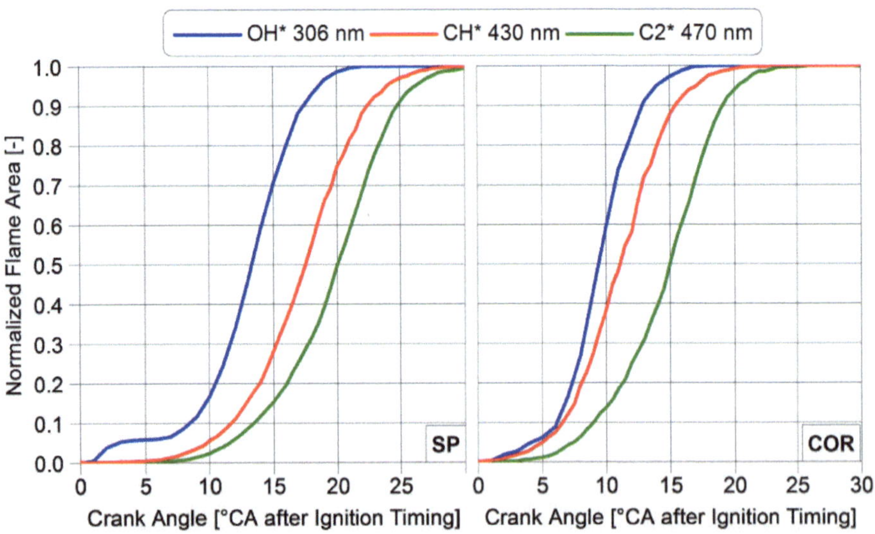

Fig. 10. Flame contour area comparing OH^*, CH^* and $C_2{}^*$ for spark plug and corona

5.3 Results for Early Engine Cycles in Transient Load Step

During the first cycles of the load step several engine parameters are adapted by control unit in order to satisfy the load demand. Prior to the load step the ignition timing is very late and changes during the first three or four cycles to its stationary value while the variable valve train is adjusted during the first ten to twelve cycles as shown in Fig. 11.

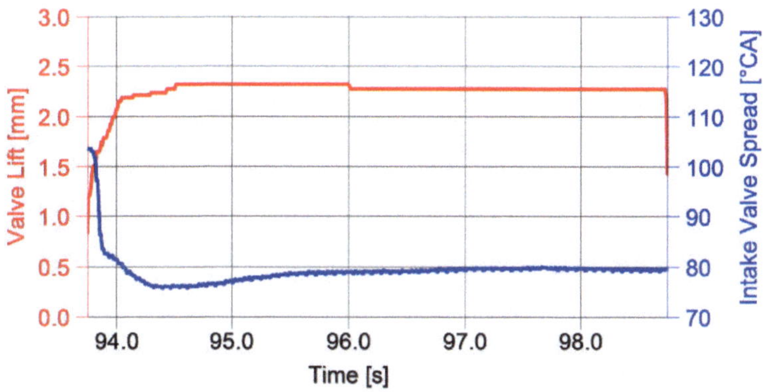

Fig. 11. Parameter change of variable valve train during load step

Since the beginning of the load step with regard to the position of cylinder 6 is not fixed, a valid comparison between different experiments for the first and second cycle is not possible in all measurements. Therefore the following comparison concentrates with the third fired cycle. Figure 12 shows the results for the area propagation of the chemiluminescence for the CH* and C_2* for both ignition systems in comparison. Hull curves and mean values are shown for fired cycle three to seven (upper diagrams) and seven to twelve (lower diagrams). The differences between the ignition systems regarding the start of combustion and early flame development as described for the stationary parts of the load step are confirmed by these early combustion cycles. The flame kernel propagates faster after ignition for the corona and the differences can again be seen for all investigated cycles.

The difference in width of the hull curves varies due two several effects in these early combustion cycles. As mentioned above the ignition timing is changing during the load step and the ignition timing of the corona is even further delayed by the experimental setup. Therefore ignition timings and heat release are not fully comparable. Comparison of the corona for both wavelengths shows a significant difference in cycle-to-cycle fluctuation. However further investigation shows that this difference is due to a difference in heat release and ignition timing and not due to the different observed wavelengths.

The cycle-to-cycle differences lessen for the later cycles. The ignition timing is already at its stationary value and the gradient of change in the parameters regarding the variable valve train is shallower compared to the cycles before. The ambient conditions at the igniter are therefore more constant and the resulting differences much smaller.

The results for the transient part of the load step confirm the observations from the more stationary part. The optical analyzation of flame area propagation is a valid method to compare the different ignition systems on a cycle-to-cycle base.

In order to quantify the difference due to species observation more experiments are needed in order to negate the stochastic nature of ignition timings in early engine cycles.

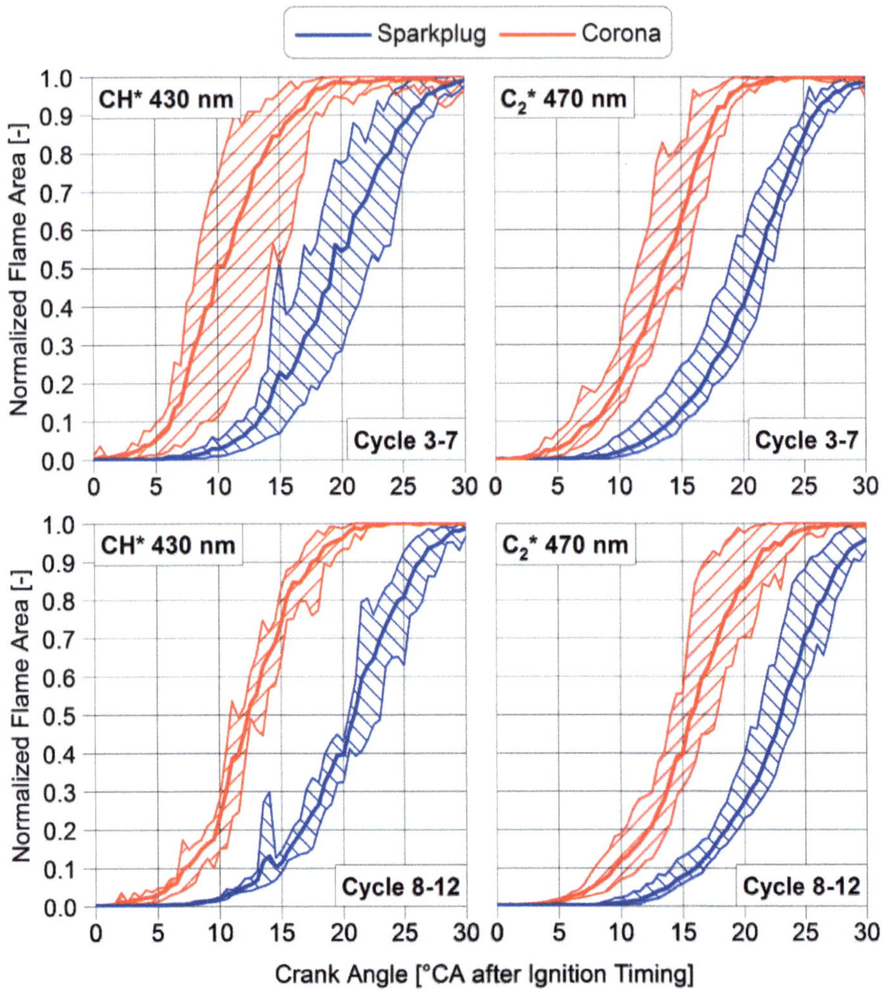

Fig. 12. Flame contour area spectral filtered to CH* and C_2* comparing spark plug and corona in early fired engine cycles of the load step

6 Summary and Discussion

Corona ignition allows several additional parameters in combustion design compared to spark ignition systems. The optical measurements show a clear difference between spark plug and corona ignition effects on flame initiation looking at the OH* chemiluminescence.

CH^* and C_2^* chemiluminescence give a clear behavior on burning duration depending on the ignition system. A correlation between optical signals and heat release is observable and can be used to further evaluate single cycle behavior of the ignition systems using the shown optical setups.

Furthermore, the ratio of CH^* and OH^* or the ratio of C_2^* and CH^* have the potential to be used as local information of the air/fuel ratio respectively the heat release. However simultaneous observation of all species would be desirable.

Using the lower ignition delay of a corona system, ignition timing can be adapted as well as mixtures ignited which need an advanced ignition timing on the other side. These facts can be used in lean mixtures, diluted mixtures or in combustion regimes with Miller timing.

Next steps in order to expand on these results have to concentrate on the very early engine cycles with adaptable ignition timings in an engine setup with corona igniter in all cylinders.

Acknowledgements. We like to thank BorgWarner Ludwigsburg GmbH providing the ignition system and application support.

References

1. Disch, C., Koch, T., Donn, C., Spicher, U.: Engine-in-the-Loop as a development tool for emissions optimisation in the hybrid context. MTZ worldwide, Oktober 2014
2. Feßler, M.: Untersuchung der Auswirkungen eines alternativen Zündverfahrens auf den Verbrennungsprozess bei einem geschichteten Brennverfahren. Logos Verlag, Berlin (2012)
3. Süß, M.: Untersuchungen zur homogenen kompressionsgezündeten Verbrennung mit Fremdzündungsunterstützung. Logos Verlag, Berlin (2012)
4. Hampe, C., Kubach, H., Spicher, U., Rixecker, G., Bohne, S.: Investigations of ignition processes using high frequency ignition. In: SAE 2013 World Congress (2013)
5. Disch, C., Pfeil, J., Kubach, H., Spicher, U., Koch, T., Thiele, O.: Experimental investigations of a DISI engine in transient operation with regard to particle and gaseous engine-out emissions. In: JSAE/SAE 2015 International Powertrains, Fuels & Lubricants Meeting (2015)
6. Gaydon, A.: The Spectroscopy of Flames. Chapman and Hall, London (1974)
7. Gaydon, A.G., Wolfhard, H.: Flames - Their Structure, Radiation and Temperature. Chapman and Hall Ltd., London (1970)
8. Kathrotia, T.: Reaction kinetics modelling of OH*, CH* and C2* chemiluminescence, Dissertation, Heidellberg (2011)
9. Ikeda, Y., Kaneko, M., Nakajima, T.: Local A/F measurement by chemiluminescence OH*, CH* and C2 in SI engine. In: SAE World Congress, Detroit (2001)
10. Nau, P., Krüger, J., Lackner, A., Letzgus, M., Bockhinke, A.: On the quantification of OH*, CH* and C2* chemiluminescence in flames. Appl. Phys. B **107**, 551–559 (2012)

Spark Control for Ion Current Sensing

Tycho Weißgerber(✉) and Gregor Kortendiek

WEISSGERBER Engineering GmbH,
Breisenbachstraße 36, 44357 Dortmund, Germany
tw@weissgerber-engineering.de

Abstract. Ion current sensing is an established method for the analysis of engine combustion. It is based on the measurement of thermally induced electrical conductivity in the combustion chamber and is used for detection of misfire, engine knocking, the air-fuel-ratio or peak pressure position. Because the spark plug electrodes are used as sensor elements, the combustion signals cannot be determined during spark duration. Particular attention therefore is drawn to the temporal transition between spark and combustion at higher speeds. In this paper, the necessary circuitry for controlling the spark will be presented for an inductive ignition system, that will maximize the measuring window as well as minimize disturbing electrical oscillations.

Keywords: Ion current sensing · Inductive ignition · Combustion analyses

1 Ion Current Sensing for Combustion Analyses

The principle of ion current measurement is based on the thermal conductivity of gases and has been studied by Irving Langmuir extensively [1, 2]. The method has been used for a long time to analyze engine combustion as well [3–6]. Simplistically a voltage is applied to the electrodes of the spark plug and the electric conductivity is measured in the region around the electrodes as a current flow via a measuring resistor (Fig. 1).

The measurable conductance between the electrodes is determined by the field determining geometry of the electrodes and the charge carriers present. According to Maxwell the space charge density follows:

$$div\, \varepsilon_0 \varepsilon_r E = \rho$$

with:

E Electrical field strength as gradient of potential distribution: $E = -\, grad \varphi$

ρ Space charges of n unit charge $\rho = e \cdot (n_+ - n_-)$

ε_0 Permittivity of free space $\varepsilon_0 = 8,854e^{-12} \frac{F}{m}$

ε_r Electrical Permittivity of air-fuel mixture $\varepsilon_r \approx 2$

Following the equation, the space charge density corresponds to the quantity of ionized species, which can be trapped by an electric field. In practice, positive DC voltages are preferred due to the higher mobility of the negative charge carriers and the resulting higher signal amplitudes. By that meaning, using alternating driving voltages

M. Günther and M. Sens (eds.), *Ignition Systems for Gasoline Engines*,
DOI 10.1007/978-3-319-45504-4_12

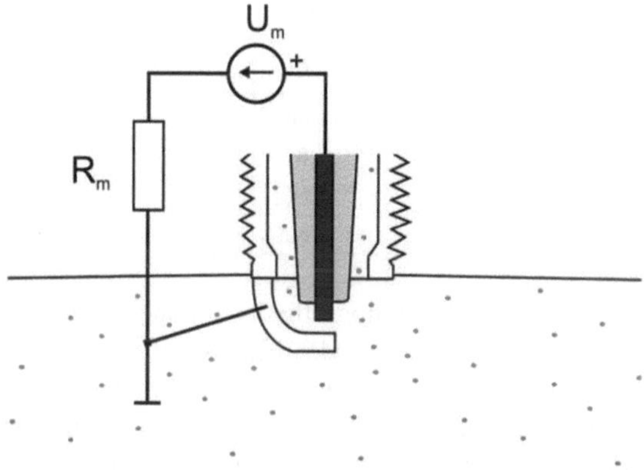

Fig. 1. Spark plug electrodes with surrounding space charges

the measurement signal appears as rectified. The arrangement of the electrodes and the position within the insulating ceramic have a significant impact on the field strength and on the amplitude of the measurement signal by that, whereas the electrode degradation by aging plays a minor role.

The degree of ionization of gas in general is described by a modified shape of the Boltzmann-distribution, that takes into account the statistical weight of the singularly ionized atom in the original state with a free electron. The Eggert-Saha equation describes the degree of ionization as:

$$\frac{n_i^2}{n_a} = \frac{2\sqrt{(2\pi m_e k T)^3}}{h^3} \exp\left(-\frac{E_i}{kT}\right)$$

with:

n_i Number of ionized molecules density
n_a Total number of molecules density
m_e Electron mass $9{,}109 \cdot 10^{-31}$ kg
k Boltzmann's constant $1{,}38 \cdot 10^{-23}$ J/K
T Temperature
H Planck's constant $6{,}6256 \cdot 10^{-34}$Js
E_i Substance's ionization energy

and the ionization energy of some selected substances:

Substance	E_i [eV]	Substance	E_i [eV]
H	13,598	H2O	12,6
H2	15,427	N	14,53
O	13,618	N2	15,576
O2	12,063	NO	9,25

(*continued*)

(*continued*)

Substance	E_i [eV]	Substance	E_i [eV]
OH	13,17	NO2	9,79
CH	10,64	K	4,3
CO	14,013	Mg	7,7
CO2	13,769	Na	5,1

Essentially, with ion current technology the local concentration of nitrogen oxides is measured around the spark plug electrodes because of the low ionization energies, where the amplitude is sensitive to the presence of alkali metals as a component of fuel additives. The ion current signals are directly dependent on the chemistry of combustion and are widely scattered, especially during the chemical phase. Nevertheless, typical characteristics for each engine operating point can be carried out, which enables cylinder-specific conclusions from single measurements or from averaging of several cycles. The ion current signal for a typical single combustion cycle is shown in Fig. 2.

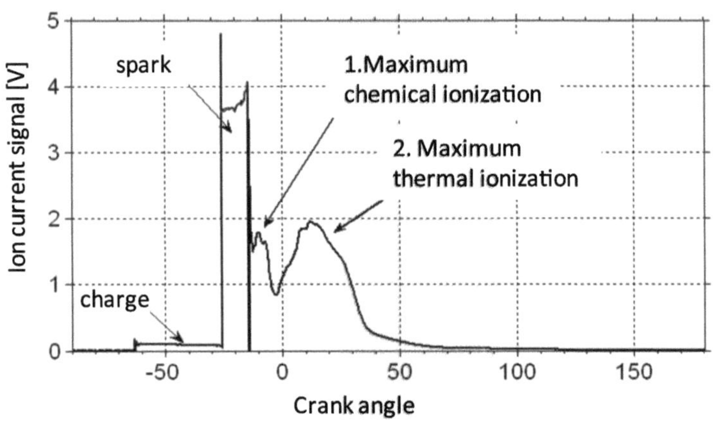

Fig. 2. Typical ion current characteristic of a regular combustion

It is measured using the later presented sensing circuit and gives information about spark and combustion. On the x-axis the time or crank angle depending progress starts with the charge phase of the transformer, continues with the ignition voltage and spark burning voltage and ends showing the ion current's signal typical characteristics with the chemical ionization in the flame front and the thermal ionization by the pressure inside the combustion chamber. Because of the different signal amplitudes, the amplitude gain factors for the ignition voltage, the spark burning voltage and the ion currents are switched respectively. At distinctive cold combustions it might also be necessary to increase the gain factor for the ion current for a robust misfire detection.

For those cases also delayed combustions with very slow impact to the power of the engine can be measured as elongated ion current signals at low level as well.

Ion currents amplitudes are determined by the cylinder internal pressure over a wide range and allow the cylinder-specific detection of knocking combustion. Figure 3 shows an example of a non-supercharged engine at 5500 rpm and medium load. Indicated on the left is the cylinder pressure, on the right the ion current. The superposition of the ion current signal by the spark duration can clearly be seen with the following two typical maxima of chemical and thermal ionization. On the decaying pressure signal the common knocking resonances are superimposed as higher frequency oscillations as well as on the ion current signal.

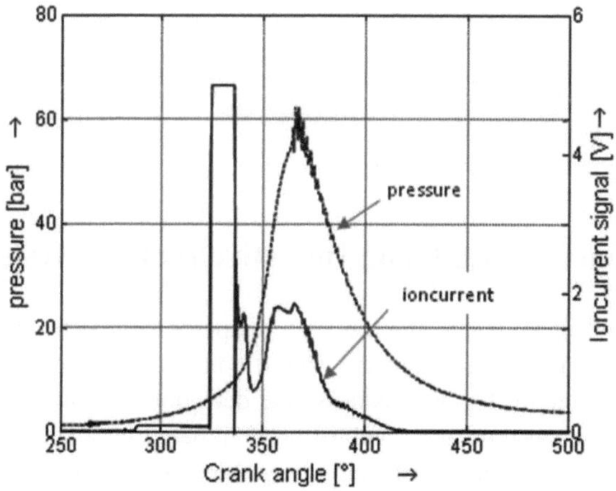

Fig. 3. Exemplary measurement of knocking combustion

By means of ion current measuring a proportional feature related to the air-fuel-ratio can be extracted, for example facilitating a cylinder balancing. Both enriching and leaning reduce the NOx concentrations including the ion current signal (Fig. 4). Therefore the amplitude gives an ambiguous information. However picture 4 also shows that with higher λ-values the combustion occurs delayed. Together both effects enable an estimation of a proportional value. The sensitivity of the amplitude caused by unknown fuel additives makes the determination of a true value impossible. Enriching and leaning are the only way of calibrating the system for known values. This is not necessary for cylinder balancing procedures as only features relative to the individual cylinders are required. One single lambda sensor is then sufficient for engine control itself.

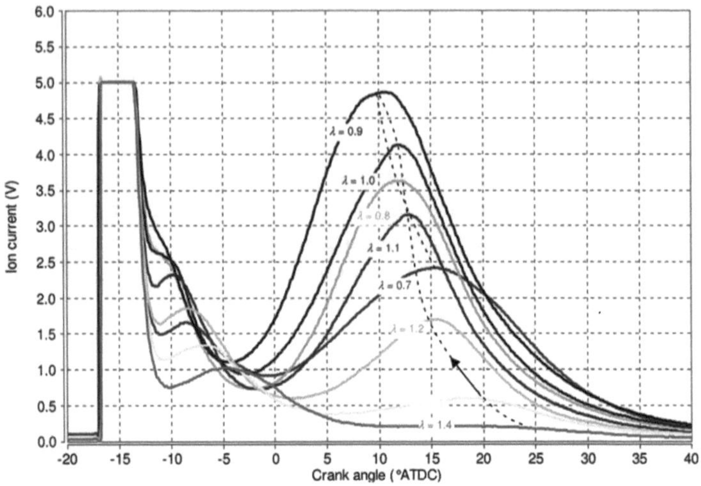

Fig. 4. Ion current signal when varying the fuel-air-ratio [6]

2 Ion Current Sensing Using Inductive Ignition Systems

Because of the low signal levels of a few microamperes for interference free measuring of the ion currents the acquisition and signal conditioning are mounted inside the ignition module. A circuit for wide band signal detection is shown in Fig. 5. As usual in inductive ignition systems, the combustion is initiated by a spark [7]. The ignition transformer therefore is integrated into the vehicle electrical system with battery voltage Ubat and cable resistance Rnet and is connected to the spark plug either directly or via an ignition cable.

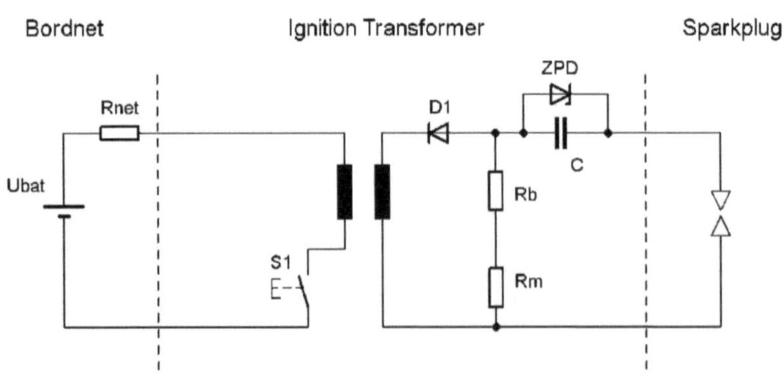

Fig. 5. Principle sketch of an ignition transformer with ion current sening circuit

By closing switch S1 the electronic control unit starts charging the ignition transformer. On opening the ignition timing is set up. Without any additional control electronics the primary current increases linear due to the differential inductance. Accomplishing the magnetic saturation the primary current increases more progressive and finally it ends with a constant value in the resistive saturation. Within the linear range the primary fed energy is controlled by the dwell time according to the current operating point. There is no current flow on the secondary side, if the spark plug electrodes are electrically open. An induced positive voltage can be measured due to

$$U = -\frac{\partial \psi}{\partial t} = -L(i_{\mathrm{Prim}}) \cdot \frac{\partial i_{\mathrm{Prim}}}{\partial t}$$

the change in the magnetic flux respectively to the change in the primary side current flow. It can achieve a few kilovolts. To prevent premature ignition, diode D1 is inserted in the secondary circuit (Fig. 6).

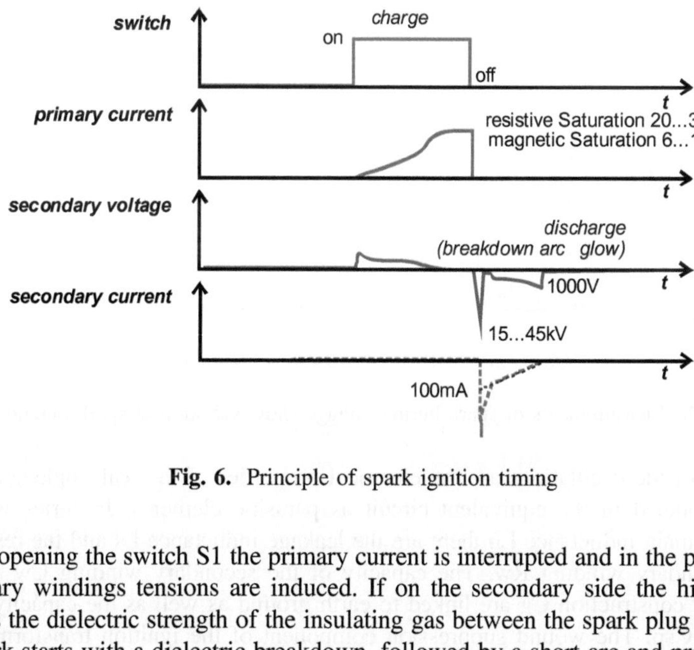

Fig. 6. Principle of spark ignition timing

On opening the switch S1 the primary current is interrupted and in the primary and secondary windings tensions are induced. If on the secondary side the high voltage exceeds the dielectric strength of the insulating gas between the spark plug electrodes, the spark starts with a dielectric breakdown, followed by a short arc and prolonged by glow discharge [8]. The essential measuring circuit for the ion current sensing include the components ZPD, C, Rb and Rm. With the Zener diode ZPD and the capacitor C, a voltage source is established, whose positive pole is applied to the center electrode of the spark plug. The ground reference is realized by the resistors Rb and Rm, wherein Rb represents a high voltage resistor for decoupling the measuring resistor Rm through which the ion current is visible as a voltage drop. The diode D for avoiding early ignitions decouples during the measurement the secondary coil, so that the ion current flows via the resistor branch.

During the spark duration no ion currents can be measured, and therefore at higher speeds or very powerful ignition transformers chemical phase is almost completely covered by the spark. In addition, the spark power determines the plasma temperature as the product of spark voltage and spark current and thus the stability of the spark plasma. Because of that the decaying end of the spark current is thermally unstable and the sparks can be blown out prematurely.

As a result, the spark duration varies significantly, which is illustrated measuring the spark burning voltage (Fig. 7, 5 V/1000 V, inverted) for several following cycles using a oscilloscope. The spark burning voltage show a couple of foot point changes, the ignition voltage is overridden and can only be used as a trigger pulse. Consequently, the beginning of the ion current measuring phase is not known. On the one hand, the end of the ignition spark of the operating state of the engine is influenced in an unknown way and on the other hand in the transition phase oscillations occur, which are superposed to the ion current signal.

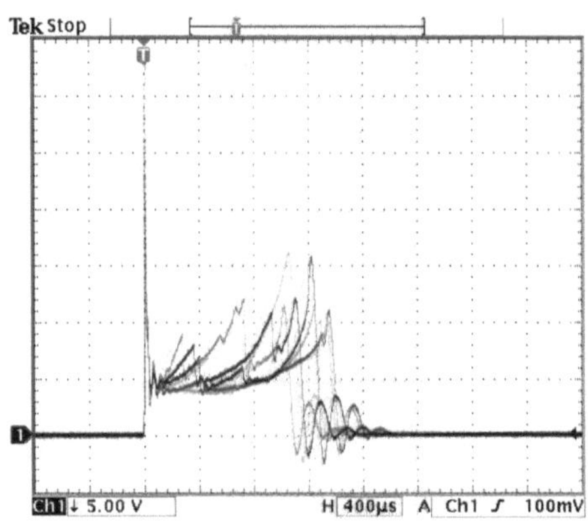

Fig. 7. Measurements of spark buring voltage show variation of spark duration

Source of the oscillations are the electrical properties of the real implementations that are modeled in the equivalent circuit as parasitic elements. In series with the differential main inductance Lh there are the leakage inductance Ls and the resistance of the secondary winding Rw. The capacity of the secondary winding Cw and the geometrical construction Cg are linked to earth ground as well as the capacity of the spark plug Csp. The wound suppression component of the ignition transformer represents an impedance of Remc, Lemc and Cemc. Inside the spark the filtering resistance Rsp is located. The resistance between the spark plug electrodes is taken into account by R(T). Depending on the operating conditions it assumes very different values. While the spark is burning, it is defined by the integration of the spark current according to Toepler's law and reaches a typical value of about 150 Ohms. Without any spark it varies depending to the temperature induced ionization between 100 k and 1000MOhm.

The inductances and capacitances with dominant Lh, Ls and Cw, Csp build a resonant circuit, in which the magnetically stored residual energy oscillates after an early termination of the spark. Diode D1 and the resistors attenuate the oscillations (Fig. 8).

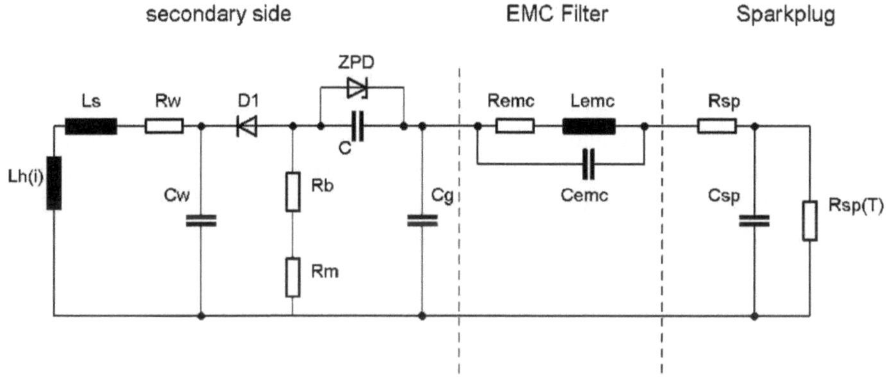

Fig. 8. Secondary side equivalent circuit with most important parasitic elements

3 Spark Control

To define the measurement window, an electronic circuit is installed to adjust the spark duration by a specific spark switch-off. Essentially, the two components Diode D2 and switch S2 are added. After charging the primary coil the opening of switch S2 initially remains without any effect. The spark continues to burn unaffected. Now if switch S1 is closed again, diode D2 takes over the transformed secondary current and commute the magnetic flux on the primary side. As a result the spark is extinguished. Another advantage of switch S2 is that the electrical system is decoupled from the measuring branch and therefore the impact of transient disturbances on the supply voltage to the measurement signal will be reduced significantly (Fig. 9).

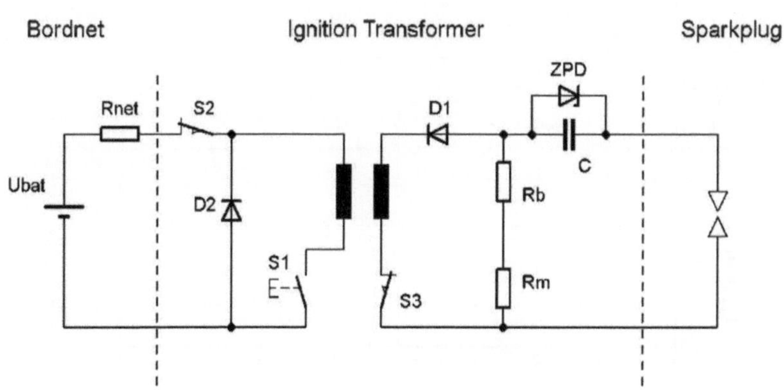

Fig. 9. Principle sketch of ignition transformer with ion current sensing an spark control

Figure 10 shows an example with three superposed cycles. The primary current Iprim is the same for all. Clearly visible is the turn-off of the ignition spark in the secondary current Isec after 300 µs or 500 µs burning time compared to the non influenced decay after about 1300 μs. The signal Uion clearly shows the overlapping of the ion current combustion signal by the spark. This is still superimposed in part by a decaying voltage which is due to the change in magnetic flux in the transformer. Optionally, this influence can be further minimized by an additional switch S3.

This methodology also allows the control of the spark energy over the duration of the ECU ignition timing with a known spark duration (Fig. 11). With increasing dwell time the primary current increases and consequently the maximum value of the spark current as well.

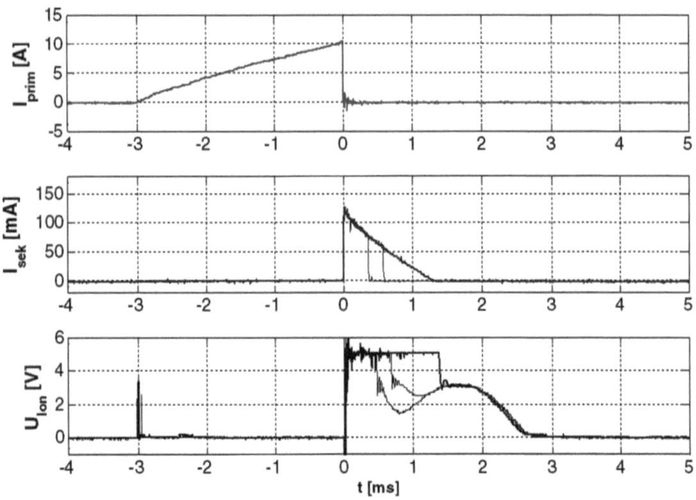

Fig. 10. Superimposed measurements of controlled spark duration

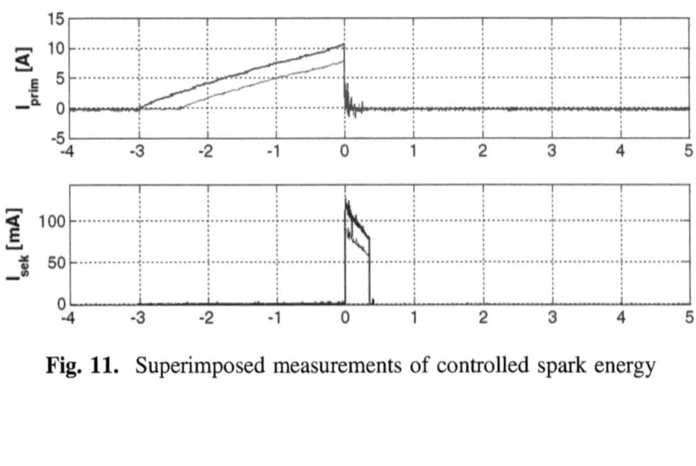

Fig. 11. Superimposed measurements of controlled spark energy

4 Summary

Starting with the physical principles, an ion current measurement circuit is presented for an inductive ignition system. It is explained that ion current analysis is based on the measurement of ionized nitrogen oxides in the combustion chamber and can be used to detect combustion misfires, knocking, the air-fuel ratio or other related information. The ion current sensing thus represents an known method for the analysis of engine combustion. The proposed circuit for the metrological detection of the ion current also offers the possibility of measuring the ignition voltage and the spark burning voltage using the resistive voltage divider. The special challenges of the ion current measuring technology are the unknown spark duration and the overlapping of the combustion signal by the spark at high speeds. In the electrical equivalent circuit the parasitic elements of the ignition transformer and the spark plug are modeled. Their impact on electrical oscillations especially in the transition region from spark to ion current sensing is pointed out. To define the measurement window for the ion current sensing, the spark duration and energy control is introduced. The principle circuitry is described, supported by exemplary measurements.

References

1. Langmuir, I., Mott-Smith, H.: The theory of collectors in gaseous discharges. Phys. Rev. **28**, 727–763 (1926)
2. Conde, L.: An introduction to Langmuir probe diagnostics of plasmas. Department of Física. ETSI Aeronáut ngenieros Aeronáuticos. Universidad Madrid (2011)
3. Wilstermann, H.: Wechselspannungszündung mit integrierter Ionenstrommessung als Sensor für die Verbrennungs- und Motorregelung. Dissertation TU Karlsruhe (1999)
4. Andersson, I.: Cylinder Pressure and Ionization Current Modeling for Spark Ignited Engines. Thesis University Linköping (S) (2002)
5. Weißgerber, T.: Ion current measurements in lean burn engines. In: 4th International CTI Conference on Emission Relevant Sensors (2011)
6. Vidmar, K.: Potential use of ion sensing for start calibration development. Graduate Thesis FH Graz (2007)
7. Abschlussbericht zum DFG-Sonderforschungsbereich SFB224 .Ottomotorische Verbrennung. RWTH Aachen (1995)
8. Maly, R.: Spark ignition: its physics and effects on the internal combustion engine. In: Hilliard, J.C., Springer, G.S. (eds.) Fuel Economy in Road Vehicles Powered by Spark Ignition Engine, pp. 91–148. Springer, New York (1984)

Combustion Processes

Ignition System Development for High Speed High Load Lean Boosted Engines

D.J. Corrigan(✉), E. Pascolini, D. Zecchetti, and F. Titus

Ferrari Gestione Sportiva, MOT, Maranello, Italy
Daire.Corrigan@ferrari.com

Abstract. The 2014 season represented a major shift in the focus of Formula One powerunit development. Prior to this date, engines were high revving naturally aspirated units where performance was primarily driven by volumetric efficiency. Thereafter, the regulations changed to a fuelflow limited formula [1]. This effectively meant that the competition was to achieve the minimum powerunit BSFC especially at 100 kg/h fuelflow and > 10500 rpm (as per FIA regulation). Turbocharging was once again permitted and engines were downsized from 2400 cm^3 to 1600 cm^3 swept volume. The focus on combustion efficiency meant that lean operation became very interesting. The move to a lean boosted engine changed significantly the requirements placed on the ignition system.

The 2014 regulations prohibited some advanced ignition systems and so development was concentrated on a conventional inductive coil and exposed spark gap plug. Early challenges were simply to cope with the very high sparking pressures and hence breakdown voltages. Once adequate robustness was obtained, the push was to have good ignitability and hence combustion stability for lean knock limited conditions.

Various aspects of spark plug geometry were explored at numerous points in the development with improvements in combustion performance of the engine often requiring tradeoffs with the ignition system to be made as sparking pressures tended to increase ever higher. As expected, the influence of the ignition system was primarily seen in early burn rates but correlations were identified between these and the optimum lambda for best knock limited efficiency. Along with significant effort on spark plug design, work also took place using an ignition coil emulator searching for even small improvements of performance through different current profiles and multisparking. Endoscopic imaging of early combustion was carried out on a single-cylinder version of the Formula One engine to aid understanding and complement conventional indication measurements.

This paper describes development that took place between 2011 and 2014.

1 Background

In 2014, the regulations of the FIA Formula One World Championship relating to the powerunit changed significantly [1]. The objective of this shift was to make Formula One more road relevant. Downsizing and hybridization had become increasingly common in the road car world and so there was little technology transfer that could take place from the high revving naturally aspirated V8 race engines used in Formula One.

M. Günther and M. Sens (eds.), *Ignition Systems for Gasoline Engines*,
DOI 10.1007/978-3-319-45504-4_13

It was decided that the best approach would be to turn Formula One into an efficiency based competition. To this end, a fuelflow limit was introduced – the maximum instantaneous fuelflow rate could not be allowed to exceed 100 kg/h. The total fuel quantity for the race would also be limited to 100 kg. These changes effectively meant that powerunit performance would be based on achievement of the minimum BSFC at this fuelflow rate.

The engine architecture also changed. The 18000 rpm V8 engine with swept volume of 2400 cm^3 was to be replaced by an engine of just 1600 cm^3. The initial proposal was for this engine to be inline with four cylinders. This engine was foreseen for the 2013 Championship year but never actually raced. The new engine regulations were postponed until the 2014 season and the layout became a V6. The fuelflow curve also changed from the 2013 proposal. The minimum speed where 100 kg/s fuelflow could be used was increased from 8500 rpm to 10500 rpm and the maximum engine speed limit was increased from 12000 rpm to 15000 rpm.

In addition to the reduced size and engine speed, the regulations also permitted turbocharging for the first time in many years. Hybridisation was also increased. Whilst the V8 engines had a Kinetic Energy Recovery System (KERS), this had a power limitation of 60 kW and an energy per lap limitation of 400 kJ. With the new regulations, the power limit was doubled and the energy limit increased by a factor of ten. Only half of the energy could be recovered kinetically and the rest would come from a motor-generator attached to the turbocharger. The knock on effect of this was that the turbine would need to recover more energy than simply the requirement to drive the compressor. This meant that exhaust pressure levels would need to be quite high.

The above changes are summarized in Table 1:

Table 1. Key regulatory changes 2009–2014

	2009-13	Provisional 2013	2014
Swept Volume, cm^3	2400	1600	1600
Maximum Engine Speed, rpm	18000	12000	15000
Number of Cylinders	8	4	6
Fuelflow, kg/h	Unlimited	100	100
Min. Speed for Max. Fuelflow, rpm	Not Applicable	8500	10500
Energy Recovery	Kinetic Only	Kinetic + Heat	Kinetic + Heat
Turbocharging	Not Permitted	Permitted	Permitted
Bore size, mm[a]	98	88	80
Bore/Stroke Ratio[a]	2.46	1.34	1.51

0[a] For 2009-13, max bore limit only, for later regulations a fixed value was imposed

Maximizing the performance of the naturally aspirated Formula One engines was mostly about maximizing the airflow rate. Valve dimensions and lift were therefore very large and a pent-roof head was used. There was little risk of knock and hence compression ratios were only limited by what could be achieved geometrically and were typically in the range 11–13. The optimum air fuel ratio was at a fuel rich lambda value of around 0.9. The intake system was designed for very high permeability and to harness unsteady gas dynamic effects. Volumetric efficiency was in excess of 120 %. Whilst the brake performance was very impressive - more than 300 CV/litre from a naturally aspirated engine – the combustion was less so. The maximum enleanment these engines could support in full load and high speed was only around lambda 1. At this lambda, 10–90 %MFB (Mass Fraction Burned) durations were around 60°CA (Crank Angle) and the required ignition advance could be in excess of 60° BTDC (Before Top Dead Centre). In addition to the work to maximize airflow, friction was also very important at these engine speeds and so there was significant development to reduce this to a minimum.

The new regulations required a completely different approach. The fuelflow would be the same for all teams and the airflow, or lambda, could be chosen with some freedom due to the presence of the turbocharger. The regulations were more road car relevant than they had been for some time, but one important difference was that emissions would be unregulated. This meant that there was no requirement to run at stoichiometric conditions to enable use of a three-way catalyst and hence lean air fuel ratios could be employed to increase efficiency.

2 Lean Boosting

The efficiency benefits of lean boosting are well known and some of these are documented in detail in [2]. They can be summarized as:

1. Increased combustion efficiency (fewer rich pockets)
2. Reduced heat losses (more dilution)
3. Improved knock limit (longer end gas autoignition delay times)
4. Higher ratio of specific heats (improved cycle efficiency)

There are some limits to how much enleanment can be used. Increased enleanment, particularly close to the spark plug at ignition timing, can lead to high cyclic variability. Main burn durations will increase making the heat release profile further from the Otto cycle ideal. At a fixed fuelflow rate, more enleanment means higher boost values. This has implications for turbocharger design and also for required intercooler heat rejection (more heat rejection means a greater aerodynamic penalty). Leaner running also lowers exhaust temperatures and so turbine power decreases relatively to compressor power. Pumping losses also increase.

Harnessing the benefits of enleanment to develop not just an optimized power unit but an optimized vehicle package is therefore far from straightforward. During the predevelopment phase, it was decided to investigate lean performance using single cylinder engines consistent with the new regulations. Sweeps of plenum temperature and exhaust pressure together with lambda were performed in order to best understand these interactions and hence to specify a well-balanced power unit. Data were constantly exchanged both with engine and car simulation groups in order to converge to an optimum package for best laptime as soon as possible.

3 2013 4-Cylinder Concept

The four-cylinder engine never made it to the track in 2013. The rules were modified to specify a V6 and postponed for one year. In spite of this, significant development work took place on this concept, particularly on single-cylinder engines. It was the first experience for many Formula One engineers of a boosted engine as such a long period had elapsed since turbochargers had last been permitted in the sport. The rate of learning was therefore arguably highest in these early stages.

3.1 Functionality Issues

The V8 normally aspirated Formula One engines had low trapped cylinder charge mass and spark advance timings of greater than 40° BTDC in full load conditions, even at slightly rich lambda values. The pressure at the plug during spark timing was therefore very low and hence also the breakdown voltage (less than 10 kV). The main requirements for the plug were to be small (M10 or even less) and to be resistant to vibrations. Plugs were typically of a semi-surface discharge type as shown in Fig. 1.

Fig. 1. Typical plug design for normally aspirated formula one engine regulations

Right from the beginning of the development of the boosted engines, it was thought that ignitability would be more important and hence an exposed spark gap with a reasonably large gap size would be advantageous. Large spark gaps mean large breakdown voltages for a given sparking pressure in accordance with Paschen's law [3]. The breakdown voltage limit of the plug depends partly on its size and the smallest plug possible was desired for combustion chamber packaging reasons. The relationship between sparking pressure and breakdown voltage needed to be investigated for candidate spark plugs to ensure adequate functionality.

With the starting spark plug configuration for the boost single-cylinder engine, issues were encountered very early in the program with misfires as shown in Fig. 2.

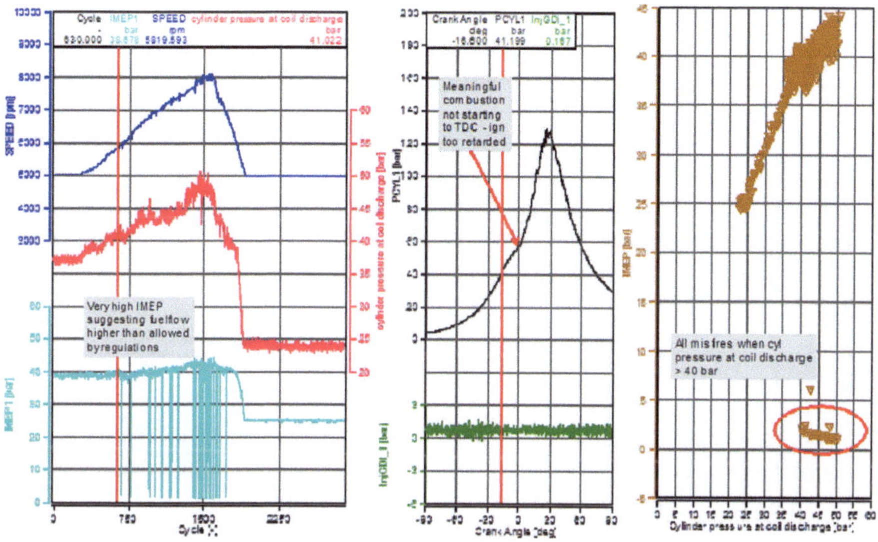

Fig. 2. Misfires encountered early in boosted single-cylinder engine running correlated with sparking pressure

Here a ramp up in engine speed and load is shown where misfires begin at an engine speed of around 6000 rpm and are very frequent by 8000 rpm. It can be seen that these misfires occurred when the cylinder pressure at the moment of attempted spark discharge (sparking pressure) was greater than 40 bar. This suggested that the voltage limit of the ignition system was being exceeded.

A dedicated test was therefore performed to understand better the capability of the available plugs. Both boost pressure and ignition retard were used to gradually increase the sparking pressure until misfire occurred. A high voltage probe was used to identify both the relationship between the pressure and the voltage and the maximum voltage where sparking reliably took place. The results are shown in Fig. 3. As would be expected for two plugs with the same gap and overall size, the maximum breakdown voltage is similar at around 32 kV. This is initially reached at around 40 bar Spark Pressure for both plugs. This is around where the misfires were observed in Fig. 2 confirming that they were due to exceeding the plug voltage limit. It is interesting that Plug 1 in Fig. 3 is able to keep sparking at greater than 40 bar with no increase in voltage, unlike Plug 2. It is likely the case that the spark is no longer taking place in the gap but rather it is tracking along the surface of the ceramic for Plug 1. Although this sparking mode should mean lower ignitability, an ignition source will still be present and this is likely the reason why this plug enabled the engine to run without misfires over a wider operating range than Plug 2 which was the starting configuration.

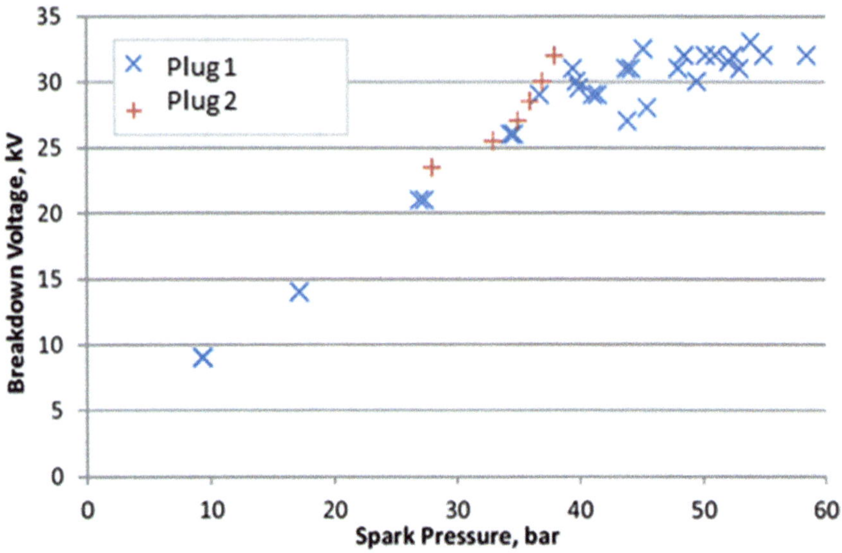

Fig. 3. Correlation between breakdown voltage and spark pressure for two 0.8 mm gap plugs

3.2 Ignitability Tradeoffs with Breakdown Voltage

The next stage in the ignition system development, having understood the starting point in terms of functionality, was to attempt a first optimisation of the plug performance within the available breakdown voltage limits. To this end, a number of plug geometries were compared over a range of plug gaps. Larger gaps may result in ignitability improvements due to a higher power breakdown phase along with a geometrically longer arc and lower chances of kernel quenching on the electrodes. For a given coil energy however, a higher breakdown voltage due to a bigger gap will also mean a shorter spark duration. This is because more of the coil's energy is transferred in the breakdown process so there is less available to support the arc and glow phases.

Figure 4 shows the different types of plug geometry that were compared. The "cheese slice" ground electrode is likely the one most resistant to vibrations of the three. This was a big concern on the previous generation of engines and it wasn't known would it be critical or not for the lower speed boosted regulation. The bulkier ground electrode may be disadvantageous from an ignitability point of view however. The J-type configuration on the other hand should be good for ignitability and is typically used in road cars. The gap protrudes somewhat more into the chamber. The slant ground electrode should be somewhere in between. It should have better vibration resistance than the J-type due to the smaller ground electrode but perhaps somewhat poorer ignitability as the spark location is less exposed. The very small ground electrode should also mean minimal quenching.

Spark timing sweeps were performed at lambda 1.4 where combustion stability was quite poor in order to highlight even small ignitability differences. Breakdown voltages were again recorded.

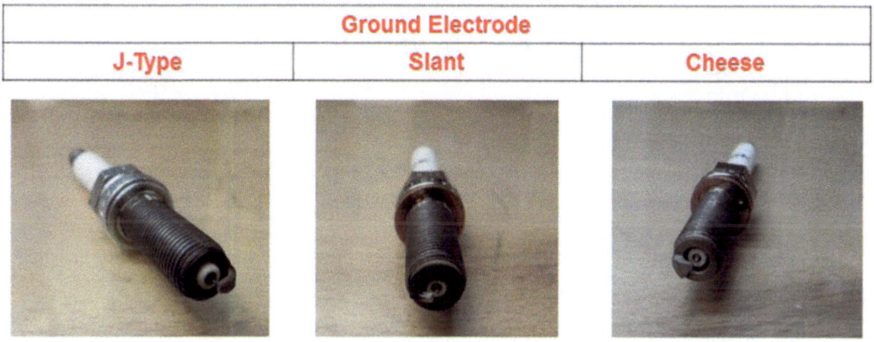

Fig. 4. Investigated plug geometries

Table 2 shows the effect of the electrode geometry on the combustion stability and breakdown voltage limit. It can be seen that a difference can be noted between the plug types. The lowest CoV of IMEP, i.e. the best combustion stability, was recorded with the J-gap plug. Note that this spark plug has a lower heat range and hence the ceramic will be hotter. This may be beneficial also for ignitability. The cheese slice electrode performed next best whilst the slant ground electrode gave the poorest combustion stability although the difference between it and the cheese slice was just 0.3 %. The maximum breakdown voltages were close to 36 kV in all cases. The results suggest that the J type design requires slightly less voltage to breakdown the same nominal gap size. This could be related to small differences in production tolerances and so may not be significant.

Table 2. Effect of electrode geometry on combustion stability

Type	Gap, mm	Heat range	Cov of IMEP, %	Max VBD, kv	Max sparking pressure, bar	Average kv/bar
J	0.8	9	2.3	35.5	53.1	0.67
Cheese slice	0.8	12	2.7	36.4	50.1	0.73
Slant	0.8	12	3.0	36.4	50.2	0.72

Table 3 shows the effect of spark plug heat range. Two heat range 10 plugs were compared to the standard heat range 12 (higher numbers mean a cooler ceramic). There was no combustion stability advantage to the hotter plug ceramic although there was also significant differences in the CoV of IMEP recorded for both heat range 10 plugs. There did seem to be a reduced voltage requirement for a given sparking pressure with the heat range 10 however. Although preignition was not observed during the test, it was thought prudent to aim for a cool ceramic (high heat range) and ground electrode and so there were concerns about the J gap design. An additional reason for discontinuing research on the J-gap plug was fear of vibration-related ground electrode failure.

Table 3. Effect of spark plug heat range

Type	Gap, mm	Heat range	Cov of IMEP, %	Max VBD, kv	Max sparking pressure, bar	Average kv/bar
Slant	0.8	10	3.2	35.7	56.4	0.63
Slant	0.8	10	3.7	36.4	58.5	0.62
Slant	0.8	12	3.0	36.4	50.2	0.72

Table 4. Effect of spark plug gap

Type	Gap, mm	Heat range	Cov of IMEP, %	Max VBD, kv	Max sparking pressure, bar	Average kv/bar	Notes
Slant	0.4	12	4.8	24	83.8	0.29	Gap manually reduced, exhaust temperature limit
Slant	0.6	12	3.5	36.8	76.9	0.48	
Slant	0.8	12	6.4	36.1	51.9	0.70	Gap manually increased

Table 4 shows the effect of changing the gap size. This was done manually starting from a native 0.6 mm slant ground electrode plug (this ground electrode type is easier to modify).

The combustion stability results suggest that 0.6 mm is best but note that the 0.8 mm CoV values are very different than the results for a similar plug with a native 0.8 mm gap, as shown in Table 3. It is likely that the ignitability of the plug has been affected by the manual gap adjustment, possibly due to unintended deformation of the central electrode tip. What is clear however, is the much lower voltage requirement as the gap size is reduced, as would be expected. Comparison of the CoV results for a native 0.8 mm slant ground electrode plug to a native 0.6 mm of the same configuration and heat range, a CoV reduction of 0.5 % is seen for the larger gap size.

The conclusions from the test were that the low heat range J type plug seemed best from an ignitability point of view, whilst the high heat range cheese type and slant ground electrode were poorer, with a slight advantage for the cheese slice. This result was not expected, considering the risk of quenching on the large ground electrode. It is also possible that the large ground electrode was conferring some shielding benefit from in-cylinder flow and hence reducing early flame kernel displacement. For plugs of the same geometry, the results in Table 3 show that the lower heat range plugs (hotter ceramics) did not give a combustion stability improvement but they had a lower voltage requirement for similar sparking pressures. Modification of the gap manually did not give the expected results for combustion stability and hence this operation may have damaged the plug. Comparison of the 0.8 mm and 0.6 mm native slant ground electrode plugs, showed a clear combustion stability benefit for the bigger gap; the CoV decreased from 3.5 to 3.0 %. However, the 0.8 mm gap had a higher voltage requirement (0.62 versus 0.48 kV/bar), which is an important factor, given that for the provisional 2013 engine concept, sparking pressures were high (in excess of 50 bar).

4 Initial 2014 Concept Testing

4.1 Impact of Regulation Change on Ignition System

The 2013 boosted engine regulations were abandoned and instead it was decided to introduce the new powertrain in 2014. The swept volume was kept the same as the 2013 proposal at 1600 cm^3 but the engine configuration was changed from an inline four cylinder to a V6. The bore size was reduced but the bore-to-stroke ratio increased from around 1.3 to about 1.5. Perhaps most significantly for the ignition system requirements, the fuel flow limit curve also changed as shown in Fig. 5. It can be seen that the fuelflow, and hence boost requirement of the 2014 regulation, was lower at all speeds until 10500 rpm. The higher speed where 100 kg/h fuel flow was permitted also meant an increase in ignition delay for this most important speed range. Sparking pressures therefore became less critical but ignitability increased in importance. In initial tests, it was observed that the single cylinder prototype of the 2014 V6 engine configuration could not run as lean as the equivalent four-cylinder concept. Spark plug investigations continued and attention also turned to the coil.

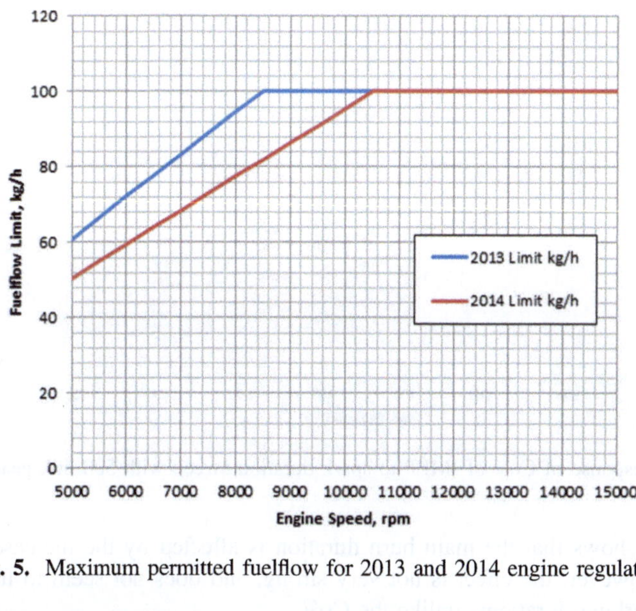

Fig. 5. Maximum permitted fuelflow for 2013 and 2014 engine regulations

4.2 Ignition Coil Emulator Testing: Round 1

Most testing during the early stages of the boosted engine development took place with a conventional 80 mJ inductive coil. It was decided to check if this could be improved upon using an ignition coil emulator. This emulator system allows the current characteristics in the spark gap to be prescribed through high speed adjustment of the voltage. It was decided to perform a sweep of spark duration at a constant peak current of 500 mA and thereafter a sweep of peak current at a constant spark duration of

330 μs. The engine was run from 10000–15000 rpm in full load at a constant lambda close to the lean limit in order to highlight differences. With the standard coil, this corresponded to a CoV of IMEP of around 7 %. The ignition timing was fixed to the standard coil values.

Figure 6 shows the response of the engine combustion stability to the spark duration sweep. It can be seen that spark durations of 100 μs result in very unstable combustion. There is clear trend for the combustion stability to improve with increased spark duration, although the benefits appear to be flattening off after around 400 μs. Note that at 12000 rpm, 1°CA corresponds to 14 μs. A spark duration of 400 μs at this speed is therefore equivalent to 29°CA. Therefore, since the CoV is strongly correlated with variations in the initial flame kernel, it is understandable that there is little effect after 400 μs.

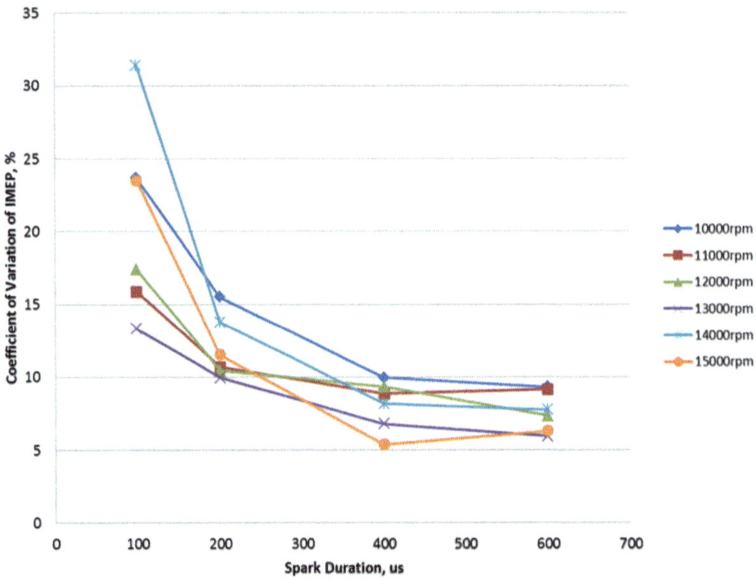

Fig. 6. Response of CoV of IMEP to spark duration sweep with 500 mA peak current

Figure 7 shows that the main burn duration is affected by the increased sparking durations. However, the effect is not very strong, and does not seem to flatten out at increased sparking durations, unlike the CoV.

In the second sweep, the spark duration was fixed at 330 μs and the peak current was varied from 100 to 600 mA. The response of combustion stability to variation in peak current is shown in Fig. 8. It can be seen that there is a high sensitivity at low peak currents, with a significant improvement in combustion stability as the peak current is increased. Above around 500 mA, the results showed no clear evidence of further improvements. The sensitivity of burn duration to peak current is shown in Fig. 9. Once again, reductions in CoV can be seen with increasing peak current. This is more significant at some speed than others.

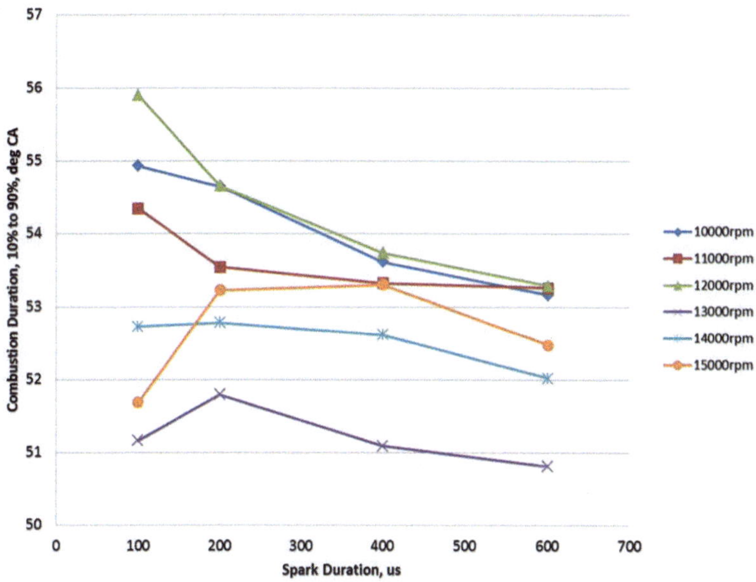

Fig. 7. Variation of 10–90 % burn duration with spark duration, 500 mA peak current

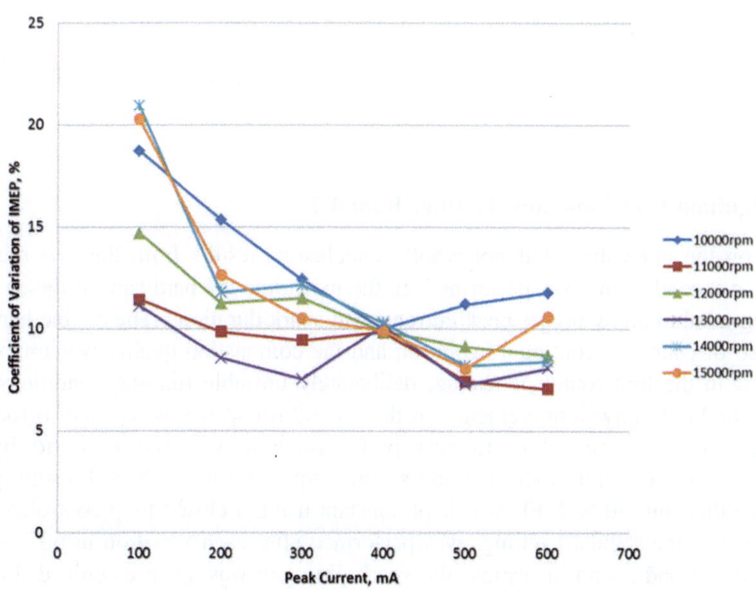

Fig. 8. Variation of CoV of IMEP with peak current, 330 μs spark duration

The results obtained with the first round of ignition coil emulator testing were interesting. An improvement in combustion stability with increasing spark duration and peak current was expected but the apparent main burn duration benefit, although small,

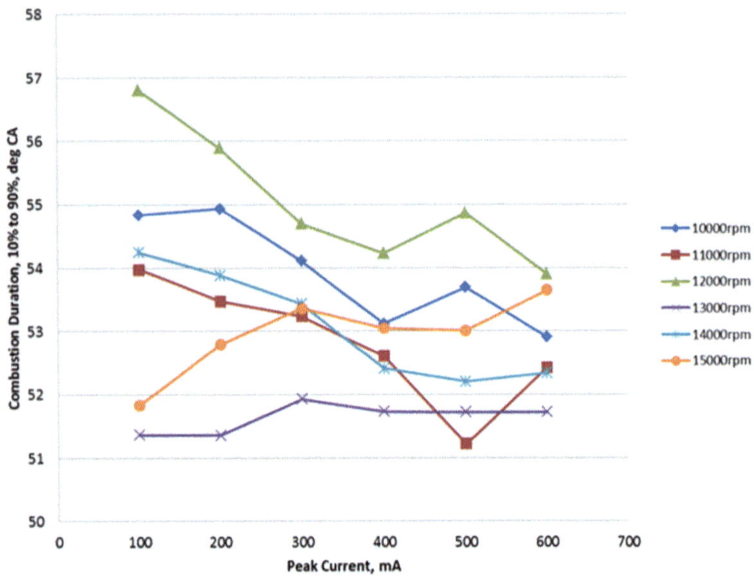

Fig. 9. Variation of 10–90 % burn duration with peak current, 330 μs spark duration

was more surprising. It should be noted that the testing methodology of fixing the ignition timing may have confounded the results - reduction of the burn delay with increasing peak current would advance the combustion phasing which would have had a knock on beneficial effect on the burn duration.

4.3 Ignition Coil Emulator Testing: Round 2

Based on the interesting, but not wholly conclusive, results from the first round of testing, a second round was performed. In the meantime the hardware of the emulator was upgraded to allow bigger peak currents and spark durations. The engine hardware was also of course in continual evolution and the combustion quality was improving. Whereas in the first round of testing, deliberately unstable running conditions were used to highlight any improvements, in the second round it was decided to focus on lambda values corresponding to best performance. Lower sensitivity to ignition parameters was therefore expected and so this required a more careful testing procedure. To this end, 50 % MFB was kept constant using a closed loop controller and a rebaseline of the standard settings was performed after each variation in the sweep.

In the second round of testing, the spark plug gap was again examined this time using three native gap sizes of 0.6 mm, 0.8 mm and 1.0 mm (as mentioned above, the 2014 regulations were less severe in terms of sparking pressure than the 2013 proposals and hence larger gap sizes could be run). All plugs were of slant ground electrode type. Sweeps of spark duration and peak current were again performed but in a two dimensional fashion over a bigger range than was possible before and with all plug types. The peak current sweep results for spark durations of 200 μs, 500 μs and 800 μs are shown in Figs. 10, 11 and 12 respectively. Two parameters were examined in detail: 0–10 % MFB (early burn delay or duration) and the standard deviation of 50 % MFB (which should be correlated to knock limit variations). The upper part of each figure shows the response of the key combustion parameters over the nine power curves that comprised each part of the test. A rebaseline with the standard settings was performed between each new test point over the sweep to check continuously for degradation during the measurements. This is sometimes referred to as “ABA” type testing. The lower part of each figure shows the correlation between the two combustion parameters examined and the peak current over the entire sweep.

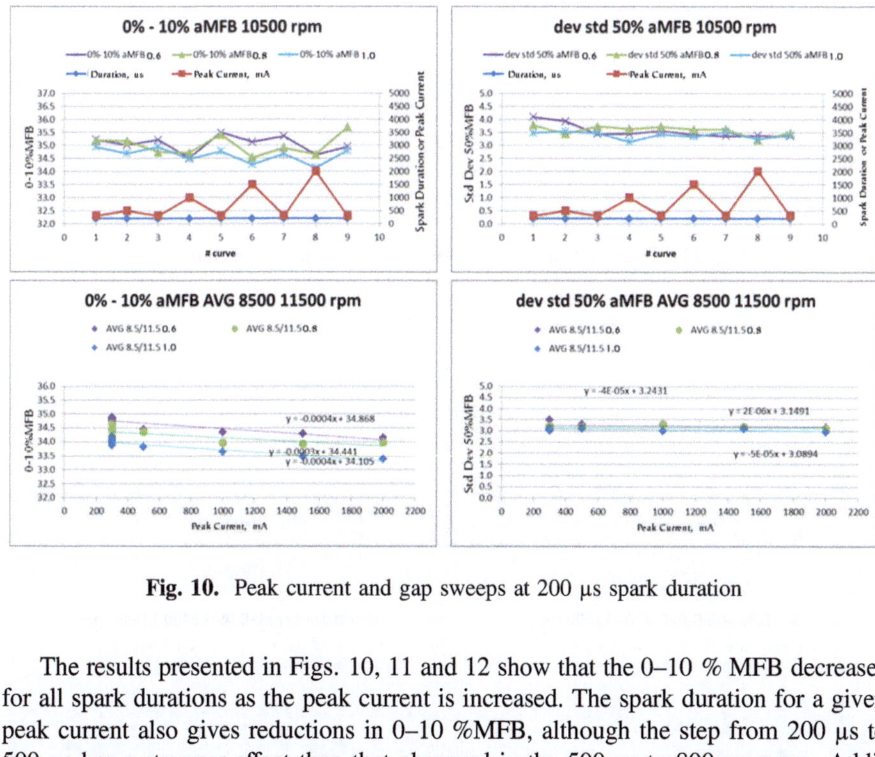

Fig. 10. Peak current and gap sweeps at 200 μs spark duration

The results presented in Figs. 10, 11 and 12 show that the 0–10 % MFB decreases for all spark durations as the peak current is increased. The spark duration for a given peak current also gives reductions in 0–10 %MFB, although the step from 200 μs to 500 μs has a stronger effect than that observed in the 500 μs to 800 μs range. Additionally, there is a clear improvement with increasing the gap size. Note that holding either spark duration constant and increasing peak current or holding peak current constant and increasing spark duration is increasing the delivered energy in both cases. Even over these quite extreme sweeps of ignition parameters, the effect of variations in peak current, spark duration and gap size is only seen clearly on the 0–10 % MFB and the magnitude of the change is in the range of 0.5 to 1.0°CA. More significant changes in the 0–10 % MFB were originally expected, based on the first round of testing. In the second round however, the base engine combustion stability was already better than in the first testing and the conditions used were more race relevant but less critical for ignitability. The influence of the ignition system for combustion was therefore diminished.

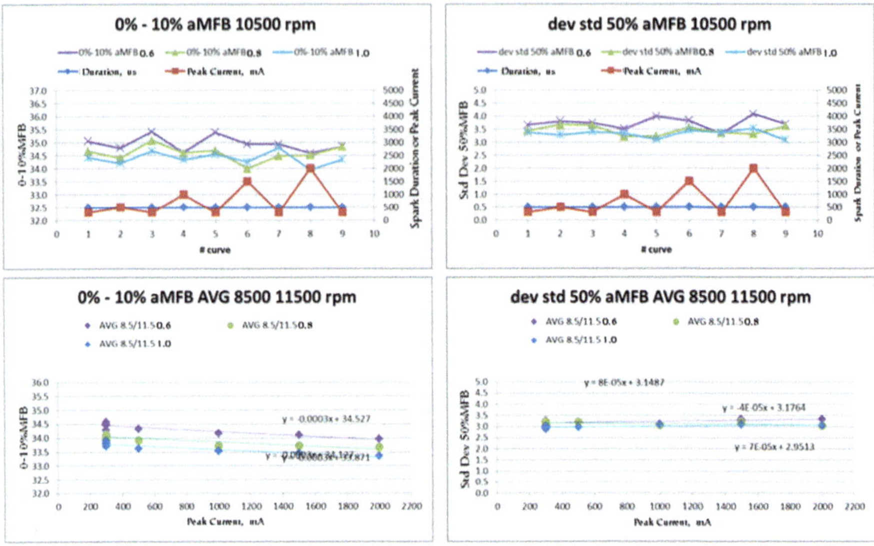

Fig. 11. Peak current and gap sweeps at 500 μs spark duration

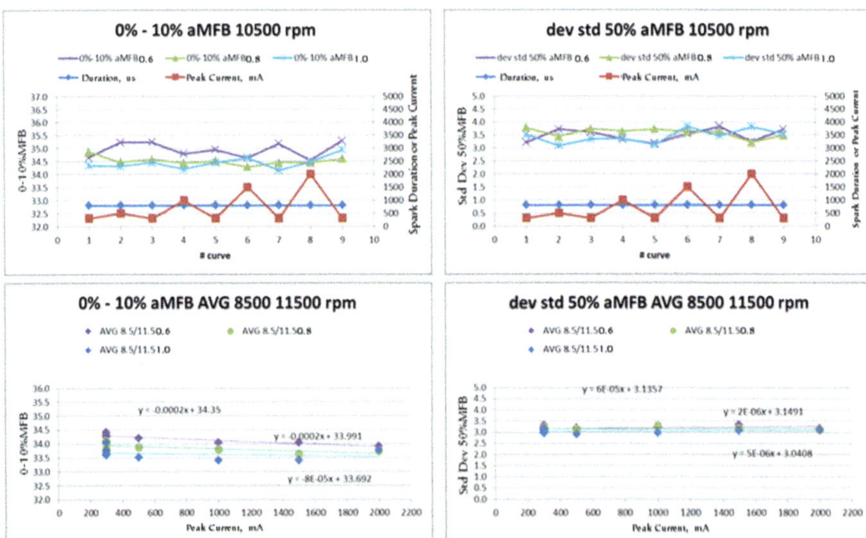

Fig. 12. Peak current and gap sweeps at 800 μs spark duration

A constant ignition energy sweep was also performed (trading off spark duration against peak current as would be more realistic for a conventional ignition system). The results are shown in Fig. 13 where it can be seen that there is no influence on the 0–10 % MFB in this test.

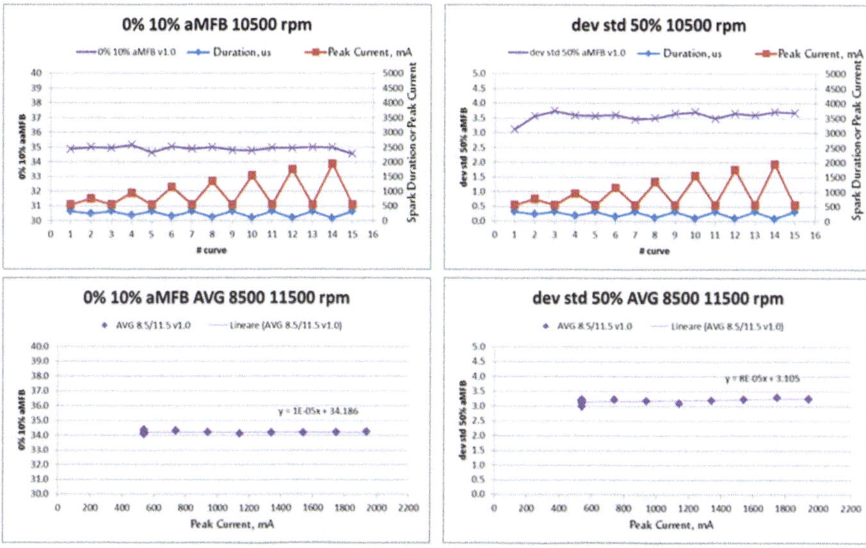

Fig. 13. Peak current and spark duration sweep at constant ignition energy

The conclusions from the second round of testing were that there was likely little to be gained from an increase in ignition coil energy significantly above that which the standard coil could supply at that time, given that the engine combustion had improved due to other factors and the test conditions were less extreme but more representative of race engine use for the 2014 configuration.

It should be noted that the ignition coil emulator is not fully representative of a real ignition coil and in particular, the voltage rise times are significantly lower. A higher energy standard inductive coil was available (100 mJ rather than 80 mJ) and this was also tried at a later date. The gain in 0–10 % MFB was somewhat larger than would be expected from the coil emulator results but was still only around 0.5° in the 0–10 % MFB duration. This was with a longer baseline 0–10 % MFB of around 44°CA, rather than the 33°CA on the hardware and test conditions of the second coil emulator test due to further combustion system modifications. It is not known if the gain was bigger because of the faster rise times or because of the longer baseline ignition delay for that round of testing. It may have been a combination of both. A small brake power gain was also observed by mapping the engine to the knock limit for both 80 mJ and 100 mJ coils but this increase was only of the order of 0.2–0.3 %, which is difficult to measure reliably. It would have been better to have performed both the coil emulator and standard coil upgrade testing on the same engine configuration but the rate of development was very high in all areas at this time and so engine configurations changed rapidly.

4.4 Further Plug Geometry Testing

An initial sweep of plug geometry types was carried out on the 2013 concept single-cylinder engine as detailed above. These plugs were of a relatively conventional exposed spark gap type which were thought to be the most appropriate solution. In an effort to make sure nothing was missed with this approach, two more radical plug types were tested on the 2014 concept single-cylinder.

The first testa part was a type of plug employed in turbocharged F1 engines during the 1980's. It is shown on the left in Fig. 14. This type should be very poor from a quenching point of view but it offers the theoretical advantage of very high shielding of the initial flame kernel from in-cylinder flow. If this quenching is a significant problem, there could be some advantages, including reduction of the breakdown voltage as arc stretching would be reduced. The gap was also a very small 0.4 mm which would limit still further the voltage requirement.

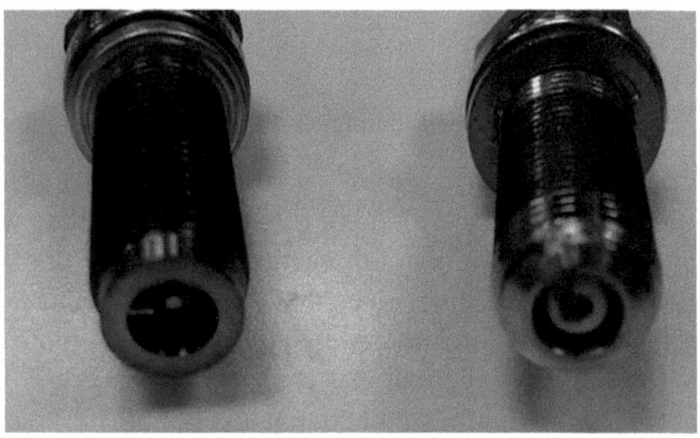

Fig. 14. Retracted electrode plug (left) and slant ground electrode (Right)

The plug was tested against a slant ground electrode plug of 0.6 mm gap. Some results are shown in Fig. 15. A very strong effect on the 0–10 % MFB can be seen, varying from a 2–7° increase depending on the engine speed. However, it is interesting to note that the CoV of IMEP is almost unchanged. The main 10–90 %MFB burn durations were quite similar, with an apparent slight advantage for the retracted design, although this could be due to imperfect measurement repeatability. Brake performance at the knock limit was also the same. It is normally the case that combustion stability is correlated with 0–10 %MFB but clearly this was not observed in this test. It may be that there is increased quenching at the initial stage of ignition but that this is quite a repeatable phenomenon and hence does not increase cyclic variation. Alternatively, increased quenching was perhaps balanced out by less kernel displacement effects during the earliest stages of combustion.

The second alternative plug type was more similar to those used on normally aspirated F1 engines. This plug type is shown in Fig. 16 and the results are presented in Fig. 17. This plug type had a positive effect on the 0–10 % MFB but only by around 1° CA or less. The combustion stability was similar to the baseline slant ground electrode

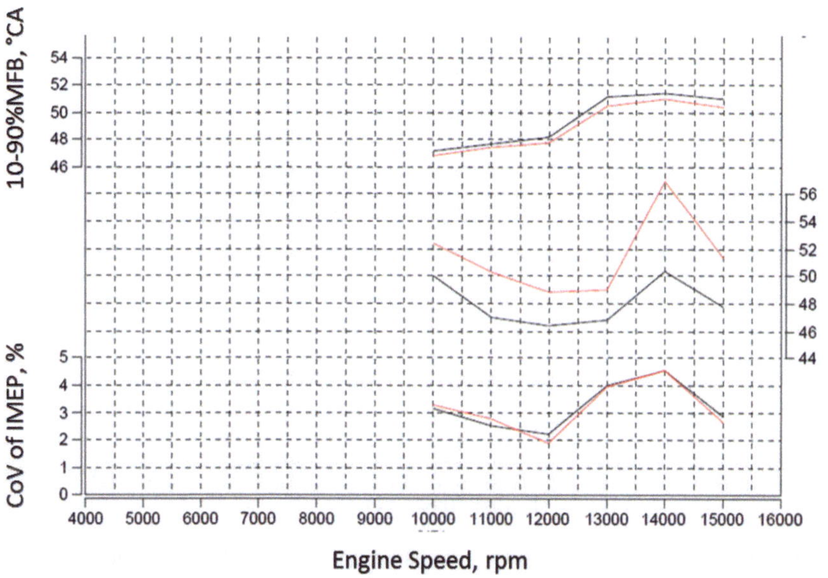

Fig. 15. Combustion data for retracted electrode plug (red) and slant ground electrode (black)

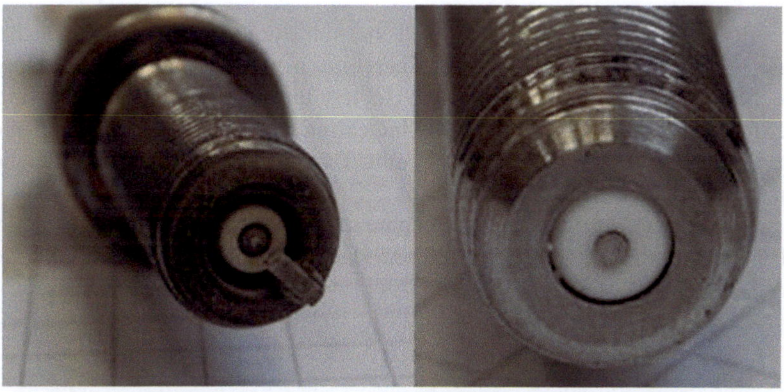

Fig. 16. Slant ground electrode (left) and semi-surface discharge plug (right)

type. However, this plug was shown to give a significant and repeatably worse knock limit in ABA (Baseline /Test /Rebaseline) testing relative to the baseline configuration.

Ultimately, although the combustion stability differences between these plugs was not significant, neither of the alternative designs gave a performance improvement on the engine.

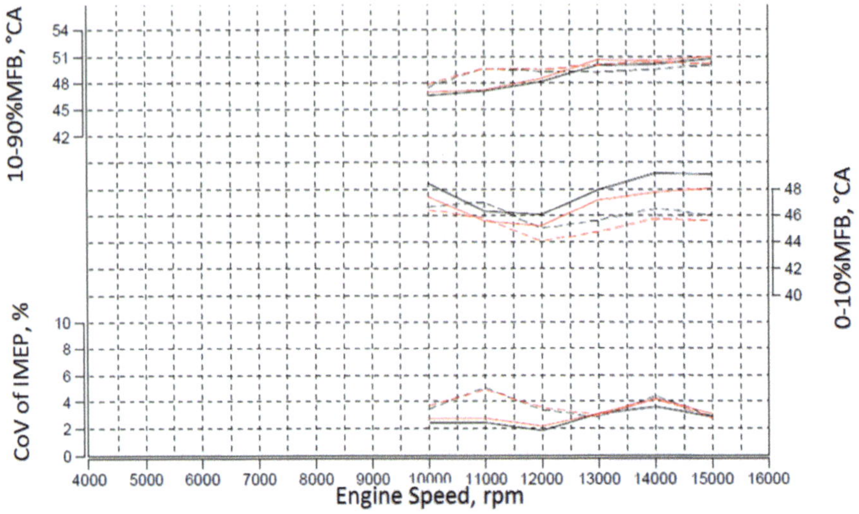

Fig. 17. Combustion data for semi-surface discharge plug (red) and slant ground electrode (black). Dashed lines refer to high exhaust pressure, solid lines to low exhaust pressure

4.5 Effect of Spark Plug Penetration

The results presented in Sect. 4.4 showed that retraction of the electrodes led to an increase in the 0–10 % MFB duration. Therefore, it was decided to try the opposite approach. The most straightforward way to test this was by simply removing the washer from under the base plug, thereby increasing the protrusion of the plug into the combustion chamber by approximately 2 mm. The effect on the early burn duration is shown in Fig. 18. The benefit is as much as 2°CA at speeds where the burn delay is the longest. On average, over the range of engine speeds tested, the decrease is around 1° CA. This testing was performed on a 0.6 mm slant ground electrode plug. It was found

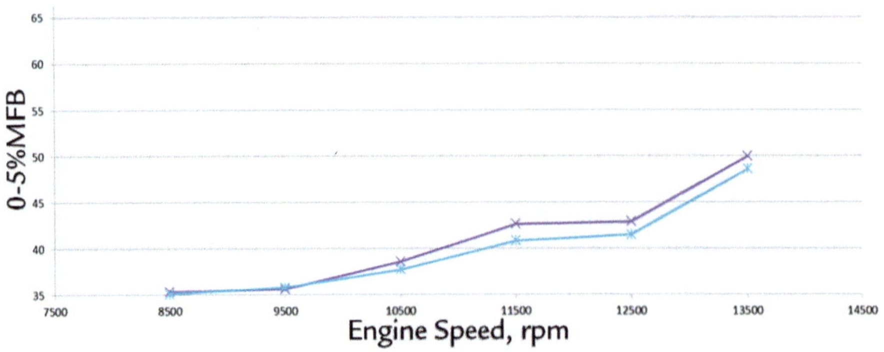

Fig. 18. Variation of 0–5 % MFB for standard plug penetration (purple) and increased penetration (blue)

that the effect of increasing the protrusion of the 0.6 mm plug was similar to running a 1.0 mm gap spark plug at the standard mounting height.

5 Endoscopic Flame Kernel Measurements

The majority of development work was carried out using cylinder pressure indication measurements in an attempt to gain insight into the early combustion behaviour. As this early behaviour is key to ensuring good combustion stability, it was thought worthwhile to examine it in more detail. Endoscopic videoscopy measurements were therefore performed using a photomultiplier camera system which is sensitive in the UV range. The endoscope was installed orthogonally to the symmetry plane of the combustion chamber. Such a system is particularly well suited to early flame kernel imaging. Especially in the early stage of a combustion, videoscopy gives significant additional information compared to cylinder pressure analysis. With flame footage, the temporal resolution of ignition delay determination is higher, since the early flame development has little effect on the cylinder pressure (there is significant uncertainty in heat-release calculations for small MFB values hence why 0–5 % MFB or 0–10 % MFB are often used rather than 0–1 % MFB which would be more correctly termed the ignition delay). Furthermore, cylinder pressure measurements cannot describe spatial behaviour of the flame. Optical measurements were performed in collaboration with MOT GmbH, a member of the APL group.

Figure 19 shows how the ignition delay is quantified optically. It can be seen that an initial light source is seen from the spark discharge itself. This light source then

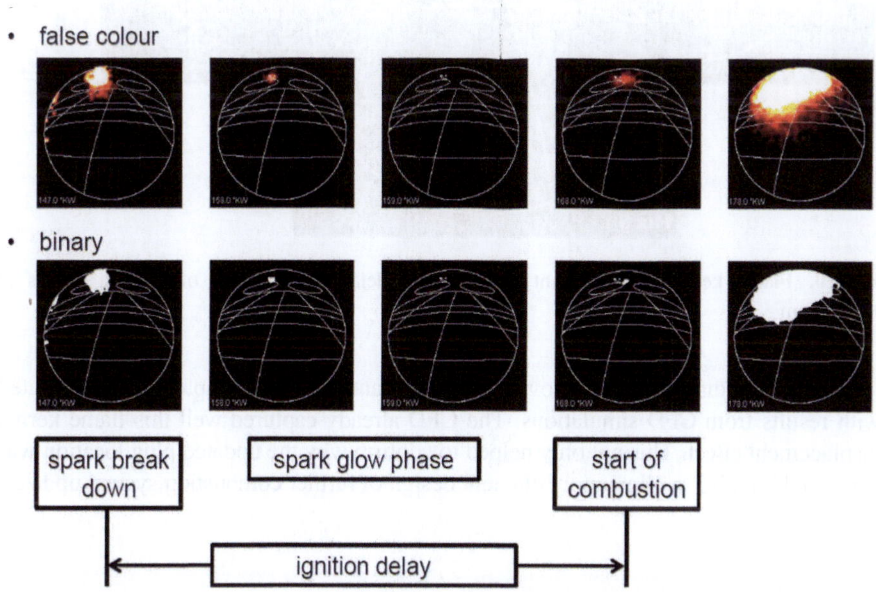

Fig. 19. Image processing details for optical ignition delay and kernel location

disappears and at a certain point the flame kernel can be seen. Thresholds are applied to the images such that the time and location of the initial flame kernel can be measured.

One useful insight that can be gained from such measurements is the cyclic repeat-ability of the initial flame kernel. Figure 20 shows how the ignition delay and flame kernel location vary over a sweep of lambda at 10500 rpm (Mix 1 to Mix 5 is progressively leaner). It can be seen that there is significant displacement of the flame kernel away from the spark plug location towards the right hand side of the images due to in-cylinder flow. Prior to performing the optical measurements, it had been noted that the bias of the spark plug location towards the left side in these figures gave a performance advantage over other locations in the combustion chamber. The optical images suggest that this is because the flame front begins growing from a location offset from that of the plug electrodes and hence the ignition source is more central with an offset plug. It can be seen in Fig. 21 that this effect is more pronounced at 13500 rpm where in-cylinder velocities are likely higher.

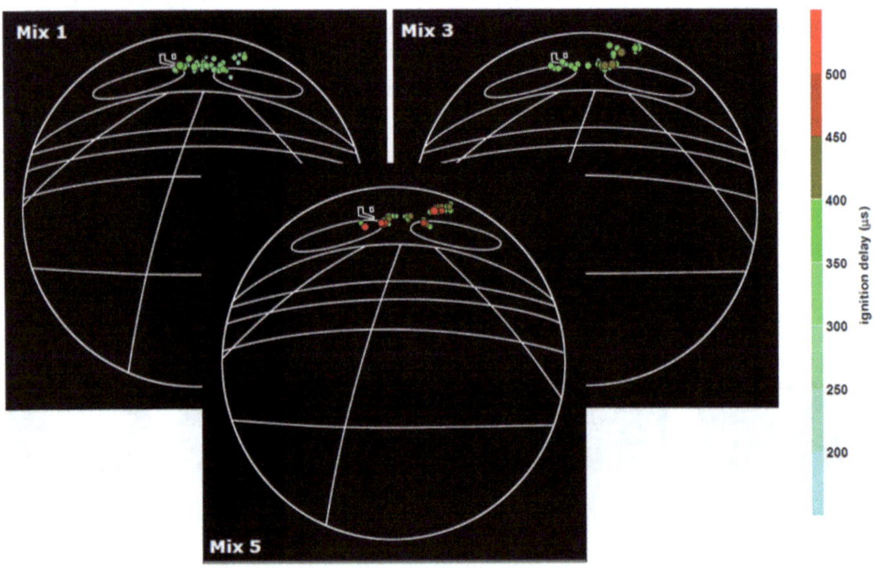

Fig. 20. Flame kernel displacement and ignition delay for a range of lambda values at 10500 rpm

The kernel displacements shown in Figs. 20 and 21 were compared and correlated with results from CFD simulations. The CFD already captured well this flame kernel displacement effect. This not only helped to confirm why the updated plug location was beneficial but also to allow more efficient design of further combustion system updates.

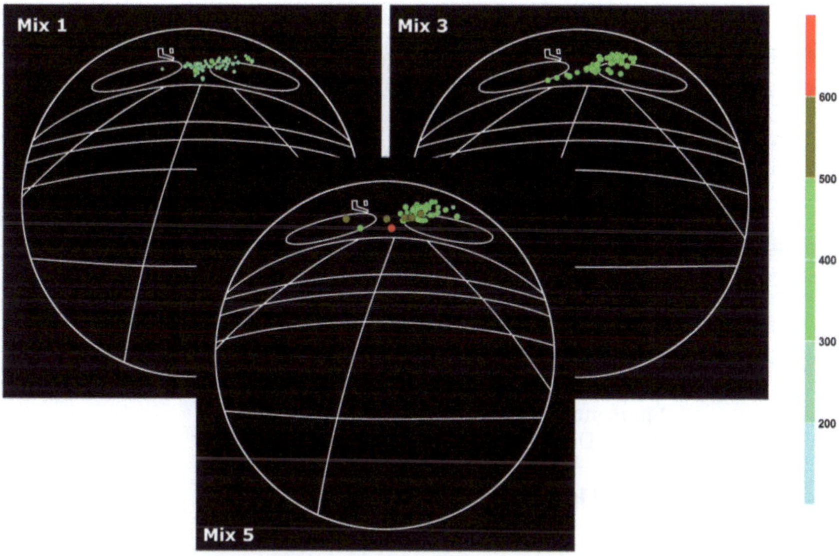

Fig. 21. Flame kernel displacement and ignition delay for a range of lambda values at 13500 rpm

6 Multispark Testing

Early testing with the ignition coil emulator had focused on peak current and spark duration. It was decided to also investigate multiple sparking events. As the combustion hardware had changed again, some prior tests were repeated. Firstly, the ignition coil emulator was compared to the standard coil by applying representative settings to the emulator unit. It was seen that the matching of results was not perfect and that the coil emulator resulted in around 1° longer 0–10 % MFB. The main burn durations and combustion stability were quite similar however. Spark duration and current were once again swept and, as expected from previous testing, it was not until values significantly less than what the standard coil produced were implemented that combustion stability became an issue and even then both a low peak current (100 mA) and short duration (40 µs) were needed together to produce a misfire. Note that this was significantly weaker than the minimum spark duration and peak current required to produce misfire in early testing of the 2013 concept. The engine was running slightly richer lambda than in the very early testing, which of course helps to improve CoV, but the combustion system had also been significantly developed. This improvement effectively led to a much less demanding ignition requirement, which is also evidenced by the shorter 0–10 % MFB values recorded in this test in comparison to what has been documented in earlier paragraphs. Using these "weak spark" settings, a sweep of the number of spark events was then performed.

Figure 22 shows a schematic of the profile used and the results obtained are presented in Fig. 23. It can be seen that it is only for the single weak spark that a misfire occurs (evidenced by the higher CoV of IMEP) and even then for one attempt this did

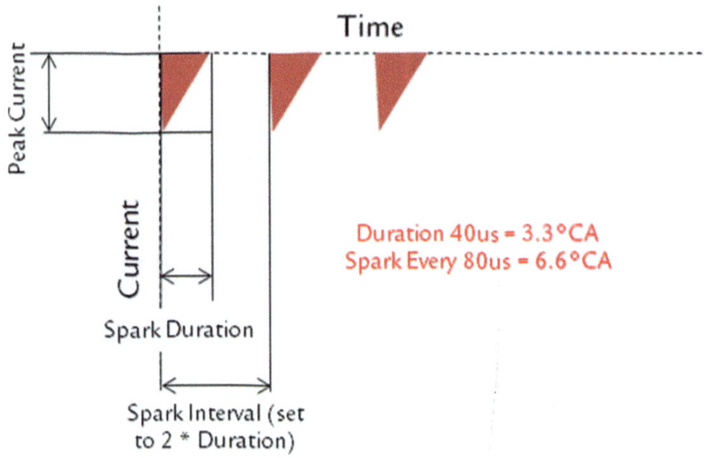

Fig. 22. Multispark current profile

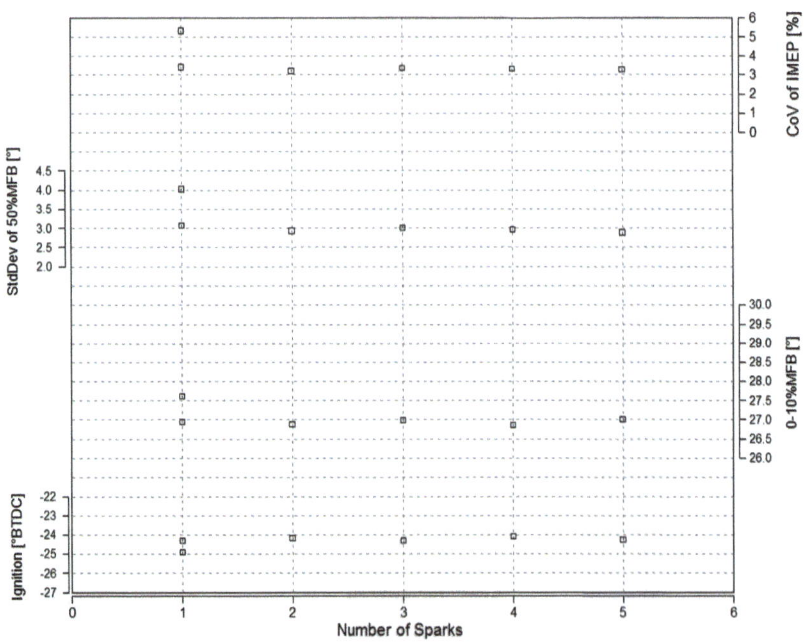

Fig. 23. Multispark results

not happen. From two sparks onwards, no misfires were seen but there was no obvious benefit arising due an increase from two until five spark events. This test once again confirmed that the existing inductive coil had significant margin of ignitability with respect to the minimum peak current and spark duration requirements of the combustion system. Multisparking did not give an advantage with spark current profiles corresponding to the standard coil values but only for much shorter, weaker sparks. On a combustion system with a higher ignitability requirement, a stronger response to multisparking is likely to be observed.

7 Importance of Early Burn Duration

The background to this paper mentioned the benefits of lean boosting. These benefits were indeed realized during the development of the boosted engines in the period 2011–2014. As expected, it was found that there was a correlation between the optimum lambda for each hardware configuration and the 0–10 % burn duration. This correlation is shown in Fig. 24. These results refer to changes not only of the ignition system but also intake ports, pistons, cams etc. It can be seen that a reduction in the 0–10 % burn duration from 45°CA to 35°CA shifts the optimum lambda for the engine 0.1 leaner. This in turn results in better BSFC and hence higher power for a given fuelflow. Figure 25 shows the effect on combustion stability judged by the standard deviation of the 50 % burn location. This parameter is closely related to the engine knock limit – higher standard deviation values result in a poorer knock limited combustion phasing. The results shown in Fig. 25 are for every point in the database and so include different lambda values as well as hardware configurations. It can be seen that shorter 0–10 % MFB are correlated with more stable 50 % MFB values.

It is more common to judge engine combustion stability by CoV of IMEP and the same dataset employed for Fig. 25 has been used to produce the CoV shown in Fig. 26. Comparison of Figs. 25 and 26 shows that the stability of 50 %MFB degrades in

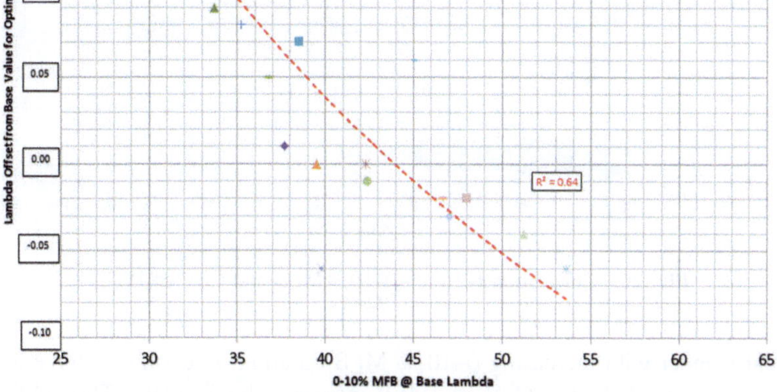

Fig. 24. Correlation between 0–10 % MFB and optimum lambda

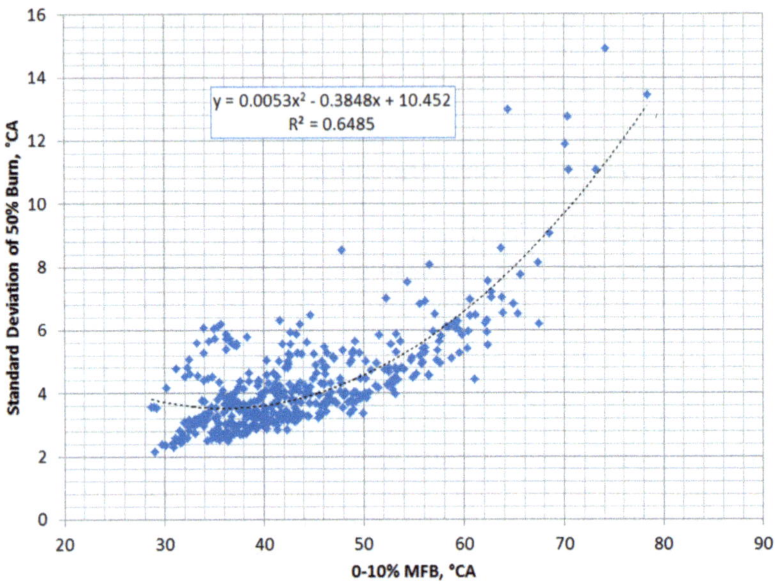

Fig. 25. Correlation between 0–10 % MFB and stability of 50 % MFB

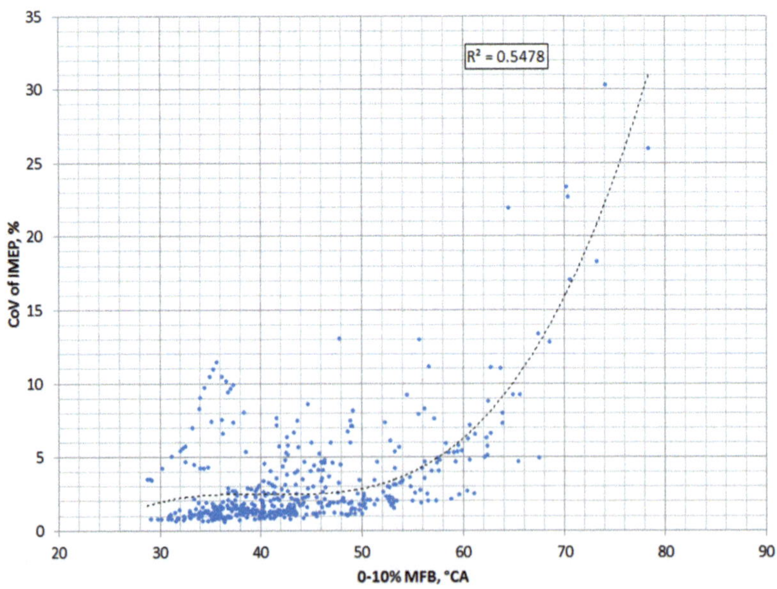

Fig. 26. Correlation between 0–10 %MFB and CoV of IMEP

a gradual manner with increasing 0–10 % MFB duration. In contrast, above a certain value of 0–10 % MFB, the CoV of IMEP increases dramatically. The "cliff point" seems to be at about 60°CA duration. This is somewhat higher than has been reported

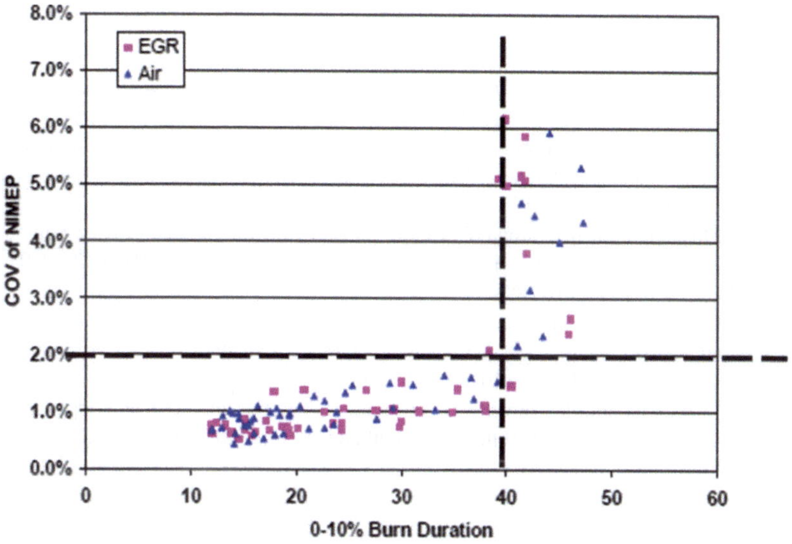

Fig. 27. Correlation between 0–10 % MFB and CoV of IMEP for Conventional SI Engines [4]

in the literature [4] for more conventional spark ignition engines as reproduced in Fig. 27, perhaps due to the different speed ranges used.

8 Conclusions

The reintroduction of boosting to Formula One and the imposition of a fuel flow limit were arguably the biggest changes ever made to the rules regarding F1 engines. The regulations implied a high speed, high load, lean boosted approach capable of supporting high exhaust pressure. When experimental activity began, the ability to reliably spark at the high cylinder pressure was not guaranteed. Changes of the engine hardware often required reinvestigations of the ignition system with different results being found in comparison to previous tests. The requirements of the ignition system in terms of ignitability and robustness were therefore constantly moving targets as engine development proceeded and they remain so to this day. Spark plug types, gap sizes, penetration and heat range were all investigated and sensitivities were found. Spark energy parameters were also looked at in some detail and optical imaging was used to good effect to add insight to standard indication measurements. It was found that the spark plug geometry and location had the strongest effect on combustion parameters providing that coil energy was above a certain threshold. The impact of changes in the ignition system was seen directly in improvements in early combustion rates but these in turn were found to correlate to combustion stability and hence enleanment capability at the knock limit. The regulations demanded a relatively conventional approach but in spite of this, highly efficient powerunits have been produced which have a BSFC in full load conditions of less than 180 g/kWh [5] due in no small part to fast, stable, lean combustion.

References

1. Federation Internationale de L'Automobile (FIA) Formula One World Championship Regulations. http://www.fia.com/regulation/category/110
2. Stokes, J., Lake, T.H., Osborne, R.J.: A Gasoline Engine Concept for Improved Fuel Economy –The Lean Boost System. Published by the Society of Automotive Engineers, Paper No. 2000-01-2902
3. Shane, T.: Investigation of Corona Threshold Voltages on International Space Station U.S. Laboratory and Airlock Orbital Replacement Units. SAE Technical Paper 2001-01-2899 (2001). doi:10.4271/2001-01-2899
4. Ayala, F.A., Gerty, M.D., Heywood, J.B.: Effects of Combustion Phasing, Relative Air-Fuel Ratio, Compression Ratio, and Load on SI Engine Efficiency. SAE Technical Paper 2006-01-0229. doi:10.4271/2006-01-0229
5. Sassi, L., Kitsopanidis, I., Lovett, G.: Evolutions in F1 engine technology: pursuing performance from today's powerunit through efficiency. In: 37th Vienna Engine Symposium (2016)

New Ignition Systems 2

Effects of Microwave-Enhanced Plasma on Laser Ignition

Jun Hayashi[1(✉)], Chen Liu[1], Fumiteru Akamatsu[1], Atsushi Nishiyama[2], Ahsa Moon[2], and Yuji Ikeda[2]

[1] Department of Mechanical Engineering, Osaka University, Yamadaoka 2-1, Suita, Osaka 565-0871, Japan
j.hayashi@mech.eng.osaka-u.ac.jp
[2] Imagineering, Inc., 7-4-4 Minatojima-Minami, Chuo, Kobe 650-0047, Japan

Abstract. The characteristics of a microwave-enhanced laser ignition of methane/air lean premixed mixture in the constant volume combustion vessel is experimentally investigated. The microwave-enhanced plasma is applied as an advanced ignition source due to the controllability of the plasma life time. The parameters of the microwave are varied for elucidating the effects of microwave-enhancement on laser ignition. Results shows that the microwave-enhanced laser ignition decrease the total time of combustion. This is because the large volume and long life time plasma decrease the time of initial development of the ignition kernel. In addition, the total energy of microwave enhanced laser ignition can be reduced effectively by applying the duty ratio of microwave.

Keywords: Laser ignition · Microwave-enhanced plasma · High-speed imaging · Constant volume combustion chamber · Methane/air premixed mixture

1 Introduction

Since the risk of depression of fuel and the amount of emissions of carbon dioxide have been concerning as the major problems in worldwide. To solve these problems, high thermal efficiency and low pollutant emissions are required for conventional internal combustion systems. The lean premixed combustion of natural gas has attracted attention as efficient and clean combustion techniques. Therefore, studies on the ignition phenomena in lean premixed conditions are becoming more important. Several innovative ignition technologies have been developed to improve ignition stability under these conditions [1–12]. This is because that the combustion systems with leaner premixed conditions can achieve higher thermal efficiency. In conventional combustion systems with spark plug ignition, the heat loss to the electrodes of the spark plug and the heat loss to the chamber wall can be the critical cause of misfire under an extremely lean condition. In order to make successful ignition with striving against that heat loss, the spark with large energy is required in the conventional ignition system. That leads the damage on electrodes.

Laser ignition is regarded as one of the innovative ignition procedures for improving the thermal efficiency and the fuel economy because the laser ignition can

M. Günther and M. Sens (eds.), *Ignition Systems for Gasoline Engines*,
DOI 10.1007/978-3-319-45504-4_14

ignite the mixture in the wide range of lean and diluted conditions [13–16]. However, the ignition under extremely lean condition and/or high turbulent condition is required as mentioned above. Since high speed of turbulent flow dismiss the plasma formed by laser induced breakdown in such conditions, it is necessary to improve ignitability of laser ignition. Here, large initial size and long life-time of the plasma are proved to be one of effective procedures for the ignition in lean premixed condition [13, 15–17]. Although a high incident energy of laser pulse increases the light emission intensity and the size of the plasma, it barely affects the plasma life-time. Ikeda and Tsuruoka conducted experimental investigation and performed the controllability of the plasma life-time by using microwave-enhanced plasma [16]. Whereas the instantaneous light emission intensity of microwave-enhance plasma was weaker than laser-induced plasma, the light emission intensity and life-time could be controlled by changing the microwave duration.

In this study, therefore, a microwave-enhanced laser ignition is proposed to achieve the successful ignition by forming long life-time plasma. Effects of microwave superposition on the laser induced breakdown and effects of the parameters of microwave on the laser ignition are experimentally investigated. To highlight the enhancement of microwave, microwave-enhanced laser ignition trials in the constant volume combustion chamber are conducted. Ignition phenomenon of premixed mixture is observed by using high speed imaging and the pressure history.

2 Experimental Setup and Method

Figure 1 shows a schematic illustration of the experimental apparatus for microwave-enhanced laser ignition which consists of a laser energy control unit, a constant volume combustion vessel (CVCC), and a focusing lens. To control the laser incident energy without changing the spatial or temporal laser profile, a combination of half-wave length plate and a polarization beam splitter utilized in this study [6]. The Q-switched Nd:YAG laser (Quanta-Ray, Spectra-Physics, wavelength 532 nm, pulse width 8 ns, initial diameter 8 mm) was used as a laser source. Two pyro-electric energy meters (Ophir Inc., PE-50) were placed in front and behind of the CVCC to measure the incident and the transmitted laser energies, respectively. A cylindrical shaped CVCC has 60 mm in diameter and 30 mm in thickness. On the side of cylindrical CVCC, two quartz windows are equipped for the laser incidence and the laser transmission. A large quartz window is equipped for direct observation and high speed imaging. The laser pulse focused at the center of the CVCC using an anti-reflective-coated convex lens of a 100 mm focal length. The pressure history in the CVCC was recorded by a pressure sensor (Kyowa Electronics, PGL-A-10Mp) at the bottom of the combustion vessel. Ignition is defined by the increasing of the pressure in the CVCC. All experimental systems were synchronized by a pulse delay generator (Stanford Research Systems Inc., DG535).

Microwaves were generated from a semiconductor device (output power 1 kW) with frequency of 2.45 GHz and transmitted by a co-axial cable. A three-stub tuner was placed in between the microwave source and antenna for adjusting the impedance of the system. A directional coupler and diode detectors were installed in the microwave

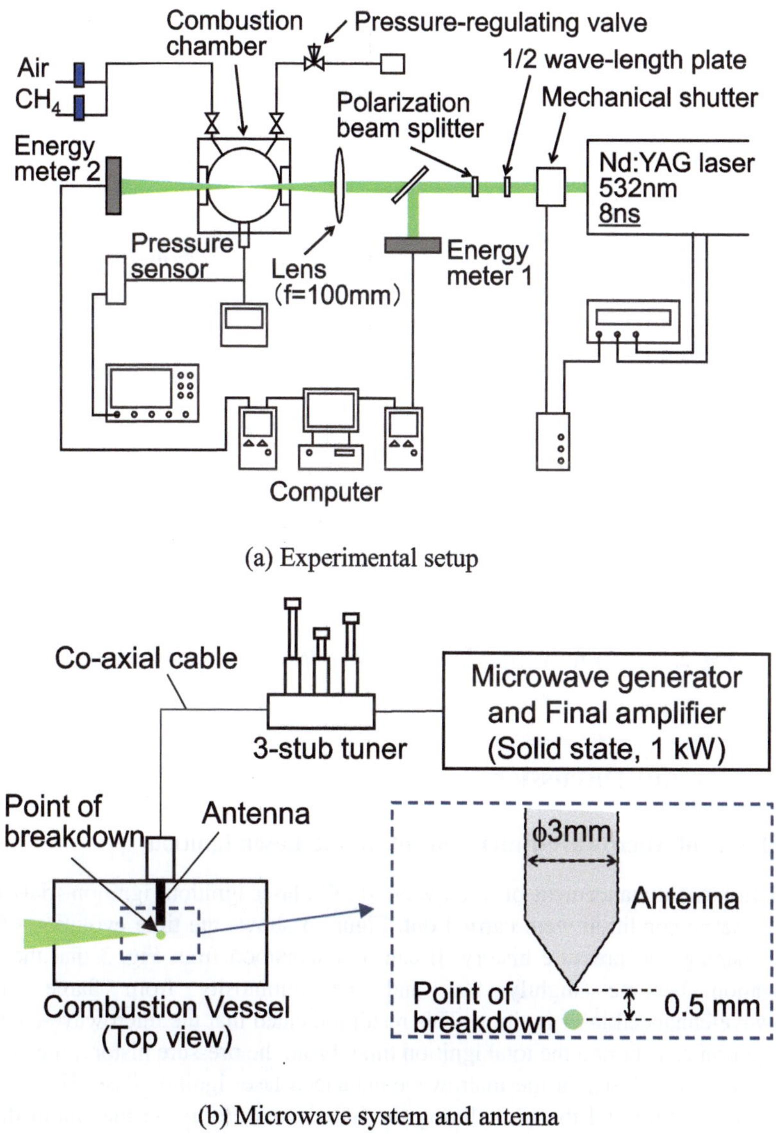

Fig. 1. Schematic illustration of (a) experimental setup and (b) microwave system and antenna

transmission line to measure the input and reflected microwave power. The antenna placed inside the combustion vessel has 3 mm in diameter with a tapered tip configuration to form locally intensified microwave field. The position of the antenna tip was 0.5 mm away from the laser breakdown point as shown in Fig. 1(b).

A schematic sample of microwave pulse pattern and laser trigger timing is shown in Fig. 2. Microwaves were injected 20 ns before laser-induced breakdown in all conditions. The total duration, duty ratio of microwave were varied. The duty ratio is defined as Eq. (1).

$$\text{Duty ratio} = \frac{t_w}{t_d} \tag{1}$$

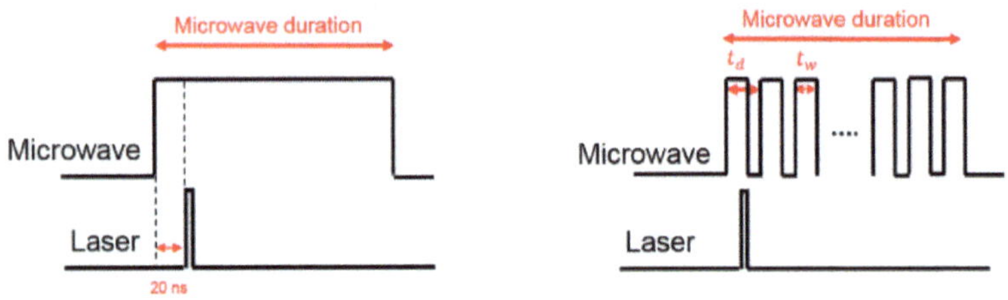

(a) Microwave pulse without duty ratio (total duration)

(b) Microwave pulse with duty ratio

Fig. 2. Schematic diagram of microwave pulses and laser indicate timing

3 Results and Discussion

3.1 Effect of Microwave-Enhancement on the Laser Ignition

To confirm the enhancement of microwave on the laser ignition, ignition trials under stoichiometric condition were carried out. Figure 3 shows the time evolution of high speed imaging and pressure history. It can be understood from Fig. 3 that the flame propagation become slightly fast and the luminosity from flame in the microwave-enhanced laser ignition. This result indicated that the microwave-enhanced laser ignition can shorten the total ignition time. From the pressure history, the pressure in the CVCC rise faster in the microwave-enhanced laser ignition case. However the maximum pressure and the time during the initial pressure to the maximum did not change. This means that the ignition source can on the initial flame kernel, but it cannot accelerate the flame propagation in the quiescent mixture because when the flame propagated away from the ignition source, the ignition source is in the burned gases in the quiescent mixture.

The effect of microwave-enhanced plasma on the minimum pulse energy for ignition was investigated. The excess air ratio was varied from 0.60 to 1.54 in this study under the initial pressure of 0.1 MPa. Figure 4 shows the relationship between the minimum pulse energy for ignition and the excess air ratio. It is clearly found from Fig. 4 that the minimum pulse energy for ignition indicated lower values in conditions of microwave-enhanced plasma. In addition, it is also found from Fig. 4 that the

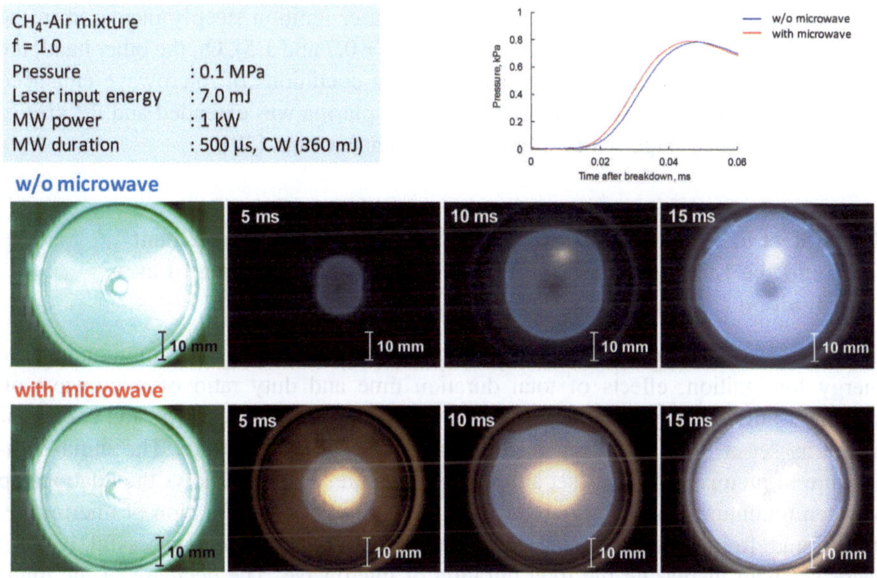

Fig. 3. Flame propagation of methane/air premixed mixture with and without microwave enhancement.

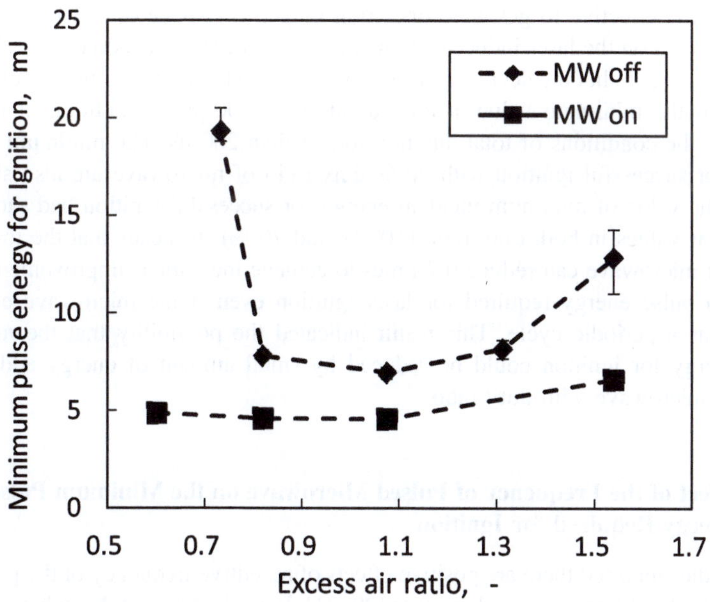

Fig. 4. Relationship between initial excess air ratio and minimum pulse energy required for the successful ignition

minimum pulse energy required for ignition in laser ignition steeply increased in the fuel rich and fuel lean conditions (excess air ratio = 0.7 and 1.5). On the other hand, the range of successful ignition was expanded in the conditions of microwave enhanced laser ignition. This is because the life time of the plasma was extended and the size of the plasma became large with the microwave-enhancement [19].

3.2 Effect of the Total Duration Time of Microwave Enhancement and the Effect of Duty Ratio on the Microwave-Enhanced Laser Ignition

To illuminate the effective conditions of microwave for reducing the minimum pulse energy for ignition, effects of total duration time and duty ratio of microwave on minimum pulse energy for successful ignitions were also investigated. The excess air ratio was set at 1.54 (fuel lean condition) at atmospheric pressure. The initial temperature of premixed gas was set at room temperature. Figure 5 shows the relationship between minimum pulse energy required for ignition and total duration of microwave. It is found from Fig. 5 that the minimum incident energy for successful ignition decreases with increasing the total duration of microwave. The decrease of the minimum pulse energy for successful ignition means that the ignition system can have possibility to add excess energy on the plasma, which is ignition source, in the microwave enhanced ignition. As mentioned above, the strong ignition source is required in the extremely lean and/or high turbulent conditions, the additional energy to the plasma is effective to get wide operation range. This is because when the larger energy applied on the laser induced plasma, the size of plasma become large. It is also found from Fig. 5 that the decrease of minimum incident energy for successful ignition reached to the minimum value at total duration of 200 μs. In addition, it does not change in the conditions of total duration longer than 200 μs. The minimum incident energy for successful ignition with 70 % duty ratio of microwave are also shown in Fig. 5. The value of minimum incident energy for successful ignition had same trend and similar values in both duty ratio (100 % and 70 %). It means that the total input energy of microwave can reduce 0.7 times to achieve the similar improvement on the minimum pulse energy required for laser ignition even if the microwave enhanced plasma has a periodic cycle. This result indicated the possibility that the minimum pulse energy for ignition could be reduced by small amount of energy addition by applying microwave with duty ratio.

3.3 Effect of the Frequency of Pulsed Microwave on the Minimum Pulse Energy Required for Ignition

Some studies reported there are positive effects of repetitive frequency of the plasma on the ignitability of lean premixed mixture [20, 21]. It can be thought that when we apply the pulsed microwave on the laser induced plasma, there would be similar effects of the frequency of pulsed microwave on the minimum pulse energy required for successful ignition. Thus, effects of the pulse frequency of microwave on the minimum pulse

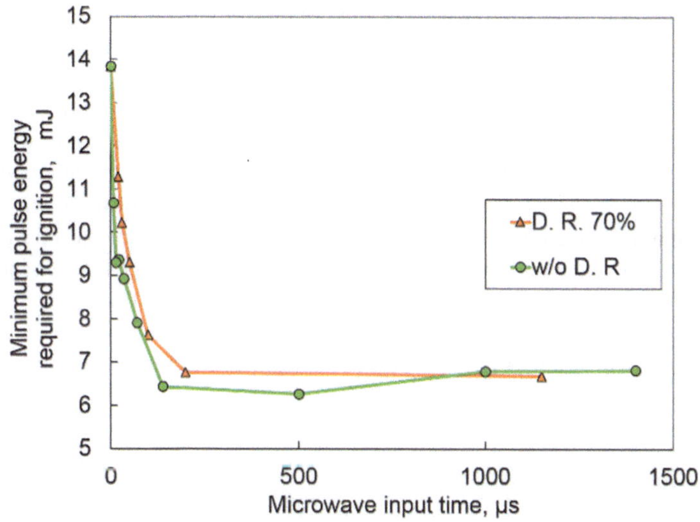

Fig. 5. Minimum ignition energy required for the successful ignition with and without Duty Ratio (D. R.)

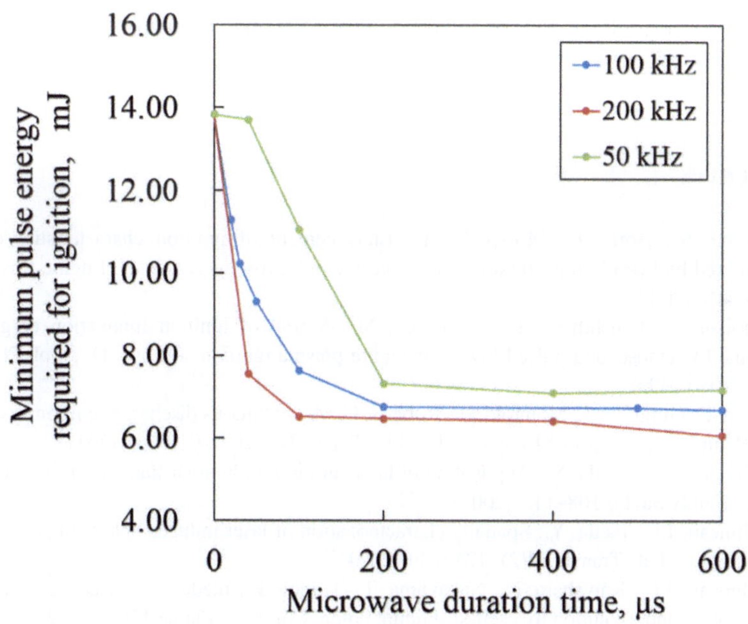

Fig. 6. Minimum laser ignition energy required for the successful ignition by Duty ratio of 70 % on various duty ratio frequency.

energy for ignition were investigated in this section. Figure 6 shows the relationship between the minimum pulse energy required for ignition and the total duration of microwave. The excess air ratio was set at 1.54 (fuel lean condition) at atmospheric pressure. The initial temperature of premixed gas was set at room temperature. In Fig. 6, the minimum pulse energy required for successful ignition were plotted with three different frequencies of microwave pulse. The duty ratio of the microwave pulse was set to 70 %. It can be found from Fig. 6 that the decrease of the minimum pulse energy for ignition take place less than 200 μs in all frequencies of pulsed microwave. Degree of improvement on the minimum pulse energy required for ignition increases in the higher frequency conditions compared with that of lower frequency. This result indicated that the pulse to pulse coupling was easy to take place in the conditions of high frequency. This kind of pulse to pulse coupling on the energy deposition was also observed in the other type of plasmas [20].

4 Conclusion

Microwave-enhanced laser ignition was proposed and the effects of microwave input pulse duty ratio on minimum ignition energy was experimentally investigated in a CVCC. Results indicated that microwave enhanced laser ignition decrease the total time of combustion. This is because the large volume and long life time plasma decrease the time of ignition phenomenon. In addition, the total energy of microwave enhanced laser ignition can be reduced effectively by applying the duty ratio of microwave.

References

1. Tanoue, K., Hotta, E., Moriyoshi, Y.: Enhancement of ignition characteristics of lean premixed hydrocarbon-air mixtures by repetitive pulse discharges. Int. J. Engine Res. **10**(6), 399–407 (2009)
2. Shiraishi, T., Urushihara, T., Gundersen, M.: A trial of ignition innovation of gasoline engine by nanosecond pulsed low temperature plasma ignition. J. Phys. D: Appl. Phys. **42**, 135–208 (2009)
3. Alger, T., Gingrich, J., Mangold, B., Roberts, C.: A continuous discharge ignition system for EGR limit extension in SI engines. In: SAE Paper, No. 2011-01-0661 (2011)
4. Beduneau, J.L., Ikeda, Y.: Application of laser ignition on laminar flame front investigation. Exp. Fluids **36**(1), 108–113 (2003)
5. Beduneau, J.L., Ikeda, Y.: Spatially characterization of laser induced sparks in air. J. Quant. Spectro. Radiat. Transf. **84**(2), 123–139 (2004)
6. Beduneau, J.L., Kawahara, N., Nakayama, T., Tomita, E., Ikeda, Y.: Laser-induced radical generation and evolution to a self-sustaining flame. Combust. Flame **156**(3), 642–656 (2009)
7. Pavel, N., Tsunekane, M., Taira, T.: Composite, all-ceramics, high-peak power Nd: YAG/Cr4 + :YAG monolithic micro-laser with multiple-beam output for engine ignition. Opt. Express **19**(10), 9378–9384 (2011)

8. Ikeda, Y., Nishiyama, A., Wachi, Y., Kaneko, M.: Research and development of microwave plasma combustion engine (Part I: Concept of Plasma Combustion and Plasma Generation Technique). In: SAE Paper No. 2009-01-1050 (2009)
9. DeFilippo, A., Saxena, S., Rapp, V., Dibble, R.W., Chen, J.Y., Nishiyama, A., Ikeda, Y.: Extending the lean limit of gasoline using a microwave-assisted spark plug. In: SAE Paper No. 2011-01-0663 (2011)
10. Nishiyama, A., Ikeda, Y.: Improvement of lean limit and fuel consumption using microwave plasma ignition technology. In: SAE Paper, No. 2012-01-1139 (2012)
11. Rapp, V., DeFilippo, A., Saxena, S., Chen, J.Y., Dibble, R.W., Nishiyama, A., Moon, A., Ikeda, Y.: Extending lean operating limit and reducing emissions of methane spark-ignited engines using a microwave-assisted spark plug. J. Combust. **2012**, Article ID 927081 (2012)
12. Wolk, B., DeFilippo, A., Chen, J.Y., Dibble, R.W., Nishiyama, A., Ikeda, Y.: Enhancement of flame development by microwave-assisted spark ignition in constant volume combustion chamber. Combust. Flame **160**(7), 1225–1234 (2013)
13. Hayashi, J., Yoneda, K., Sugeta, K., Liu, C., Akamatsu, F.: Characteristic of microwave-enhanced laser ignition in methane/air premixed mixture. In: Proceeding of the 2nd Laser Ignition Conference, LIC6-2 (2014)
14. Yanagisawa, K., Saito, T., Furutani, H.: Gasoline engine performance with laser-induced breakdown ignition under inert gas dilution condition. J. Combust. Soc. Jpn. **56**(178), 346–354 (2014)
15. Nishiyama, A., Moon, A., Ikeda, Y., Hayashi, J., Akamatsu, F.: Ignition characteristics of methane/air premixed mixture by microwave-enhanced laser-induced breakdown plasma. Opt. Express **21-S6**, A1094–A1101 (2013)
16. Ikeda, Y., Tsuruoka, R.: Characteristics of microwave plasma induced by lasers and sparks. Appl. Opt. **51**(7), B183–B191 (2012)
17. Ikeda, Y., Nishiyama, A.: Combustion improvement of gasoline engine by non-thermal plasma. J. Plasma Fusion Res. **89**(4), 234–237 (2013)
18. Hayashi, J., Nakatsuka, N., Sawanaka, Y., Okada, T., Furui, K., Akamatsu, F.: Effects of laser profiles on laser ignition of methane/air premixed gas. In: Proceedings of the 1st Laser Ignition Conference, vol. 1, pp. LIC8–LIC6 (2013)
19. Hwang, J., Bae, C., Park, J., Choe, W., Cha, J., Woo, S.: Microwave-assisted plasma ignition in a constant volume combustion Chamber. Combust. Flame **167**, 86–96 (2016)
20. Lovascio, S., Ombrello, T., Hayashi, J., Stephanyan, S., Stancu, G., Laux, C.: Effects of pulsation frequency and energy deposition on ignition using nanosecond repetitively pulsed discharges. In: Proceedings of the Combustion Institute (2016) (In Print)
21. Stephanyan, S., Hayashi, J., Laux, C.: Large-volume excitation of air, argon, nitrogen and combustible mixtures by thermal jets produced by nanosecond spark discharges. AIAA Sci. Tech. (Accepted)

Pulse Train Ignition with Passively Q-Switched Laser Spark Plugs Under Engine-like Conditions

Sebastian Lorenz, Mark Bärwinkel(✉), Wolfgang Mühlbauer, and Dieter Brüggemann

Chair of Engineering Thermodynamics Und Transport Processes (LTTT), Bayreuth Engine Research Center (BERC), University of Bayreuth, Bayreuth, Germany
lttt@uni-bayreuth.de

Abstract. Lean combustion and downsizing are two concepts to increase fuel efficiency and to reduce emissions of engines. However, the requirements on the ignition system increase consequently due to increased flow velocity and pressure at the time of ignition. In this context, the application of miniaturized passively q-switched laser spark plugs with pulse train operation provides an alternative to the conventional spark plug. To exploit the full potential of the pulse trains, ignition and combustion processes induced by passively q-switched laser spark plugs are investigated in this study.

Keywords: Laser ignition · Pulse train ignition · Flame kernel · Laser-induced plasma

1 Introduction

To meet future exhaust-gas legislations, engine developers apply lean combustion and downsizing. This leads to growing challenges regarding a reliable ignition, as these measures are accompanied by increased flow velocities and grade of turbulence as well as a higher pressure at the time of ignition. In this context, the laser ignition is considered to be a promising alternative to conventional spark plugs [1–5].

During the laser ignition, a short laser pulse (nanosecond timescale) is focused and a plasma breakdown occurs. The ambient conditions, the laser properties as well as the focusing parameters influence the plasma formation and propagation [6–8]. A shock wave is emitted because of high temperatures and pressures within the plasma. After the shock wave a hot gas ball remains. A flame kernel with a donut-like structure with a third lobe is developed due to gas dynamic effects [9–11]

The first ignition of an engine by a pulse laser was studied in 1976 [12]. However, laser ignition has been unable to achieve commercial success because of high demands in practical operations and high cost of mostly used actively q-switched lasers. The application of miniaturized passively q-switched laser (pQL) spark plugs is regarded to be a promising solution, since these systems are smaller, more robust and more cost efficient in comparison with actively q-switched lasers [6]. The successful tests of

M. Günther and M. Sens (eds.), *Ignition Systems for Gasoline Engines*,
DOI 10.1007/978-3-319-45504-4_15

VCSEL-pumped laser spark plugs support the development of economical laser ignition systems [13–15].

An example for a pQL spark plug can be seen in Fig. 1. The laser spark plug system is end-pumped by a diode laser. By increasing the pump energy or pump duration, such a pQL can emit pulse trains with a temporal pulse spacing from 60 μs to 250 μs. Thus, the overall ignition energy can be delivered by multiple laser pulses with an energy of a few millijoule per pulse. During a pulse train ignition, the local ambient conditions for the plasma development of consecutive laser pulses depend strongly on the previous ignition process [6].

In this study, ignition and combustion processes induced by prototypes of pQL spark plugs provided by the Robert Bosch GmbH are investigated to exploit the potential of the laser induced pulse train ignition. Therefore, the influence of the laser pulse profile on the flame kernel formation is examined. Furthermore, a constant volume combustion chamber was used to investigate the influence of the temporal pulse spacing, the pulse number on the ignition and combustion process of methane/air-mixtures. Flow velocity and ignition pressure variations can be investigated simultaneously at a specially designed test bench [6, 16].

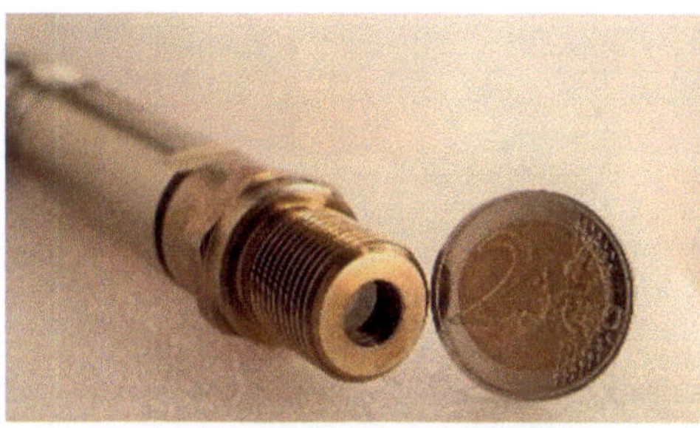

Fig. 1. Prototype of a laser spark plug provided by the robert bosch GmbH

2 Influence of Laser Pulse Profile on Flame Kernel Formation

In a first step, the temporal and spatial energy distribution within the laser pulses of different pQL spark plugs was measured [17]. It can be shown that the distribution depends on the laser design and influences the individual steps of the entire laser ignition process as well as the early flame kernel structure [6, 17]. Different early flame kernel structures are mainly affected by a change of the hot gas flow dynamics arising in the hot gas before the early flame kernel. The gas flow dynamic itself mainly results from a spatial plasma temperature during the plasma-cooling phase [16].

In Fig. 2 the general flame kernel structure formation is illustrated. To model these 3D structures, Schlieren images were used as a template. The hot gas ball at 3 μs after the breakdown remains after the laser induced breakdown. Due to gas dynamic effects a donut-like structure is formed (10 μs to 60 μs). In the further process, the gas dynamic mostly presses hot gas out of the donut-like structure and a so-called third lobe occurs [6].

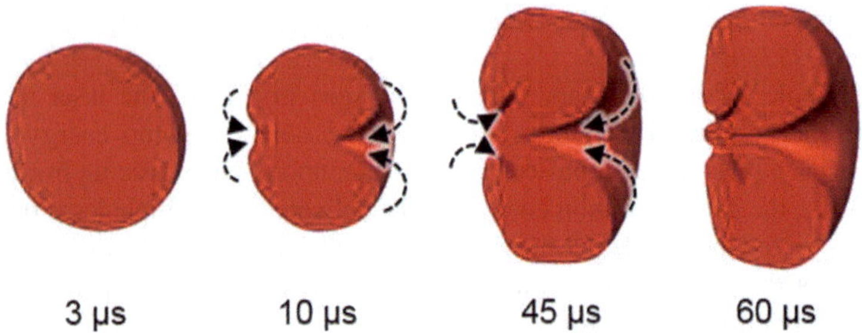

Fig. 2. Formation of a laser induced flame kernel [16]

The effect of different laser pulse profiles on the flame kernel development can be seen in the Schlieren images shown in Fig. 3. Depending mostly on the duration of the laser pulse, the third lobe is generated to a greater or lower extent. In the case of long pulses (A), a significant third lobe generation can be observed. In contrast, a short laser pulse (B) leads to little or no third lobe generation [6, 16].

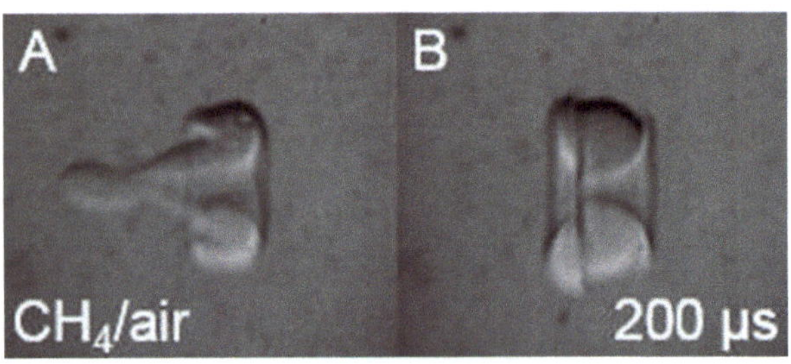

Fig. 3. Schlieren images of flame kernel structures generated by two different lasers [16]

3 Influence of Pulse Trains on Flame Kernel Formation

In the case of a pulse train ignition, consecutive laser pulses and their flame kernels interact with the flame kernel structure induced by previous laser pulses. Three different scenarios of interaction can be divided depending on the flow velocity of the ambient gas. In the first scenario (slow flow velocity) the initial flame kernel is hardly moved by the flow at the time of the second laser pulse. Thus, unburned gas with sufficient density is provided by the third lobe effect at the focus point in the spatial center of the existing flame kernel. The second laser pulse induces a second plasma and a flame kernel in the center of the first flame kernel. In Fig. 4 a 3D model calculated from a Schlieren-image of this case is shown 2 µs after the second laser pulse. The structure (1) is the flame kernel of the first laser pulse. The hot gas ball of the second laser pulse (2) develops in the center of the first flame kernel. The shock wave of the second laser pulse (3) propagates faster along the shape of the first flame kernel due to its lower density in comparison to the surrounding gas [6, 16].

The second scenario occurs if the first flame kernel is moved by the flow and the consecutive laser pulse is focused in its flame structure (structure 1 in Fig. 4). Due to a low density, no or an insufficient plasma breakdown takes place. The consecutive pulse does not contribute to the ignition in this case [6, 16].

At high flow velocities and sufficient temporal pulse spacing consecutive plasmas and flame kernels are induced outside of the previous flame kernel by consecutive laser pulses (third scenario). The individual flame kernels merge to one single flame kernel during the flame kernel growth (Fig. 5) [6, 16].

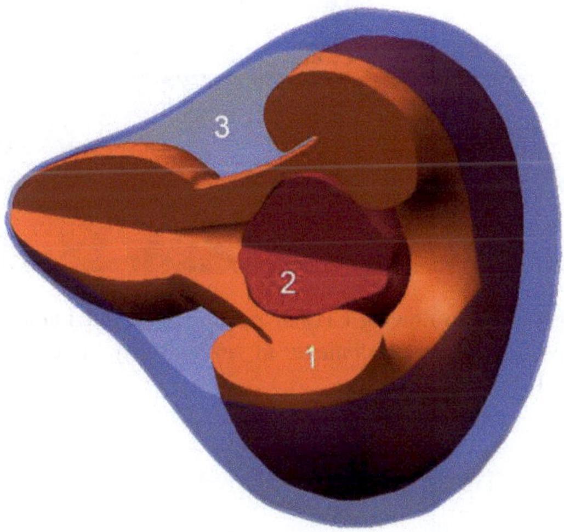

Fig. 4. Calculated 3D-modell: flame kernel of a first laser pulse (1) and hot gas ball (2) and shock wave (2) of a second laser pulse [16]

4 Ignition and Combustion Process After Pulse Train Ignition

To investigate the scenarios introduced in Chap. 3 under engine-like conditions, a constant volume combustion chamber test bench was developed. Ignitions at defined fuel/air-ratios, flow velocities and at elevated pressures can be realized (Fig. 5, left). The combustion chamber is optically accessible. In consequence, the flow and the flame kernel development can be recorded by the High-Speed-Schlieren technique. In Fig. 5 (middle and right) a pulse train ignition with five laser pulses is shown. It can be seen that each laser pulse creates an individual flame kernel. In the further process the individual flame kernels merge. Pressure course analyses show a more efficient combustion of lean mixtures in the case of appropriate temporal pulse spacing [16].

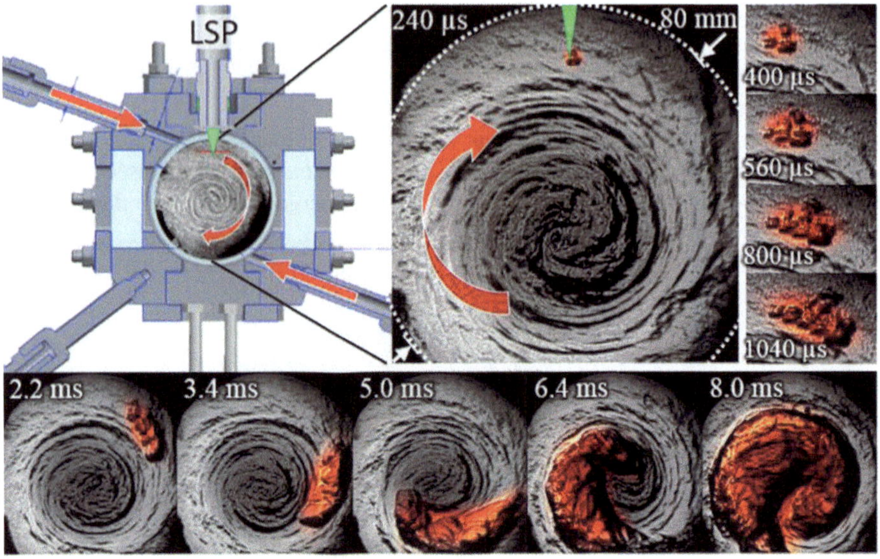

Fig. 5. Pulse train ignition of a flowing CH_4/air mixture at elevated ambient pressure. Images recorded by high-speed Schlieren technique in an optically accessible constant volume combustion chamber [16]

5 Conclusion

In this study, the ignition with passively q-switched laser spark plugs was investigated. The analysis of early flame kernel structure formation and pulse train ignition formed the primary focus. It can be shown that the temporal-spatial energy distribution of laser pulses of passively q-switched laser spark plugs mainly influences the early flame kernel structure. By increasing the pump energy or pump duration, the pQL emits pulse trains. Three different scenarios occur how consecutive laser pulses interact with the flame kernel structure of their predecessors at different flow velocities. It is shown that pulse train ignition can enhance the ignition and combustion of lean mixtures in the case of increased flow velocities and lean fuel/air-mixture conditions.

References

1. Dearden, G., Shenton, T.: Laser ignited engines: progress, challenges and prospects. Opt. Express **21**, A1113–A1125 (2013)
2. Fuchs, J., Leitner, A., Tinschmann, G., Trapp, C.: Concept for high-performance direct ignition gas engines. MTZ Worldwide **74**, 18–23 (2013)
3. Tauer, J., Kofler, H., Wintner, E.: Laser-initiated ignition. Laser Photonics Rev. **4**, 99–122 (2010)

4. Albin, E., Nawroth, H., Göke, S., D'Angelo, Y., Paschereit, C.O.: Experimental investigation of burning velocities of ultra-wet methane-air-steam mixtures. Fuel Process. Technol. **107**, 27–35 (2013)
5. Mastorakos, E.: Ignition of turbulent non-premixed flames. Prog. Energ. Combust. **35**, 57–97 (2009)
6. Lorenz, S., Bärwinkel, M., Stäglich, R., Mühlbauer, W., Brüggemann, D.: Pulse train ignition with passively Q-switched laser spark plugs. Int. J. Engine Res. **17**, 139–150 (2016)
7. Schwarz, E., Gross, S., Fischer, B., Muri, I., Tauer, J., Kofler, H., Wintner, E.: Laser-induced optical breakdown applied for laser spark ignition. Laser Part. Beams **28**, 109–119 (2010)
8. Bärwinkel, M., Lorenz, S., Stäglich, R., Brüggemann, D.: Influence of focal point properties on energy transfer and plasma evolution during laser ignition process with a passively Q-switched laser. Opt. Express **24**, 15189–15203 (2016)
9. Morsy, M.H.: Review and recent developments of laser ignition for internal combustion engines applications. Renew. Sust. Energ. Rev. **16**, 4849–4875 (2012)
10. Phuoc, T.X.: Laser-induced spark ignition fundamental and applications. Opt. Laser Eng. **44**, 351–397 (2006)
11. Bradley, D., Sheppard, C.G.W., Suardjaja, I.M., Woolley, R.: Fundamentals of high-energy spark ignition with lasers. Combust. Flame **138**, 55–77 (2004)
12. Dale, J.D., Smy, P.R., Clements, R.M.: Laser Ignited Internal Combustion Engine – An Experimental Study. SAE Technical Paper 780329 (1978)
13. Schwarz, J., Stoppel, K., Nübel, K.-H., Engelhardt, J.: Pumping concepts for laser spark plugs – requirements, options, solutions. In: OPIC 2nd Laser Ignition Conference, Yokohama, Japan, 21–24 April 2014
14. Gupta, S., Bihari, B., Biruduganti, M.: Performance benefits of laser ignition in a natural gas 6-cylinder engine. In: OPIC 4th Laser Ignition Conference, Yokohama, Japan, 18–20 May 2016
15. Suzudo, T., Hagita, K., Ikeo, T., Izumiya, K. Jikutani, N., Higashi, Y., Taira, T.: Total design of high power VCSEL pumped passively Q-switched micro-lasers for laser ignition. In: OPIC 4th Laser Ignition Conference, Yokohama, Japan, 18–20 May 2016
16. Lorenz, S.: Impulskettenzündung Mit Passiv Gütegeschalteten Laserzündkerzen Unter Motorischen Bedingungen. Logos Verlag, Berlin (2016)
17. Lorenz, S., Bärwinkel, M., Heinz, P., Lehmann, S., Mühlbauer, W., Brüggemann, D.: Characterization of energy transfer for passively Q-switched laser ignition. Opt. Express **23**, 1647–1659 (2015)

Advanced Plasma Ignition (API): A Simple Corona and Spark Ignition System

Paulus Krüger(✉) and Barend Visser

North-West University, Potchefstroom, South Africa
Paulus.Kruger@NWU.ac.za

Abstract. Due to stringent emission and fuel consumption requirements, several innovative ignition systems have been proposed to address the shortcomings of current spark ignition systems. Two of the most promising candidates are high-energy spark ignition systems and corona ignition systems. However, both these systems have some fundamental drawbacks. Spark ignition systems are very mature and relatively inexpensive, but the performance of these systems is limited by the small electrode gap, which results in a small ignition volume. Corona systems, on the other hand, have a much larger ignition volume, but have been slow to enter the market due to their complexity and cost.

These drawbacks are overcome by the Advanced Plasma Ignition (API) system which is described in this paper. API-corona is a very simple, reliable and inexpensive corona system that is directly compatible with most modern engines. API-spark is a high-energy spark ignition system with a ten times longer spark length that a normal spark system.

Keywords: Corona ignition · High-frequency ignition · High-energy spark ignition · Automotive ignition system

1 Introduction

There is a large variety of ignition technologies that deliver better performance than conventional transistor coil ignition (TCI) systems. However, the low cost, reliability and simplicity of TCI systems still make them the preferred system in modern engines. Various ways have been investigated to improve TCI systems. However, the small gap size of a spark plug severely limits the performance of any TCI system. One of the most promising new technologies is corona ignition, but most corona systems available on the market are much more complex and expensive than TCI systems and require engine and ECU modifications.

The Advanced Plasma Ignition (API) systems [1–4] have been developed to address these problems. The API systems can either be operated as a corona ignition system (API-corona) or as a spark ignition system (API-spark). API-corona, as a corona ignition system, is the simplest, cheapest and probably the most reliable corona system currently available. API-spark, as a spark ignition system, has a spark length that is an order of magnitude longer than a conventional spark.

M. Günther and M. Sens (eds.), *Ignition Systems for Gasoline Engines*,
DOI 10.1007/978-3-319-45504-4_16

The purpose of this paper is to describe the technical aspects of the API systems. The performance of the systems on various engines will be described elsewhere.

2 General Description

Both the API-spark and API-corona systems make use of a patented plasma plug [4]. In the API-spark system, the plug is configured so that a spark is generated from the central electrode to the ground electrode, through the combustion chamber, as shown in Fig. 1a. The spark length is about 8 mm, which is much longer than a conventional spark of about 0.8 mm. After the spark is formed, it can be sustained and heated by a current (of about 100 mA) for several milliseconds. Should the spark be quenched due to turbulence, a new spark is formed and sustained.

In the API-corona system, the plasma plug is configured to generate a corona into the combustion chamber as shown in Fig. 1b.

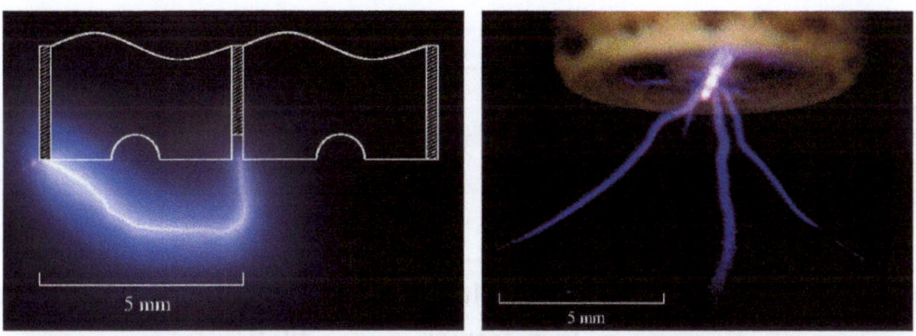

(a) API-spark (b) API-corona

Fig. 1. Photo of a spark and corona generated by the API system in a pressure chamber (in dry air at room temperature and 10 bar pressure).

2.1 Compatibility

The API systems are designed to be directly compatible with engines and ECUs designed for TCI systems:

- They consist of a unit that is screwed into the engine, replacing both the spark-plug and ignition coil.
- They make use of the battery voltage of the vehicle.
- They expect a conventional digital signal from the ECU, consisting of a dwell-time before the ignition event, with ignition occurring at the end of the dwell-time signal.
- They are compatible with ECUs that generate multiple signals for multiple sparks.

2.2 Safety

Modern ignition systems can be dangerous due to the high voltages and powers used in these systems. API is designed so that there is no dangerous high voltages anywhere in the system. Even the high voltage on the plasma plug is safe, due to its high frequency.

3 Technical Description

The API systems consist of four main components as shown in Fig. 2:

- A current source,
- An oscillator PCB,
- A high-voltage transformer and
- A plasma plug.

The novelty of the system lies in the plasma plug, the oscillator and the high-voltage transformer that is used.

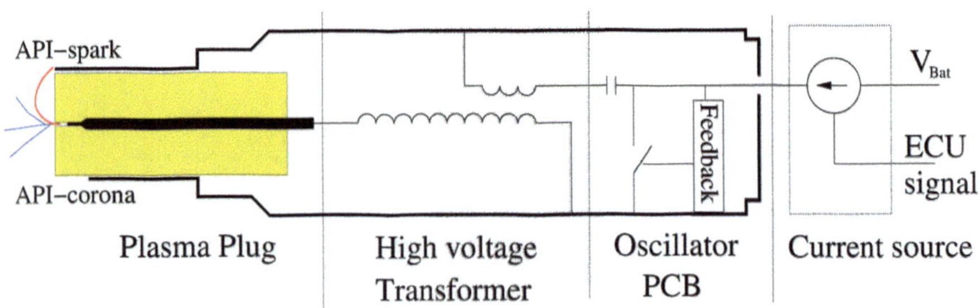

Fig. 2. The components of the API systems.

3.1 Plasma Plug

The plasma plug consists of a central electrode that ends in a small bore inside the ceramic (see [2,4] for more detail). When driven by a high-frequency ($> 1\,\text{MHz}$), high voltage ($> 20\,\text{kV}$) signal, a corona is formed inside the bore. The plug is designed so that even at high gas densities, a corona can be generated without excessively large voltage. The electrode tip also stays cold and does not become blunt with use.

Once initiated, the corona grows out of the bore into the combustion chamber. Specially designed grooves on the ceramic keep the corona away from the surface.

The difference between the API-corona and API-spark plasma plug is that the API-spark plug has a ground electrode that extends down to the edge of the ceramic insulator, whereas the ground electrode of the API-corona plug is further away, as shown on the left in Fig. 2.

In API-corona the corona grows away from the plug into the combustion chamber. Several corona 'fingers' are formed, as shown in Fig. 1b, resulting in a large volume of gas being ignitied. To limit engine corrosion due to the corona, a DC offset voltage can be applied to the central electrode.

In API-spark, the corona grows towards the ground electrode. Once the ground electrode is reached, the conducting plasma channel forms a spark discharge. The spark ignites a large volume of gas away from the cold ceramic, resulting in a short ignition delay and fast flame front.

3.2 Oscillator

The oscillators used in most corona ignition systems make use of a voltage power source to drive the resonator and make use of current feedback to lock on the resonance frequency. The drawback with these systems is that the output power inherently increases when a spark is formed and sophisticated control techniques are required to detect sparking and to keep the power in control. The API oscillator uses a current power source to drive the resonator and voltage feedback to lock on the resonance frequency as shown in Fig. 2. This has the advantage of the output power automatically decreasing when a spark is formed, so that no additional control is required.

In API, the plasma plug and high-voltage transformer is driven at its resonant frequency in a high power mode. Direct feedback from the primary is used to lock on the secondary side resonance frequency. When the secondary side becomes non-resonant due to a spark discharge, it inherently follows a more efficient lower energy transfer mode. In this low-power mode it is still self-oscillating and the high-voltage transformer becomes an AC current source ($> 100\,\text{mA}$) that sustains the spark. When the secondary side becomes resonant again, it returns to the efficient, high power, resonant mode.

To keep the system simple, only one inexpensive switching device is used. By switching at the right phase, the switching circuit is highly efficient with very little switching loss. Only direct (analog) feedback is used and no digital control or integrated circuits (ICs) are used. This makes the oscilator circuit simple and reliable.

The ten electronic components used on the API PCB are carefully selected to be able to reliably operate at under-hood conditions such as high temperatures, thermal cycling and vibrations.

The oscillator is designed so that the maximum DC voltage of the current source is about 50 V, so that no additional electrical safety precautions have to be taken.

3.3 Current Supply

Normal corona ignition systems require a voltage that is higher than the automotive voltage, and make use of a DC-to-DC converter. For API any such DC-to-DC converter can easily be adapted to a current source (these converters are normally voltage-to-current-to-voltage converters anyway). However, to make API even simpler and compatible with normal ECU signals, a simple inductor and switch can be used (as in a TCI system). During the dwell time, the inductor is charged to a predetermined current, after which it acts as the current supply for API.

3.4 High-Voltage Transformer

The high voltage transformer consists of a primary and secondary winding with weak magnetic coupling and operates very similarly to a Tesla coil. However, in a normal Tesla-coil configuration, the magnetic field of the secondary side inductively heats the primary windings, resulting in loss. This loss is minimised by constructing the transformer in such a way that the secondary side has a very high quality factor (Q) when driven at resonance.

4 Conclusion

API is made possible by the combination of three novel concepts:

- A plasma plug that can either generate a corona or a long spark at high gas densities.
- A simple and efficient current oscillator that uses direct feedback to drive the transformer at high power to generate a corona and at a lower power level to sustain a spark.
- A low-loss, high-voltage, high-frequency transformer.

API-corona uses a corona plasma plug to give a simple, inexpensive and reliable corona system that is compatible with current engines and ECUs. API-spark uses a spark plasma plug to give a spark length that is almost an order of magnitude longer than a conventional spark. The improved ignition performance of both API-corona and API-spark are currently being demonstrated on different engines. This will be published separately.

References

1. Visser, B., Krüger, P.P.: Ignition system. US Patent 8191540 B2 (2012)
2. Krüger, P.P., Visser, B., Mackenzie, J.: Advanced plasma and variable spark ignition system. In: IAV 2nd Ignition Converence, Berlin (2014)
3. Krüger, P.P., Visser, B., Mackenzie, J.: Advanced plasma and variable spark ignition system. In: International Conference: SIA Powertrain - Versailles (2015)
4. Krüger, P.P., Visser, B.: Ignition Plug. PCT patent application, PCT/IB2015/058299 (2016)

Alternative Ignition Systems

Analytical and Experimental Optimization of the Advanced Corona Ignition System

John Burrows[1] and Kristapher Mixell[2(✉)]

[1] Federal-Mogul Limited, Manchester, UK
John.burrows@federalmogul.com
[2] Federal-Mogul Corporation, Plymouth, MI, USA
Kristapher.mixell@federalmogul.com

Abstract. Benefits in performance and emissions can be achieved by the use of a properly designed corona ignition system, but these can only be realised in a cost effective manner if the designer is able to assess the likely impact of design changes to both the ignition system and the combustion chamber in which it operates.

Finite element modelling is able to capture the processes with a high degree of fidelity, but at the expense of high computational cost and long run times.

Methods described here allow evaluation of design changes to be achieved with reduced computational effort, allowing for a wider array of investigations to be completed. It is shown how the performance of the total system may be improved using these methods.

1 Introduction

The benefits of the corona ignition system are well documented and understood and will not be discussed in this paper.

Corona ignition systems operate and are governed by principles well understood in other fields of electrical engineering and application. Methods for generating, insulating, controlling and/or eliminating the formation of corona have been studied and applied in diverse fields such as power generation and transmission; telecommunications; high altitude and space travel; and fundamental research of the electrical breakdown phenomenon to name a few.

However, study related to ignition and more specifically its application in an automotive environment are limited. Considerations related to packaging, power consumption, efficiency, economic, duty and lifetime effects idiosyncratic to automotive systems are seldom published and therefore merit focused study. Careful selection of system architecture, components and materials are essential to the successful execution of a commercially viable system.

Many corona ignition systems comprise of a number of basic blocks, configured as a series-resonant R-L-C circuit, as shown in Fig. 1.

The system is comprised of:

- An electronic system configured to provide high frequency power in such a way as to produce corona in the combustion chamber. This system will consume power

M. Günther and M. Sens (eds.), *Ignition Systems for Gasoline Engines*,
DOI 10.1007/978-3-319-45504-4_17

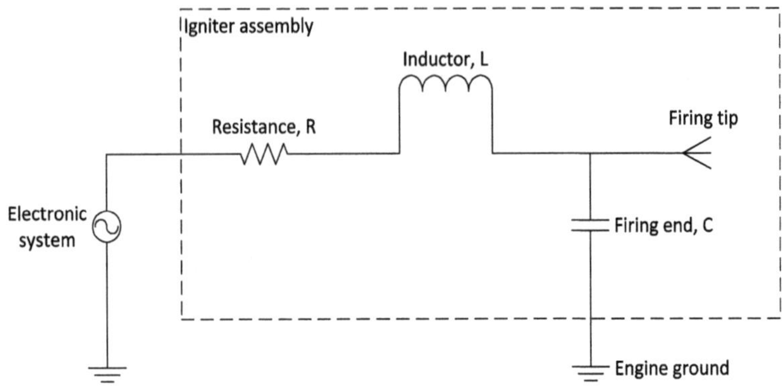

Fig. 1. Simplified equivalent circuit of corona ignition components

from the vehicle system and use data about engine operation to provide power to the igniter assembly with the proper timing, intensity and duration to give optimal combustion.

- An igniter assembly which takes this power and uses it to generate corona in the combustion chamber. This will usually involve some transformation from a lower voltage outside the chamber to the high voltage required to produce corona inside the chamber. This igniter assembly further comprises:
  - An inductor. The "coil-like" part which receives the input power from the electronic system and helps to transform this to the high voltage required. In the figure this is described as a pure inductive component L, but it additionally has some resistance modelled separately as R. This value of R must be evaluated under operating conditions where it will be many times higher than the value measured using a DC source. It includes resistance representing ohmic loss, skin and proximity effects in the inductor, tanδ losses in all dielectrics, parasitic corona formation and corona formation in the combustion chamber. As these are additive, they can be represented by a single value of R, although this may change with operating condition.
  - A firing end. The "spark plug-like" part which is installed into the head of the combustion chamber and from which corona formation occurs. This typically contributes the majority of the capacitance C of the igniter assembly. The firing end will usually have a firing tip, adapted to project into the combustion chamber and which will typically have an electrode having one or more relatively sharp tips which are designed to produce a locally high electric field strength from which corona formation can begin.
  - Any components and connections between the inductor and the firing end must be considered. These are critical as they are exposed to the same high voltages as the firing end and can contribute significantly system performance. Additional capacitance is contributed by these connection components; this is shown here lumped into a single C value as the effect is of capacitors in parallel and hence additive.

– Other system capacitance at the input side of the inductor may be neglected as the large system gain makes the energy stored in these elements very small: typically 1000 times smaller than at the output.

Analysis of the system performance may be completed using this simplified model and relying on a combination of calculated and measured values for these circuit elements.

2 Energy Audit

2.1 Effect of Operating Frequency

Selection of operating frequency has wide-ranging effects on the performance of the system as a whole and on energy requirements in particular. There are many factors to consider against one another when selecting an operating range.

By using higher frequencies:

- Corona tends towards increasing branching with a reduced tendency to arc [1].
- Magnetic components may be reduced in size and mass.
- Switching losses in the electronic system increase.
- Dielectric losses generally increase.
- Higher current is required in the igniter to achieve the same output voltage.

By using lower frequencies:

- Corona formation tends towards longer streamers with less branching but an increased tendency to arc [1].
- Larger magnetic components are required.
- Switching losses in the electronic system decrease.
- For a given voltage at the igniter, a lower current is required.

It is well known that this series resonant circuit will produce a maximum output at a frequency given by:

$$f = \frac{1}{2\pi\sqrt{L.C.}} \tag{1}$$

Where:

f = Resonant frequency [s^{-1}, Hz]
L = Igniter inductance [H]
C = Equivalent capacitance [F]

And at this frequency the igniter will be purely resistive with resistance R. System efficiency is dominated by the quality factor Q of the igniter assembly (firing end, inductor and any connection between) as this represents not only the available voltage gain at the igniter but is also a direct measure of its efficiency. At resonance, it may be shown that Q is given by:

$$Q = \frac{1}{R}\sqrt{\frac{L}{C}} \tag{2}$$

Where:

Q = Quality factor of the igniter assembly
R = Equivalent resistance of the igniter assembly [Ω]

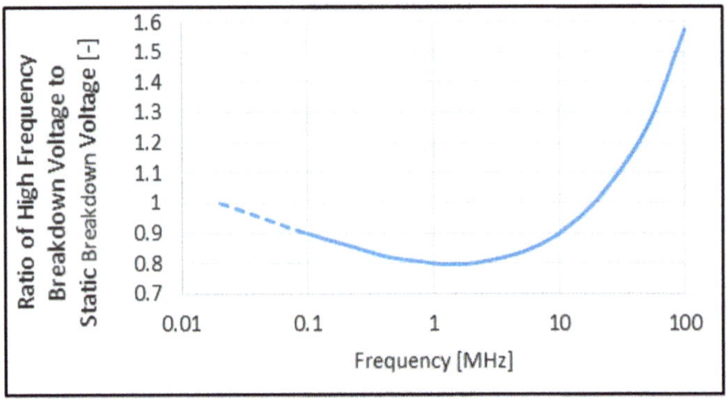

Fig. 2. Ratio of high-frequency breakdown voltage to static breakdown voltage as a function of frequency for an air Gap [2, 3]

Now it is often the case that the equivalent capacitance is difficult for the designer to control. Geometric constraints are usually enforced by the engine design: thread diameter and length, minimum projection into the combustion chamber, distance from the inductor to igniter and so on. A ceramic material, alumina in common practice, must be selected in order to withstand the operating conditions of high pressure, temperature and electric field. Combining these constraints means that capacitance of the firing end can only be controlled over a limited range: typically 15 to 30pF depending on plug type. Considering only the efficiency of the firing end assembly, and given that C cannot be freely varied, Eq. (2) shows that L should be maximized and R minimized for best efficiency. Reference to Eq. (1) shows that this would lead to the lowest possible operating frequency.

Clearly there are other constraints to consider, with one important factor being the effect of frequency on the voltage needed to generate corona. Figure 2 shows how the voltage required to generate breakdown in air changes with frequency [2, 3], where it can be seen that values reach a minimum in the range 1-5 MHz. Operating in this region has many benefits as it allows a lower voltage to be generated at the igniter tip whilst maintaining similar performance. This lower voltage is also applied to all the components connected to the tip: the ceramic insulator, the output end of the inductor, any connecting components and all the interfaces between. Clearly, using a lower voltage here leads to less opportunity for electrical failures and helps to reduce tanδ losses in all the insulating components.

Using this curve as an indicator of the voltage required at each frequency, we may use these data to help evaluate the effect of changing frequency on a selection of parameters. Especially important amongst these are the size of the magnetic components and the power required by the system.

Size of the magnetic parts directly impacts the ability to package the igniter assembly, as well as affecting the weight and cost of the system. For magnetics using a core, lower frequencies require larger magnetic cores due to the increasing volt-second product according to:

$$\Delta B = \frac{V.t}{A_e.N} \tag{3}$$

Where:

B = magnetic field [T]
V = applied voltage [V]
t = time for which the voltage is applied [s]
A_e = equivalent cross section of magnetic core [m^2]
N = number of turns on inductor [-]

Lower frequency gives a longer time in each half-cycle of the resonating circuit and hence larger values of ΔB which can lead to saturation of the core material, giving high losses and poor performance. This situation typically requires an increase in A_e and hence larger mass and volume of the magnetic component.

Similarly for air-cored inductors, where L α N^2, the volume increases approximately as the root of inductance, although difficulty in maintaining perfect magnetic coupling between the turns of a higher-value inductor leads to a volume which increases above this ideal. Again, lower frequencies require a significantly larger volume.

Another important factor is the power requirement of the system and this also changes with frequency. In order to reach a specified voltage at the igniter tip, we must apply a current into the igniter given by:

$$\bar{I} = 4.f.C.V_{peak} \tag{4}$$

Where:

$\bar{I}$ = average current between zero and peak value, $\bar{I} = \frac{1}{T}\int_0^T |i(t)|dt$ [A]
V_{peak} = peak voltage at igniter tip [V]

Maintaining our assumption of approximately constant capacitance, this implies that current consumed varies directly with f; in reality there are additional losses at higher frequency leading to a stronger relationship. This is due in part to increasing losses as the current rises, but additionally because of the requirement to reduce inductance L in order to raise frequency, remembering our approximately-fixed capacitance C, causing a drop in Q (see Eqs. 1 and 2).

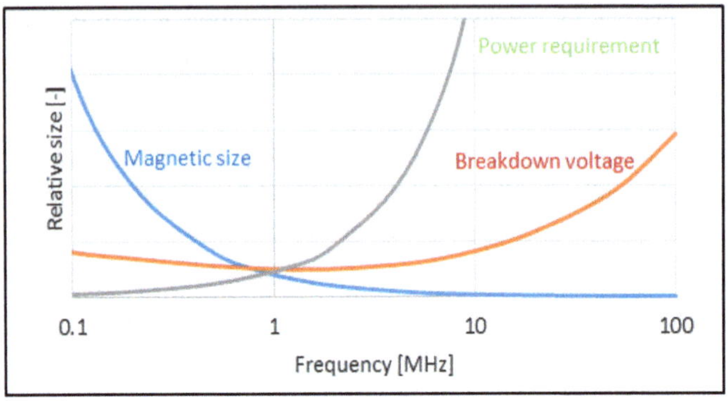

Fig. 3. Variation of key parameters with frequency. Arbitrary vertical scale.

Figure 3 shows how these parameters vary over a wide frequency range, from which we can deduce:

- The lowest possible operating frequency in the configuration described here, if we consider the system capacitance approximately fixed, is in the region of 700 kHz. Below this range it is no longer possible to make the inductors small enough to be easily packaged on an automotive engine.
- Optimum frequency range to give the lowest voltage requirement is 1-5 MHz. Selecting a frequency in this range will reduce losses and protect the system from electrical failure.
- Power requirement tends to rise with increasing frequency, although this increase is slightly offset in the range up to 2 MHz by the fall demand voltage. There is also a drop in resistance of the inductor; higher frequency requires lower inductance which means less turns and a shorter winding. However, these effects are far too small to counteract the rapid rise in current due to Eqs. (2) and (4).

Based on these factors, the system of this paper was designed to run in the range 800 kHz to 1.8 MHz to meet the conflicting requirements of voltage requirement, igniter assembly size and current/power requirements.

2.2 Analysis of Power Distribution

The total input power to the ACIS system when operating the engine is easily measured, and the power delivered to the igniters can also be measured to find the efficiency of the electronic system. Whilst power losses in an electronic system are well understood and design procedures are in place to deal with these, distribution of losses in the igniter assembly is not well understood and has a critical impact on the thermal and electrical design. A program of analyses and testing has been completed to determine the magnitude and distribution of losses, along with a determination of the factors contributing to these losses.

Analysis Method. An analysis of the firing end assembly was carried out using the HFSS module of the ANSYS software suite, which allows analysis of the high frequency behavior of the system, and which numerical engine is particularly suited to address narrow-band problems like high-Q resonance. Figure 4 shows the modelled domain for the investigation and includes the effects of:

- Ohmic losses in the coil winding, using an experimentally derived model for the stranded cable used in the actual design,
- turn-to-turn and turn-to-ground parasitic capacitance and associate losses,
- tanδ losses in all dielectric materials,
- eddy current losses in the shield,
- radiated loss to the model boundary.

Considering the measured efficiency of the drive electronics and using a fixed power input to the model, we can allocate losses according to the results of the HFSS model, as shown in Table 1. The model assumes that the pressure in the combustion chamber is such that no corona formation occurs; this leads to the highest current in the igniter assembly and hence the highest losses, and is therefore a worst-case analysis. It is also a case which is easily and repeatedly duplicated in actual testing for empirical verification.

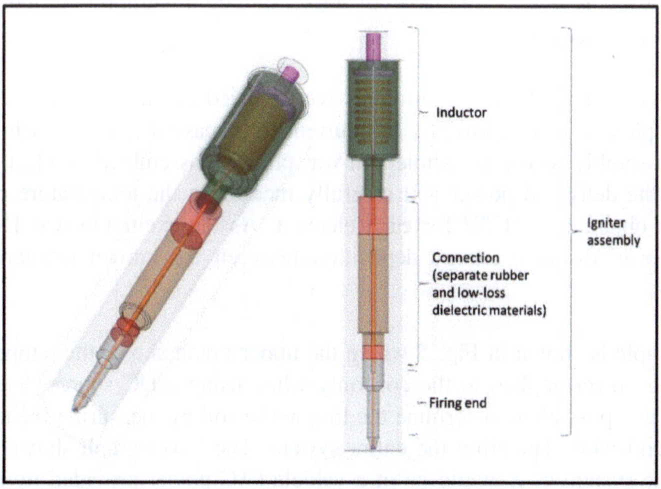

Fig. 4. Modelled domain for the high frequency analyses of losses in the igniter assembly

From this table, we can make the following observations:

- Of the energy delivered to the igniter, a high proportion is dissipated in the coil. This requires careful thermal design in the final system.

Table 1. FEA investigation of the distribution of power losses

	Electronics	Coil		Connection		Firing end	
		Joule	Dielectric	Rubber dielectric	Insulator dielectric	Dielectric	Corona
Percentage of power from battery	20 %	40 %	13 %	1 %	1 %	24 %	0 %
	20 %	53 %		2 %		24 %	0 %
Percentage of power into igniter	n/a	51 %	16 %	1 %	1 %	30 %	0 %
	n/a	67 %		3 %		30 %	0 %

- The loss in the connection components is negligible. Although still important in the total design as they contribute to system capacitance, thermal and power loss considerations are trivial.
- A significant proportion of the total power is dissipated as heat in the dielectric materials of the firing end. This is due to restricted geometrical envelope in this area and also the necessity of using a ceramic insulator, with its relatively high dielectric constant, in order to withstand the combustion environment. Fortunately, this also has the effect that these losses may easily be accommodated by the firing end.
- No energy is transferred to the cylinder mixture through corona. This is an underlying assumption in the model and is made to give a worst-case analysis.
- Radiation losses to the boundaries of the model are a negligible proportion of the whole at this frequency.

Experimental Verification. Experiments were carried out to verify this energy balance. A simple test setup allowed a DC current to be passed through each element of the igniter assembly whilst the whole unit was properly assembled in a test fixture. By controlling the delivered power and carefully measuring the temperature rise, a calibration was obtained in °C/W for each element. As the excitation was DC it was a trivial matter to evaluate energy deposition between the known resistances of the components.

An example is shown in Fig. 5 where the upper graph shows the temperature rise with varying power applied to the coil only when using a DC source. Knowing this calibration, it is possible to determine the loss in the coil by measuring the temperature rise in the coil while operating the entire system. The lower graph shows that, when operating the system as it would be in a vehicle (AC power provided by the Corona Control Unit, CCU), the temperature rise in the inductor is 2.7°C/W measured against the power into the total system. This means that the proportion of total system input power which appears as loss in the coil must be $2.7 \div 4.8$ or 56 %. Note that this includes all heating factors in the AC case: joule losses, dielectric losses and eddy current losses, but without identifying the contribution of each. Notice also that the components were modified by removal of all cooling solutions, such as heatsinks and mounting brackets, to give an increased thermal sensitivity in the test. A similar procedure gives the values for connection and firing end.

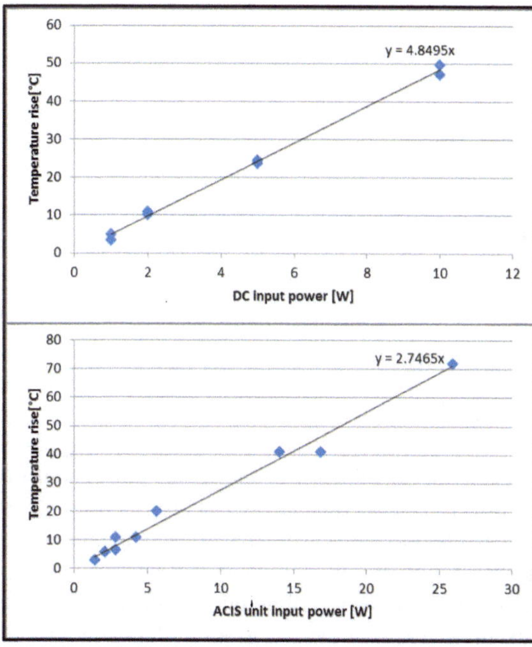

Fig. 5. Measured temperature rise in the inductor with DC excitation to the coil only (top) compared to AC excitation of the entire system (bottom)

Table 2 shows how the measured values compare with those calculated from FEA results. The firing end was installed into a suitable tests chamber and subjected to a gas pressure high enough to suppress all corona; this allowed direct comparison with FEA results. Experimental results agree very closely with those predicted by FEA. Note that

Table 2. Calculated losses vs measured losses, without and with corona formation

	Electronics	Coil		Connection		Firing end	
		Joule	Dielectric	Rubber dielectric	Insulator dielectric	Dielectric	Corona
Percentage of power from battery (by FEA)	20 %	40 %	13 %	1 %	1 %	24 %	0 %
	20 %	53 %		2 %		24 %	0 %
Percentage of power from battery (by measurement **without corona formation**)	18 %	56 %		1 %		25 %	0 %
Percentage of power from battery (by measurement **with corona formation**)	18 %	43 %		2 %		20 %	**17 %**

an additional row in the table shows the measured energy balance when the test chamber pressure is reduced to allow the formation of corona. Corona is calculated as the balance after losses are excluded: in this case about 17 % of the power input from the battery is deposited as corona in the chamber. Of course, the percentage of other losses fall commensurately.

2.3 Conclusions

A simplified model of the corona igniter has been combined with published and experimental data to evaluate the effect of changing system parameters on a variety of important system characteristics including size and weight, efficiency and losses, power requirements and dielectric withstand requirements. System operating parameters have been selected to balance these conflicting requirements.

FEA modelling has been completed using a relatively simple representation of the ACIS igniter to evaluate the distribution of the losses throughout the system and this investigation has been verified experimentally. Data from this analysis has been used to make an accurate estimate for the power lost in each component at each operating condition, allowing the designer to take thermal and electrical loads into account. The highest loss was noted in the inductor, although significant losses are expected in the firing end. In this example, 17 % of the energy from the battery is delivered as corona. As we can expect 60 % of this energy to be transferred to the gas [4], this compares favorably with conventional ignitions systems (inductive and CDI) where under 1 % of the spark energy is actually delivered to the gas [4].

Using the selected design, it has been shown possible to operate a four cylinder automotive engine with homogeneous mixture (port injected or GDI) for a total power draw of less than 100 W across all operating conditions. More challenging combustion systems with high levels of EGR and lean operation can be run while consuming less than 200 W power for a four cylinder unit.

3 Igniter Optimization

Like all ignition systems, corona igniters must face the competing constraints of thermal, electrical and mechanical requirements, whilst satisfying general requirements for package, manufacturability and cost. The corona igniter faces some challenges unique to this type of system.

Firstly, it is desirable to have a very high electric field as the initiation site for corona; far higher than the average field in the vicinity of the igniter. Electric field throughout the igniter must be controlled such that corona forms preferentially at the electrode tip. This usually requires one or more sharp tips on the igniter electrode which must maintain their sharpness throughout the life of the product, giving additional thermal and material challenges. The electrode must also be at a location relatively far from the grounded metal surfaces of the combustion chamber, leading to a greater projection than would be normal for a spark plug insulator and hence to increased thermal sensitivity.

Secondly, the high-frequency supply to the corona igniter is capable of driving current through the body of any insulating component which results in significant losses due to dielectric effects, as noted in the sections above. In addition, care must be taken to avoid the formation of parasitic corona which may easily cause additional loss and even component damage which can occur throughout the firing end assembly unless careful design practice is followed.

Described below are some of the methods used to optimize corona igniters for electrical and thermal performance. Mechanical analysis is identical to the commonly-used methods applied to conventional spark plugs and is not discussed here.

3.1 Electrical Design

Analysis Method. FEA may be used to examine the electrical field in the corona igniter, modelling the behavior at high frequency and the initiation and propagation of the corona streamers [5]. However, most optimization of the design may be carried out using a greatly simplified model: an electrostatic model used to analyze the electrical fields at the moment of corona inception. In this case a quasi-static analysis is used which does not consider the time-varying current in the corona nor development of corona streamers. This approach to the electric field calculations if justified as the characteristic wavelength exceeds the system dimensions by well over 100 times. It leads to a huge reduction in computational requirement and yet can still be used to evaluate designs for a number of important parameters:

- Relative magnitude of the electric field at all locations in the igniter. This can be used to verify that corona formation will occur at the desired locations and evaluate the possibility of formation in undesirable locations.
- Evaluation of the relative peak and average electric field across each of the insulators in the igniter, allowing verification of adequate dielectric strength in all areas.
- Investigation of voltage profiles over the insulator surfaces, joint surfaces and so on, especially with respect to the effect this has on propagation of corona over these surfaces.

If the analysis is carried out at the point of corona inception and the material properties are linear with voltage, then the results are fully scalable. As the applied voltage does not change the shape of the result, only the magnitude, it becomes convenient to analyze with one volt applied in all cases, making all results directly comparable. This model cannot consider the changes in electric field due to the formation and propagation of corona streamers, but is nevertheless useful when undertaking an initial igniter design.

An additional consideration is the effect of mesh size on electric field evaluation. As a general rule, satisfactory results may be achieved if the element size is approximately half the size of the smallest feature to be modelled. Unfortunately, there are many geometrical arrangements where this would lead to an unacceptably high mesh density and some geometries where this limit does not apply and the mesh must, theoretically, be infinitely dense [6]. Although there are methods to deal with these limitations the simplest is to use a fixed mesh size in all areas of high field, the same size throughout the model. This method is followed here, as it allows results for different areas to be directly compared, along with results from different models.

Firing Tip Optimization. Figure 6 shows a model of the tip of a typical corona igniter. This model was placed into a virtual test chamber of infinite size, filled with air and having all boundaries grounded. The igniter shell was grounded and 1 V applied to the central electrode. The lower left image shows how the voltage applied to the central electrode is distributed throughout the tip, where there are materials of very different permittivity. The lower right image shows a prediction for the electric field in this same region. It can be seen that this example is designed to maximize the electric field at the igniter tip whilst minimizing field elsewhere, the objective being to guarantee that corona formation occurs at the tip where it is fully exposed to the fuel-air mixture in the combustion chamber.

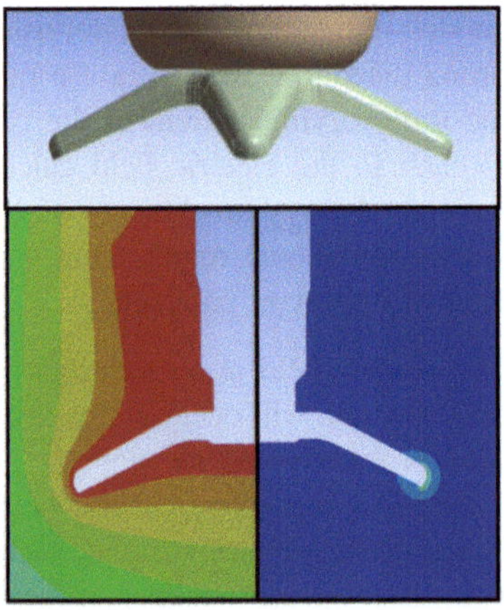

Fig. 6. Quasi-static electric field analysis of typical corona igniter tip. Left: Voltage [V] and right: Electric field [V/m] with voltage applied to the central electrode, other boundaries ground

The shape of the insulator and electrode components is critical to the formation and propagation of the corona, and small design changes can have a significant effect on system performance. In the example of Fig. 6 the igniter tip first stamped from a sheet and then formed into a dished shape to direct the corona as desired. In another example of Fig. 7 we see the same igniter tip with a modified manufacturing method: the dish is formed during the stamping process giving a small change in geometry at the end of the firing tips. There is no change on overall thickness or diameter, all features are the same and the radii of all cut edges is unchanged. In this case the electric field appears similar but close inspection shows that the modified tip produces a peak electric field over 30 % higher, as shown in Fig. 8.

This highlights the importance of detailed design of the igniter geometry: a very small geometry change gives a significant change in performance with lower required

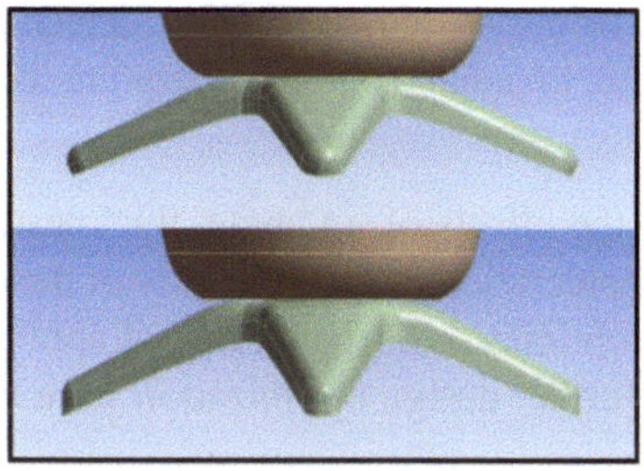

Fig. 7. Original (upper) and modified (lower) corona tip.

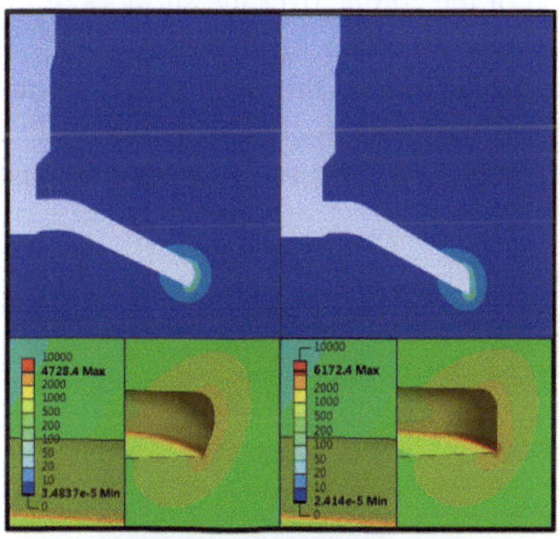

Fig. 8. Comparison of peak electric field, showing effect of minor modification. Electric field [V/m] with voltage applied to the central electrode, other boundaries ground. Linear scale (upper) and logarithmic scale (lower).

voltage and hence reduced power requirement. Higher local field without increasing bulk field also allows for a wider voltage range over which corona is formed in preference to an arc [7]. Note that this high sensitivity to detailed geometry changes also points to a challenge of igniter design: the need to maintain system performance over lifetime. This sensitivity suggests that corrosion and erosion of the firing tip could have a significant impact on igniter performance. This can be addressed using a combined approach:

- Design of the igniter geometry such that these detail changes have the minimum possible impact on performance. For example, ensure that the highest electric field is always at the igniter tip even when the maximum expected material loss has occurred.
- The low current in the corona coupled with features of the electronic unit used in this system are able to mitigate much of the electrical erosion normally expected in conventional ignition. However, mechanical design of the igniter must ensure that remaining material loss is within acceptable limits. For example, control of operating temperature and the use of high melting-point materials.

- Monitoring by the electronic system is able to detect parameters relating to the corona effectiveness and to adjust corona operating conditions to minimize the effect of igniter ageing.

Similar analyses allow the effects of tip angle, thickness, diameter and so on to be quickly studied without consideration of the geometry of the combustion chamber. This allows the firing end to be optimized independently of application to meet the required design criteria. Of course, further design optimization for a target combustion chamber may also be required.

Insulator Optimization. The tendency to produce corona from any given location in the plug may be evaluated as described above, but after corona inception it is important to predict the propagation of the corona over the surfaces of the igniter components. It is usually undesirable to allow corona to propagate over the surfaces between the igniter tip and any grounded surface, as this leads to a low-impedance conductive path which limits the volume of the discharge and can adversely affect ignition [14]. An initial evaluation of the tendency for this to occur may be made using the quasi-static analyses described above.

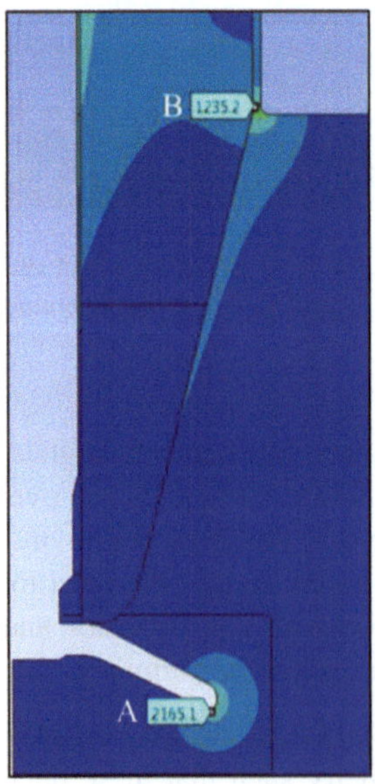

Fig. 9. Cross section of corona igniter firing end, showing electric field in insulator corenose and surrounding air. Electric field [V/m] with voltage applied to the central electrode, other boundaries ground.

Consider the corona igniter of Fig. 9, showing a cross section of the components exposed to the combustion chamber. It can be seen that this igniter is well-designed in respect of electric field, as the field at the firing tip [A] is almost double the field at any other location: there is a very high probability of corona formation preferentially at the desired location. However, at higher voltages, it is still possible for additional corona formation at the secondary location between the insulator and shell [B] and in this case it is important to evaluate how this corona will propagate.

Consider the case where the central electrode is positive with respect to the structure of the engine. In this case the gas at [B] may become ionized by the application of a high electric field, creating a region of both positive and negative charge. The positively charged particles may pass immediately to the shell of the plug which is, at this instant, acting as a cathode. The negatively charged particles cannot pass through the insulator to the central electrode but will instead propagate over the surface of the insulator towards the firing tip. This propagation will proceed so long as the potential gradient along this path is positive i.e. the corona will continue in a direction of rising voltage. If the corona reaches the positive electrode, a low-impedance path will be formed which will shunt all the corona energy into an arc with an associated drop in ignition quality [14].

Figure 10 shows the voltage at the surface of the insulator in this same condition. It can be seen that voltage rises steadily from points 1 to 2 (shell to firing tip). Corona formation at location B, the location of peak E-field, may propagate unimpeded to the igniter tip providing sufficient voltage is applied for sufficient time. This is often an undesirable situation.

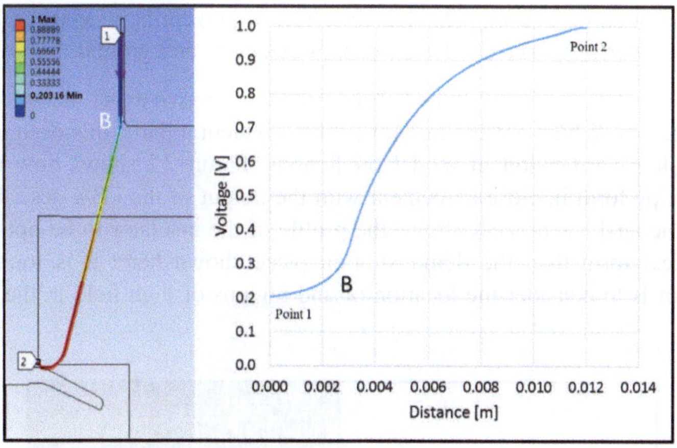

Fig. 10. Voltage evaluated over the insulator surface, 1 V applied to central electrode

If the propagating corona streamer encounters a region where the field is reversed, i.e. voltage is falling in the direction of propagation, then this situation may be avoided. Adding a feature to the insulator to break up this steady voltage gradient can be a useful modification; Fig. 11 shows one such example. In this case a rib is placed onto the insulator profile which has the following benefits:

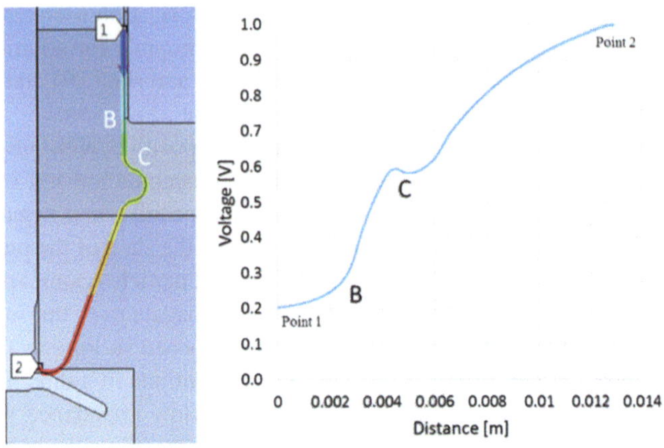

Fig. 11. Voltage evaluated over the insulator surface, 1 V applied to central electrode, rib added to insulator profile

- Distance over the surface of the insulator is increased (flashover length) which reduces the probability of an arc forming over this surface between the igniter tip and ground [15].
- The angle of the flank of the rib is such that there is a reversal in voltage gradient on the insulator surface at [C]. This has the effect of preventing propagation of corona streamers over the surface between central electrode and ground.

Corona igniters of this design have been produced to give experimental verification of these analyses. Whilst not optimal for actual production parts, this design allows an easy qualitative verification of model predictions. Figure 12 shows how the experimental parts perform in close agreement with the output of the FEA described above.

This same analysis process allows the profile of the insulator to be optimized to a more practical form than the demonstration piece shown here; it is found that the critical factor is to consider the location of and regions of high field in the air around

Fig. 12. Experimental insulator profile (left) showing how corona propagation is arrested by rib feature (right)

the firing end, along with the voltage profile over the surface of the insulator. By consideration of these factors, along with appropriate thermal and mechanical analyses, the optimum design of the insulator may be approached.

3.2 Thermal Design

Analysis Method. FEA and direct measurements show that bulk temperature changes around a combustion cycle are relatively small in a corona igniter. The method applied here therefore depends on a cycle-averaged heat load which greatly reduces computational requirement whist predicting the same average temperature as a full cycle analysis. This method has proved reliable for many years in the analysis of spark plug performance. Heat transfer to the exposed components in the combustion chamber is given by:

$$\dot{Q}_{wall} = A_{wall}(\theta) * h(\theta)\left(T_{gas}(\theta) - T_{wall}(\theta)\right) \tag{5}$$

Where:

$\dot{Q}_{wall}$ = Power transferred to the wall [W]
$A_{wall}(\theta)$ = surface area of the wall at crank angle θ [m^2]
$h(\theta)$ = surface heat transfer coefficient at crank angle θ [$W/(m^2.K)$]
$T_{gas}(\theta)$ = gas temperature at crank angle θ [K]
$T_{wall}(\theta)$ = wall temperature at crank angle θ [K]

For a cycle-averaged heat load, we may simplify this to:

$$\dot{Q}_{wall} = \overline{A_{wall}} * \bar{h}\left(\overline{T_{gas}} - T_{wall}\right) \tag{6}$$

Where cycle average heat transfer coefficient and gas temperature are defined by:

$$\bar{h} = \frac{1}{720}\int_0^{720} h(\theta).d\theta \tag{7}$$

$$\overline{T_{gas}} = \frac{\int_0^{720} h(\theta).T_{gas}(\theta).d\theta}{\int_0^{720} h(\theta).d\theta} \tag{8}$$

Notice that this gas temperature is weighted towards parts of the cycle with higher heat transfer coefficient; it is not the average temperature around the cycle on a crank angle or time basis.

Now the area of the corona igniter compared to the entire combustion is very small meaning that the bulk heat transfer is almost insensitive to the igniter design, allowing different igniters to be evaluated without changing the global heat transfer models. The average temperature of the chamber wall as a whole, T_{wall}, may be considered constant at constant operating condition, although different values may be appropriate in different areas of the combustion chamber (cylinder head, valves, liner, and piston crown). Relationships to estimate the instantaneous values for heat transfer coefficient h are widely employed [9–11]. Engine brake power can be used, along with measured or assumed energy balance figures, to calculate the average heat flux through the chamber wall for a given load point. A suitable combination of these inputs allows the Eqs. (6), (7), (8) to be closed by providing:

- Metal temperatures around the plug.
- Average metal temperature of the chamber walls (which is required to correctly evaluate radiation effects).
- Average gas temperature.
- Average heat transfer coefficient at the wall.

Model Optimization. Design of a typical sparkplug will be aiming to achieve three primary thermal goals:

1. An adequately high insulator temperature at low load conditions to ensure the removal of fuel and combustion deposits, called the self-cleaning temperature. This typically may be in the region of 350-450°C [12, 13].
2. Control of the maximum insulator temperature to avoid the possibility of any hot spot forming on in the insulator surface which may lead to preignition and possible engine damage. This may typically be in the region of 800-850°C.
3. Control of the electrode temperatures both for the avoidance of preignition and to extend service life, as corrosion and erosion of the spark surfaces accelerates rapidly with increasing temperature. Temperatures in the range up to 900–950°C are usually acceptable, with lower temperatures giving better plug life.

Figure 13 shows the thermal field for a typical spark plug in a modern automotive environment, along with two example corona igniters. Here, a generic heat load is

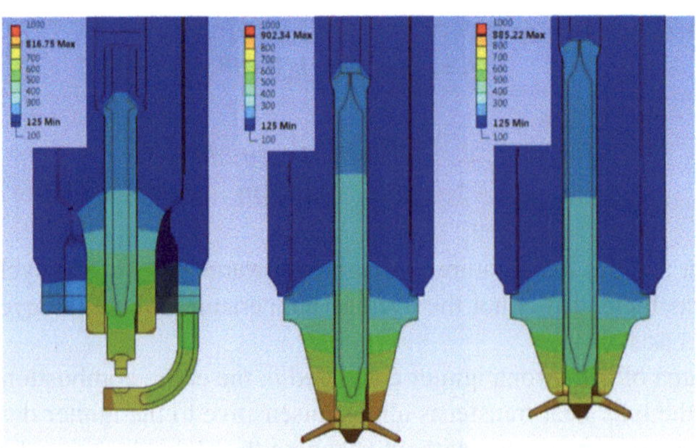

Fig. 13. Thermal field for a typical modern sparkplug compared to example corona igniters of two differing corenose lengths, with the same generic loading applied to each

applied for comparison purposes; analyses for a specific engine would use customized metal temperatures and gas loads following the process above. The left-hand image shows a conventional spark plug where, in this example, the ceramic corenose is predicted to be a little over 800°C and ground electrode tip is similar. The central electrode is cooler at about 750° and this is typical of almost all sparkplugs: heat is removed from the ceramic corenose by flowing through the central electrode.

Consider now the corona igniter shown in the central image. This plug has the same corenose length as the standard plug and comparison between the two allows us to make some important observations:

- Prior FEA work has led to the design of a construction method which optimizes the corona igniter such that the heat path from the insulator to the shell is maximized, in comparison to the conventional internal gasket seen in the conventional plug. This optimization results in the insulator inside the shell being close to the metal temperatures and helps to mitigate the temperature rise which would otherwise occur.
- With the same length of corenose, the projection of the corona igniter is much greater, and in this case greater than the ground electrode of the typical standard plug. This has the benefit of placing the corona initiation far from the grounded cylinder head, but leads to a high heat flux on the surface exposed to combustion gas. This gives a higher insulator temperature in the corona igniter (about 850°C instead of 800°C) despite the optimization described above.
- This high projection, in conjunction with the long exposed igniter tip of large aspect ratio, leads to higher temperatures at the tips of the electrode (about 900°C instead of 750°C).
- A shorter corenose length (right image) allows the corenose temperature to be reduced, becoming very similar to the standard plug. Central electrode temperatures are also reduced to the values typical in a solid ground electrode, although not to the levels in the copper-cored electrode shown here.

3.3 Conclusions

It has been shown that a simplified model for electric field may be employed to examine many of the parameters controlling corona igniter design. This model concentrates on analysis of the electric field at the instant of corona inception and may be used to evaluate the dielectric safety of components, the likely locations of corona inception and the likely propagation of corona after inception. This simplified FEA approach has been employed to demonstrate how the corona igniter may be modified to inhibit undesirable arc formation, and experimental samples confirm these numerical results. This has allowed development of an insulator profile suitable for engine use and mass production.

Thermal optimization of the corona igniter has also been considered by applying a cycle-averaged heat load model. Generic loads may be used to compare between similar designs in a non-specific engine application, while engine-specific loads may be developed to verify operation of corona igniters in a known target environment. These methods allow features to be modified in order to mitigate some of the problems faced uniquely by corona igniters which are not encountered in conventional sparkplug technology.

4 Combustion Chamber Optimization

4.1 Simplified Treatment in FEA

Analyses described above have been concerned with the corona igniter in isolation, without direct consideration of their installation into a combustion chamber. Clearly, the environment in which the igniter operates can have a significant effect on its design. Thermal and mechanical effects on the igniter may be treated using the well-known techniques of sparkplug design. However, the analysis of corona formation and propagation is unique and requires a specialized approach. A simplified analysis technique is described here which follows the methods described above, evaluating the electrical field conditions at the moment of corona inception and using these as an indicator of corona behavior. No effort is made to model the physics of corona, evaluate the changes in electric field due to corona formation, the gas pressure and flow field around the igniter nor the random nature of collisions and ionization which shapes the path and branching of real-world corona streamers. Instead, a visualization is developed which gives indication of the likely location and length of corona.

Analysis Method. Analyses were carried out using the ANSYS fluid dynamics module CFX, by using its ability to process additional variables. The CFX model is extended by the addition of the following:

- $E_{potential}$: electrical potential (scalar) [V]
- $\vec{E}_{field}$: electric field (vector) [V/m]
- ε_0 = Permittivity of free space [F/m]
- ε_r = Relative permittivity of material [−]

When:

$$\vec{E}_{field} = -\frac{dE_{potential}}{dx}\hat{\imath} - \frac{dE_{potential}}{dy}\hat{\jmath} - \frac{dE_{potential}}{dz}\hat{k} \tag{9}$$

Where $\hat{\imath}$, $\hat{\jmath}$, $\hat{k}$ are normal Cartesian vectors of unit length. Now using Gauss's Law and considering each element as having a Gaussian surface S:

$$\oiint_S \vec{E}_{field} . dA = \frac{Q}{\varepsilon_0 . \varepsilon_r} \tag{10}$$

Where Q is the net charge inside the element bounded by surface area S and A is an infinitesimally small area of this surface. By the divergence theorem, this may be reformulated as:

$$\nabla \cdot \vec{E}_{field} = \frac{\rho}{\varepsilon_0.\varepsilon_r} \tag{11}$$

Where ρ is charge per unit volume [C/m^3]. Now for models where the voltage is applied as a boundary condition in $E_{potential}$ only, Q = 0 everywhere inside the model and hence ρ = 0. Therefore:

$$\nabla \cdot \vec{E}_{field}.\varepsilon_0.\varepsilon_r = 0 \tag{12}$$

Solution of Eqs. (9) and (12) allow the potential and electric field at every point in the model to be evaluated. This approach has some benefits in the ANSYS environment:

- The mesh used may be shared with CFD, thermal and mechanical analyses.
- Inflation controls allowed in CFD meshing give a precise method for maintaining a regular mesh of uniform size to allow direct comparison between results (see discussion in 3.1 above).
- Flow visualization tools allow the extent and direction of corona propagation to be estimated as a post-processing operation.

However, the treatment here also has some limitations, most notably that the electric field is quasi-static which means that the shape of the field is not changed due to the formation and propagation of corona streamers throughout the model space. In practice the conductivity of the corona means that the local potential in and near the plasma would be expected to influence the surrounding field and these changes should penetrate the model E-field as the corona grows. These effects are the subject of a future development. However, experimental observations confirm that this "static" analysis of corona propagation in sufficiently accurate to allow general trends to be seen in corona location, size and direction.

Initiation of Corona. Initiation of corona has been discussed earlier in Sect. 3.1 and this method relies on the same type of analysis: a fixed voltage is applied (here 1 V) and corona inception is assumed to occur at the location of highest electrical field, where the field everywhere is evaluated using a fixed element size. In order to assess the likely formation and propagation of streamers, the following method is used with reference to Fig. 14 below which shows a close-up view of one of the firing end tips:

- With a fixed voltage applied, evaluate the electric field in 3D space (top left). Mesh size is kept constant at all air gaps, and at conductor and insulator surfaces throughout the model.
- The location of the highest electric field is identified (top right). However, if the single highest point is taken for the inception of corona, then the model will predict only a single streamer. In order to give a "cloud" of streamers showing the most likely locations for corona formation, a range of field values are taken: here, the highest 20 %.
- A number of model streamers are attached to this region, randomly located. The direction and velocity of streamer propagation is then determined by the gradient of electrical potential (bottom).

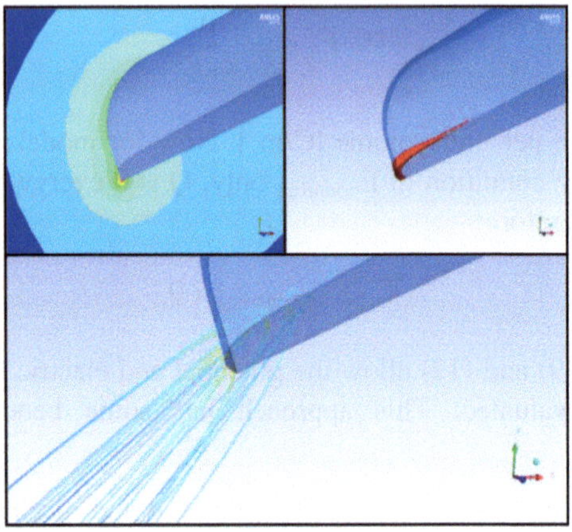

Fig. 14. Evaluation of corona inception location

- The ultimate length of the streamers is identified as where the first streamer trace reaches a potential of zero. This represents the point at which the corona would form a conductive path between the igniter tip and ground and is therefore the largest corona which could be produced in this configuration.

Corona Propagation. Once the initiation sites are identified, propagation of the corona through the chamber may be evaluated. This allows the location and design of the corona igniter to be optimized and can suggest changes to the combustion chamber which could help in the integration of a corona igniter into the overall combustion system. Using the post-processing capabilities of the CFD program, it is possible to estimate the direction and relative speed of corona propagation. The speed estimate is not a physically-based model but instead is based on a virtual particle being seeded into the electric field map and allowed to "flow" in the direction of that field at a velocity related to field strength. As such, the actual speed and length is not relevant; however, the results are directly comparable between models, allowing designs to be contrasted.

The results of a sample analysis are shown in Fig. 15, where a model corona igniter is placed into a generic combustion chamber and corona formation is modelled for various piston locations corresponding to varying ignition timing. At TDC (top) the piston is close to the igniter tip and arcing to the piston is likely due to its dominant effect on corona trajectory. The size of the corona volume shown here is calculated at the instant the corona streamers contact the piston; we can normalize the results by selecting this corona volume to be assumed 100 %. At 30°BTDC (middle) the corona is seen to spread approximately in the center of the air space between piston and head, with the ultimate length of the streamers being limited when the right-hand streamers touch the piston. The corona is influenced approximately equally by the grounded surface at both the head and the piston, giving a corona volume at this point over 1600 % of the value at TDC. Finally (bottom) when the piston is far from the igniter, the effect of the cylinder head dominates and the corona curves upwards giving corona

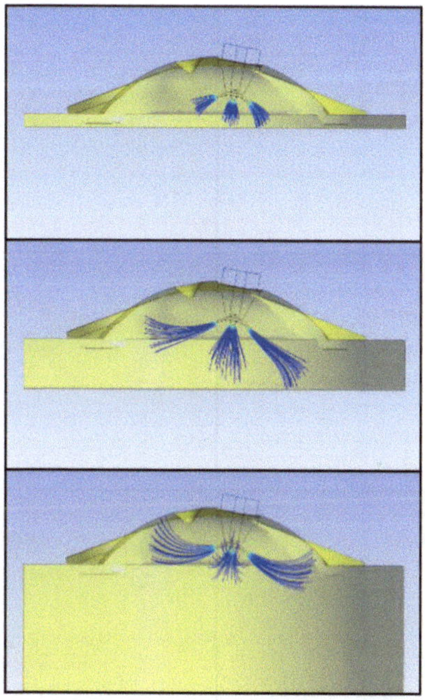

Fig. 15. Effect of ignition timing on corona propagation, overhead injector and igniter. TDC, 30°CA BTDC, > 60°CA BTDC

volume predicted to be about 1200 % of the TDC case which is smaller than at 30° BTDC. This is an important design point: corona volume may be optimized when the influence of grounded surfaces in directions normal to the desired corona growth is balanced as far as possible. The design shown here is close to optimal in this chamber for ignition timing around 30°BTDC. A longer corenose would be suitable where the piston was more distant, for example at earlier ignition timing or where the piston crown had a bowl under the igniter.

Another important variable which has a large impact on the effectiveness of the corona ignition system is the design of the combustion chamber. Figure 16 shows chambers with three different layouts, evaluated with the same corona igniter at TDC and 30°BTDC. All these chambers have the same compression ratio.

- The top chamber has an overhead injector and igniter, and each is mounted at a slight angle to accommodate them into the design. This is the design discussed previously above.
- The middle chamber had a central overhead igniter and a side injector. This design allows significantly greater corona volume at all modelled conditions, allowing 50 % more corona volume both at TDC and 30°BTDC, despite the relatively small change in igniter location. This is most favorable configuration for the corona system.
- The lower chamber shows a central injector and hence has the igniter to one side. This is the most unfavorable location for the igniter, giving only half the corona

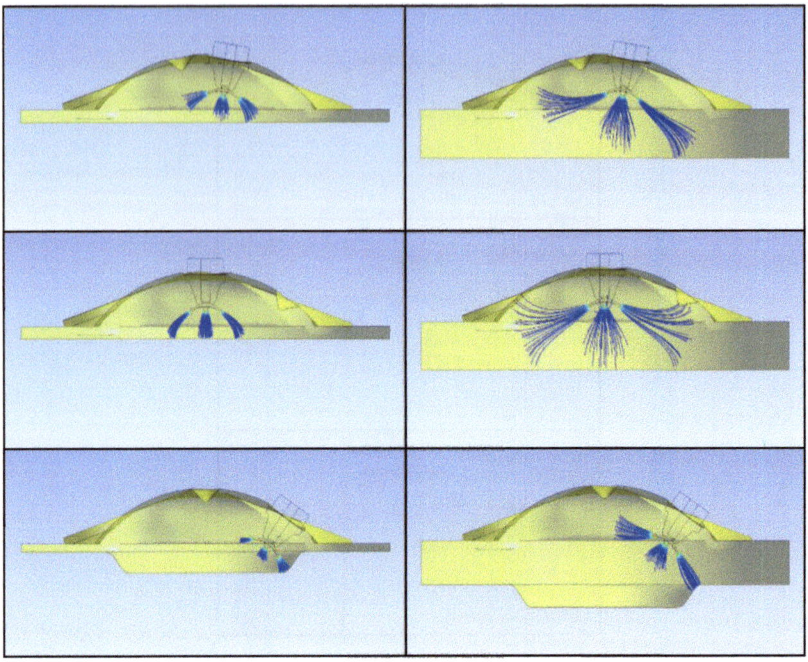

Fig. 16. Effect of chamber design on corona propagation at TDC (left) and 30°BTDC (right)

volume before the onset of arcing. Changing corenose length has only a minor effect in this design due to the close proximity of the piston to one of the igniter tips. This configuration may require a modified igniter which is biased away from corona production in this direction.

4.2 Conclusions

A solution is presented which allows the effects of corona igniter and combustion chamber geometry to be evaluated with respect to corona inception and propagation. This method is based on a quasi-static electric field analysis and as such does not attempt to capture all the physical phenomena involved; instead it gives an indication of the most likely location, trajectory and possible volume of the corona.

The importance of the correct corona initiation location with respect to the combustion chamber components is demonstrated, along with the effect of geometry changes due to piston motion. The igniter firing tip should be placed equidistant to the piston and cylinder head at the time of ignition for best corona performance; this may be achieved by correctly combining corenose length and projection, firing tip diameter and angle.

The most favorable igniter location is centrally overhead, with no nearby protrusions from the cylinder head if possible. A small deviation from a central location can have a significant effect on corona volume in the model presented here. An igniter location far from the chamber centerline or which rotates the igniter axis away from the chamber axis is undesirable, although measures may be taken in the design of the igniter to alleviate the associated problems.

5 Summary

A corona ignition system can give substantial benefits in ignition, especially in difficult combustion conditions. This translates directly into an extension of the possible operating regimes which can be accommodated by the ignition source and hence offers the promise of improved engine performance. However, development of a successful corona ignition system requires consideration of many factors not normally included in the design of conventional systems. Shown here are a selection of the FEA methods developed to allow analysis of designs in a timely manner, based on a simplified treatment of the relevant physics.

An electrical topology has been selected which allows a corona system to run a conventional 4-cylinder engine across the full operating range while consuming under 100 W from the vehicle. Alternative combustion strategies, having far more severe ignition requirements, can be operated for under 200 W consumed power. With analysis of high-frequency losses in all system elements, designers are able to accurately predict thermal and electrical stresses in all components early in the design stage.

The firing end of the system has been analyzed using thermal loads derived from cycle-averaged data and design features added to address the challenges unique to corona igniters. These thermal data feed into mechanical models and design choices in the usual way which are not discussed here.

The electric field of the igniter and surrounding combustion chamber are analyzed in detail using quasi-static models to predict the formation and propagation of corona both over insulator surfaces and through the volume of the combustion chamber. These data may be used to make design choices about many aspects of the igniter including the location of corona inception, the shape of the insulating elements and relationship of the igniter to the combustion chamber. The importance of apparently small design changes is highlighted by this modelling work.

Finite Element Analysis, supported by confirmatory experimental work, allows development of the corona ignition system in a way that would not otherwise be possible, giving insight into the details of design which would be difficult or impossible to achieve using experimental methods. The simplified treatments described here allow these analyses to be completed in a timely manner, giving engineers feedback more quickly and allowing faster design iteration. Known limits to the existing models mean that this effort is constantly ongoing, working to towards a higher fidelity with the aim of fewer iterations and still faster development.

References

1. Loeb, L.B.: Electrical Coronas: Their Basic Physical Mechanisms. University of California Press, Berkley (1965)
2. Chu, T.D.: High Frequency Breakdown Voltage. Superconducting Super Collider Laboratory; SSCL-539, March 1992
3. Effects of high frequency voltage stress on air insulation and solid insulation. In: 2010 IEEE Product Safety Engineering Symposium
4. Ronney, P.D., Gundersen, M.: Transient plasma discharge ignition for natural gas fueled internal combustion engines. In: 4th Annual Advanced Stationary Reciprocating Engines Conference, Downey, 18 September 2007
5. Breden, D., Raja, L.L., Idicheria, C.A., Najt, P.M., Mahadevan, S.: A numerical study of high-pressure non-equilibrium streamers for combustion ignition application. J. Appl. Phy. **114**, 083302 (2013)
6. Bayer, C.F., Baer, E., Waltrich, U., Malipaard, D., Schletz, A.: Simulation of the electric field strength in the vicinity of metallization edges on dielectric substrates. Fraunhofer institute for integrated circuits and device technology. IEEE Trans. Dielectr. Electr. Insul. **22** (1), February 2015
7. Javadi, H., Farzaneh, M., Peyda, A.: Determination of electric field at inception based upon current-voltage characteristics of AC corona in rod-plane gaps. Iran. J. Electr. Electron. Eng. **6**(2), 119–128 (2010)
8. Heywood, J.B.: Internal Combustion Engine Fundamentals. McGraw-Hill, New York (1988)
9. Eichelberg, G.: Some new investigations on old combustion engine problems. Engineering **148**(463–446), 547–560 (1939)
10. Woschni, G.: A universally applicable equation for the instantaneous heat transfer coefficient in the internal combustion engine. SAE Paper 670931
11. Annand, W.J.D., Ma, T.H.: Instantaneous heat transfer rates to the cylinder head surface of a small compression-ignition engine. Proc. Instn. Mech. Engrs. **185**, 976–987 (1970-1971)
12. Snow, C.W., Wallace, D.R., Daniell, A.E.: The carbon black – oxygen reaction: oxidation promotors. In: 8th Biennial Conference American Carbon Society (1967)
13. Automotive Handbook 6th edn. (2004). ISBN 0-7680-1513-8
14. Burrows, J.A., Mixell, K.I., Reinicke, P-B., Riess, M., Sens, M.: Corona ignition: assessment of physical effects by pressure chamber, rapid compression machine, and single cylinder engine testing. In: 2nd International Conference on Ignition Systems for Gasoline Engines (2014)
15. Gorur, R., Shenoi, I.: Computer modeling evaluated performance of station posts in AC systems. INMR Magazine, 13 March 2015

Comparative Optical and Thermodynamic Investigations of High Frequency Corona- and Spark-Ignition on a CV Natural Gas Research Engine Operated with Charge Dilution by Exhaust Gas Recirculation

Fabian Marko[1(✉)], Gerhard König[1], Tobias Schöffler[1], Steffen Bohne[2], and Friedrich Dinkelacker[3]

[1] Daimler AG, Mercedesstraße 137, 70546 Stuttgart, Germany
{fabian.marko, gerhard.koenig, tobias.schoeffler}@daimler.com
[2] BorgWarner Ludwigsburg GmbH, Moerikestr. 155, 71636 Ludwigsburg, Germany
s.bohne@borgwarner.com
[3] Institut für Technische Verbrennung, Leibniz Universität Hannover, Welfengarten 1A, 30167 Hannover, Germany
dinkelacker@itv.uni-hannover.de

Abstract. Investigations are presented comparing the conventional spark ignition system with a high frequency corona ignition system at a single cylinder heavy duty truck type engine with optical access to the combustion chamber through a large piston window. The focus of the thermodynamic as well as the optical studies lies on the evaluation of the ignition system with regard to an increase of the maximum EGR Rate as well as on a reduction of cyclical variations. Further considerations are the influence of specific parameters of the corona ignition system, such as ignition duration and ignition voltage on the initial flame kernel formation. For the underlying combustion concept, the corona ignition system was found to offer a high potential for a significant increase of the EGR compatibility.

Keywords: Corona ignition · Spark ignition · Flame kernel development · Cyclical variations · Commercial vehicle · Truck engine · EGR · Natural gas · Swirl · Optical engine

1 Introduction and Motivation

In addition to the use of highly efficient diesel engines, due to a favorable H/C ratio the use of natural gas offers a huge potential to reduce CO_2 emissions in the commercial vehicle sector. The Daimler AG successfully launched a monovalent Euro VI gas engine for medium-duty commercial vehicles (M936G) in the market in 2015. This engine is based on the diesel engine platform (OM936). It uses a stoichiometric combustion process with charge dilution by cooled exhaust gas recirculation (EGR) as

M. Günther and M. Sens (eds.), *Ignition Systems for Gasoline Engines*,
DOI 10.1007/978-3-319-45504-4_18

well as a conventional spark ignition system with the spark plug arranged in the center of the cylinder head [1]. Compared to the basic diesel engine, this concept has the advantage of a less complicated exhaust gas after treatment, very low pollutant emissions and reduced combustion noise. Stoichiometric combustion is enabled using charge dilution with high EGR so that critical component temperatures are avoided. Additionally this yields advantages in thermodynamic efficiency due to lower wall heat losses as well as de-throttling at part load. Compared to the basic diesel engine, up to 20 % CO_2 savings are attained in the steady-state operating map [1]. A further increase of the maximum EGR-rate and thus efficiency as well as power density is limited by the inflammation characteristics of the conventional ignition system. Consequently this results in the need for an evaluation of alternative ignition systems, such as the corona ignition system EcoFlash®, in order to achieve a more reliable ignition of highly diluted EGR mixtures. The corona ignition system shows advantages with regard to inflammation of natural gas-mixtures on passenger car engines [2]. Further publications for the gasoline operation of passenger car engines (e.g. [3, 4]) show advantages with charge dilution by EGR or excess air as well as for full load operation which was limited by knocking combustion. The reduction of the variation of the mean effective pressure is shown in [5] for the lean operation of a large stationary gas engine.

However, details of the ignition and early flame propagation are only rarely known, especially with regard to charge dilution with cooled EGR for the characteristics of the present gas-engine combustion concept. This provides the motivation for comparative thermodynamic as well as optical investigations of ignition systems using a single cylinder truck type engine with full optical access to the combustion chamber.

2 Basics

2.1 Ignition Systems

The conventional spark ignition system consists of two main components: a spark plug and an ignition coil. Due to the breakdown of the of the primary coil′s magnetic field, a high voltage of up to 40 kV is induced in the secondary coil. The shape of the discharge is characterized in the consecutive phases of breakthrough, arc discharge and glow discharge [6]. A thermal plasma channel with temperatures in the breakthrough phase of up to 60.000 K is created between the ground electrode and the bow shaped electrode. During the subsequent longer lasting phase of the glow discharge, still plasma temperatures of 3.000 K are attained [7]. The required breakdown voltage is dependent on pressure, temperature, mixture composition and the gap between the spark electrodes, as described by Paschen [8].

For this analysis, the corona ignition system EcoFlash®[1] is used. It consists of an ignition control unit (high-frequency generator and power electronics) which can supply up to six ignition modules (called igniter). The electrical set-up of the system is described in detail in [9, 10]. Schematically the corona ignition system forms a tip-plate arrangement in the combustion chamber with the igniter tip as one electrode and the

[1] manufactured by BorgWarner Ludwigsburg GmbH.

surrounding metal surfaces in the combustion chamber as the counter electrode. The gas volume between the electrodes represents an almost perfect isolator. When supplying the igniter with a high-frequency and high alternating-voltage, a strongly inhomogeneous electrical field is generated inside the combustion chamber. Due to the influence of the electric field free electrons start drifting. Further electrons can be generated by ionizing collisions when the kinetic energy of the electrons, absorbed in the electric field, becomes higher than the ionizing energy of the gas molecules. The newly generated electrons again absorb energy in the electric field and ionize new atoms via collisions; hence there is a snowballing increase in the number of charge carriers. These processes have been modelled by Townsend [10]. This process generates non-thermal plasma channels (called streamers) from the igniter tips into the combustion chamber. When the ignition energy is too high, a complete flashover with high plasma temperatures could occur. The required high frequency alternating voltage in the single digit MHz range is generated by the controller out of the onboard electrical power supply.

The igniter works as a transformer to generate the high-voltage at the igniter tips It contains an electrical series resonant circuit consisting of a coil and a capacitor. For laboratory use a PC tool allows the adjustment of all necessary settings, such as the duration of ignition and the ignition voltage (Buck-voltage).

2.2 Initial Phase of Combustion

The initial flame kernel formation is a complex process which depends on various factors. These includes especially the flow and turbulence conditions, the ignition processes as well as the mixture composition and the reaction kinetics. The initial flame kernel development can be divided into several phases (described for example by Herweg and Maly [7] or Bradley and Lung [11] for conventional spark ignition): the first phase up to approx. 300 μs after spark breakthrough is influenced by the spark plasma. In this initial phase, high spark energy yields to a thermal support of early flame propagation. By means of a directed flow at the spark plug, the so-called flame holder effect can be generated. A slightly larger volume of the initial mixture can be inflamed when fresh gas is passing the plasma channel.

In the following transition phase, the spark plasma is extinguished and the flame propagation is dominated by thermo-chemical reactions. The flame front is wrinkled by small-scale turbulence; a corresponding turbulent burning velocity can be attained. In this critical phase, the flame kernel can be extinguished or long time delays in flame kernel development may occur. The factors which are especially crucial in this phase are low mixture densities, high heat losses to the electrodes, as well as low laminar burning velocities in cases of high residual gas content. In the subsequent flame propagation the flame front is wrinkled more and more by larger turbulence eddies and the flame propagation velocity is therefore increased.

These known phases of the initial spark-ignited flame propagation can be transferred to the corona ignition system. Differences between the ignition systems concerning flame kernel formation arise mainly through the five ignition areas of the corona ignition system. In combination with the long streamers during this first phase, a significantly larger volume of the initial mixture compared to the spark ignition system is captured. This generates a very large initial flame area and consequently a much larger active flame front surface. This overcompensates the compared to the thermal plasma of the spark plug significantly lower plasma temperature. A further benefit for the propagation of the individual flame kernels is a significantly reduced heat loss to surrounding metal components due to thin igniter tips and a large distance to the cylinder head. In addition, the impact of turbulent fluctuations on the initial flame zone formation is lowered by the five separate flame kernels.

Characteristic for the natural light emission of premixed hydrocarbon combustion is a bluish flame color, which is based on emission spectra of different molecules resulting from chemiluminescent reactions. The characteristic spectra are the CH* radical chemiluminescence at 430 nm and the CO_2 wide-band radiation from 300–500 nm. The hydroxyl (OH*) radical, which is having its maximum spectral emission at 310 nm [12], can be used as an indicator for the heat release.

2.3 Exhaust-Gas Recirculation

There are three ways to increase the heat capacity of the in-cylinder charge: first the operation with excess air (lean-burn engine, $\lambda > 1$), second the stoichiometric operation ($\lambda = 1$) with exhaust gas recirculation (EGR) and third the operation with charge dilution by a combination of excess air and EGR. The exhaust gas recirculation can be realized either by valve overlap, or, as applied here, by the external recirculation of cooled exhaust gas. The EGR rate of mixture aspirating engines is defined as:

$$X_{AGR} = \frac{\dot{m}_{AGR}}{\dot{m}_L + \dot{m}_K + \dot{m}_{AGR}} * 100 \tag{1}$$

The use of recirculated cooled exhaust gas for charge dilution was investigated in [13] for the same engine concept as utilized here. It has the following advantages:

- improved combustion efficiency due to a lower combustion temperature and therefore reduced wall heat loss.
- an increase in power density due to the significant reduction of exhaust and cylinder head temperature.
- slightly reduced knock sensitivity due to an increased chemical ignition delay.
- possibility of using a simple exhaust gas after-treatment by means of a three-way catalytic converter for ultra-low NO_x emissions- in contrast to an engine operated with excess air.

Challenging for increasing the EGR rate is however a significantly reduced burning rate [14]. As described in [13], one way to extend the EGR tolerance limit to attain higher EGR rates is to optimize of the in-cylinder flow by means of an optimized piston geometry. Higher EGR rates can be obtained with the help of an increased turbulence level leading to faster combustion after the transition from laminar flame kernel development to turbulent flame propagation. In addition to that, an alternative ignition system can offer further potential to enhance the maximum EGR tolerance by accelerating and stabilizing the initial phase of combustion.

3 Experimental Set-up

The basic set-up of the test rig, as well as the ignition systems are described in the following chapter. All tests are being performed in the research laboratories of Daimler AG, Stuttgart using a heavy-duty 4-stroke single-cylinder optical research engine. The engine provides optical access to the combustion chamber by means of a large quartz-glass piston window. The combustion concept of the engine is derived from the spark ignited natural gas engine combustion concept which is applied to the M936G series natural gas engine of the Daimler AG.

3.1 Research Engine

In the development of the optical engine, for a good comparability of the results, focus has been placed on using as many identical parts as possible and on matching the optical engine geometrically to the conventional single-cylinder engine. Important technical data regarding the optical engine are listed in the following Table 1:

Table 1. specifications of the optical research engine

	Unit	Value/range
Mixture formation	–	external gas injection into intake manifold
Piston geometry		deep bowl with large squish zone
In cylinder flow		swirl with central swirl axis
Spark plug position		central
Displacement	cm^3	2473
Bore	mm	139
Stroke	mm	163
Geometric compression ratio		13.1:1
Number of valves		4
Valve timing		intake with early valve closing
Fuel		methane 99.9 %

Optical access to the combustion chamber for visualization of ignition and flame propagation is achieved through a large piston window made from high-purity quartz glass. The combustion chamber surface is shaped identically to the conventional metal piston. The viewing direction is through the bottom of the bowl onto the cylinder head. The principal set-up is described in [15].

The contours of the piston and the cylinder head, as seen with the high-speed camera, are outlined in the following Fig. 1. Here also the mounting position of the corona igniter and the piston bowl geometry is schematically shown.

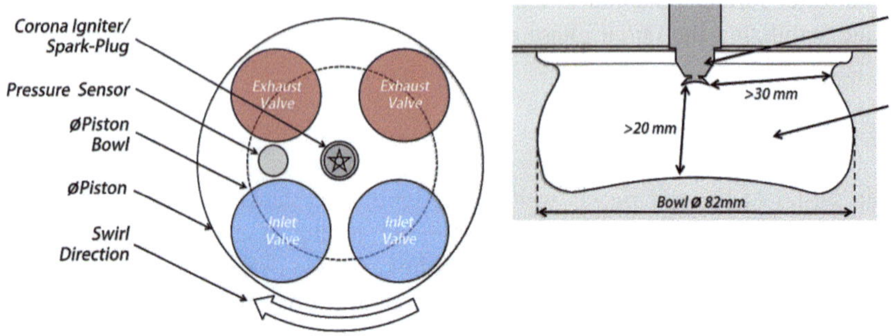

Fig. 1. Left: Schematic representation of the combustion chamber as seen with the camera; right: Sectional view - piston geometry and igniter arrangement in the combustion chamber.

The mounting position of the corona igniter can be described as nearly optimal. For instance in reference [16] a similar bowl type combustion chamber is recommended. The minimum distance from the corona igniter to the piston is 24 mm. As a result, high ignition energies without flashover to ground can be achieved.

3.2 Measurement Instrumentation

The optical engine is identical to the conventional single cylinder engine with regard to pressure and temperature measurement as well as pressure indication measurement technology. Air and gas mass flux (deviation from rated value ≤ 0.1 %) are determined with Coriolis mass flow meters. The cylinder pressure indication uses a piezoelectric pressure sensor which is installed in the cylinder head flush-mounted directly to the combustion chamber.

A high-speed camera (Photron SA-X2) is used to record the flame emission of combustion. The camera offers a very good sensitivity in the blue wavelength range. Three recording modes, with frame rates from 13,200 to 120,000 fps, are used. The traces of chemiluminescent light emission of the OH* radical are measured by means of a Hamamatsu miniature photomultiplier unit. Spectral filtering of the spectral lines of OH* radical is done with an optical band-pass filter with maximum transmission at 308 nm.

3.3 Specifications of the Ignition Systems

The conventional ignition system consists of a transistor coil arrangement with a commercially available M14 × 1.25 J-gap spark plug. In order to keep the heat losses low, an iridium center electrode with a slim shape design is used. High cylinder pressures at ignition timing, especially with late ignition timing at high load points, require a narrow electrode gap, which is adjusted for the test series to 0.35 mm. The maximum spark energy is 70 ± 3 % mJ for a maximum secondary voltage of

38 ± 3 kV. For all operating points, the ignition coil is fully charged. The position of the spark gap in the combustion chamber, measured from the cylinder head surface to the middle of the spark gap is $z_F = 4.7$ mm. The orientation of the spark plug (with the open side of the bracket pointing to the right in the combustion images) is identical for all tests. According to [13] no influence of the spark plug position could be found in this engine concept.

For the analysis of the corona ignition system a specifically manufactured M14 × 1,25 igniter with five igniter tips (prongs) is used (Fig. 2). The igniter shaft is extended in this design for the application in the very deep spark plug duct of the single-cylinder engine. Compared to the conventional spark plug, the location of the igniter prongs (where the corona streamers emerge) can be positioned deeper in the combustion chamber at a value of $z_F = 9.9$ mm. A previous investigation in a pressurized chamber verifies that there is no difference in the streamer formation if the piston consists of quartz glass or of aluminum. A further result of this study is that a flashover discharge of a streamer always goes to the cylinder head and not to the piston.

Fig. 2. Extended corona igniter M14 with 5-prongs

4 Test Procedure

4.1 Operating Points

Two representative load levels (called A50 & A75) with mean indicated pressures of IMEP = 11 and 17 bar at an engine speed n = 1100 rpm form the basis for studies on the commercial vehicle single-cylinder optical engine. The operating point data are listed in Table 2. For all operating points a stoichiometric air fuel ratio $\lambda = 1$ is set. A variation of the EGR rate is performed, based on these load levels. The ignition timing is adjusted iteratively for each operating point to achieve a constant position of the center of combustion (MFB_{50}) at 8 deg aTDC. When approaching the misfire limit, the EGR rate is increased in 1 % steps up to the misfire limit.

The A50 load point forms the basis for the analysis of the two corona ignition system parameters: duration of ignition and ignition voltage (buck-voltage). Based on this load point, the ignition settings are varied for two EGR rates ($X_{EGR} = 0$ % and 17 %). The ignition timing is set for each particular EGR rate in order to attain the MFB_{50} at 9 deg aTDC with a buck-voltage of 120 V. The buck-voltage is varied with $\Delta U_{BV} = 20$ V in the range of $U_{BV} = 100$–180 V until corona streamer flashover to ground starts occurring. The ignition duration is kept constant at $t_I = 0.4$ ms.

Table 2. Operating points EGR variation

Description	Load	EGR	Ignition timing	P_{Cyl} bei IT	Buck-voltage	Ignition duration
	IMEP	X_{AGR}	IT		U_{BV}	t_I
	[bar]	[%]	[°aTDC]	[bar]	[V]	[ms]
EGR variation A50	11	0-Max	MFB_{50} konst.	Var	Max	0.4
EGR variation A75	17	17-Max	MFB_{50} konst.	Var	Max	0.4
Buck-voltage variation	11	0	−9.5	30	100–180	0.4
		17	−16.3	30		
	17	17	−17	45		
Ignition duration variation	11	0	−9.5	30	110 & 180	0.1–1.2
		17	−16.3	30		
	17	17	−17	45		

The ignition duration is varied in the range of t_I = 0.1–1.2 ms for two ignition voltages U_{BV} = 110 V and 180 V.

4.2 Single Cylinder Optical Engine Operating Mode

The set-up and the operation mode of the optical engine are intended to achieve a good match of its thermodynamic results with those of the conventional single-cylinder research engine. In addition to the geometrical consistency described in chapter 3.1, an optimal operating mode is worked out. In order to determinate the basic parameters and to validate the results of the optical engine, several variations of operating parameters are investigated with the conventional single-cylinder engine.

The main difference between the operation modes of both engines is the limited number of fired cycles. For component protection reasons, the number of fired cycles is limited to 60. Identical gaseous fuel and air mass flows to those set on the single cylinder engine are used. The operating point is relatively stable after 30 combustion cycles so that the subsequent 30 combustion cycles are representative for the chosen operating point. The EGR is realized at the optical engine with a synthetic mixture of CO_2 and N_2 with the mass fractions w_{N2} = 76 % and w_{CO2} = 24 % [14]. Compared to the conventional single-cylinder engine, the EGR mass flow m_{EGR} is slightly increased to compensate for the missing water content in the synthetic gas mixture. The reason behind this is to achieve the same overall heat capacity C_P of the inert gas mass in the combustion chamber. This results in a slightly higher demand for charging pressure at the optical engine. So a very good match of the heat release rate dQ between the cylinder engine (SCE) and optical engine (OE) is achieved (Fig. 3):

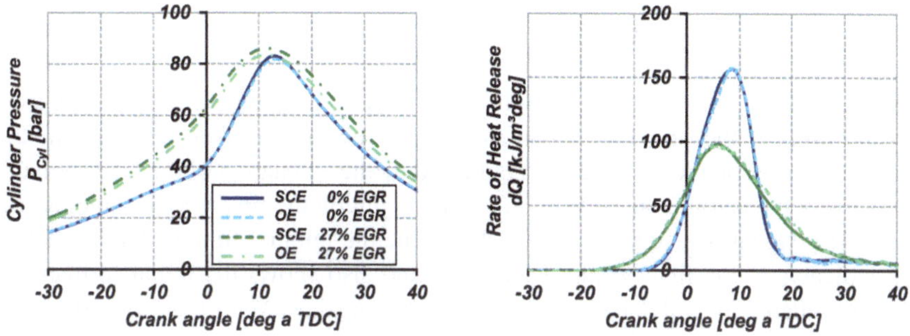

Fig. 3. Thermodynamic comparison of single cylinder engine and optical engine for A50 load points (0 % and 27 % EGR), averaged values.

4.3 Data Analysis

All characteristic thermodynamic and optically determined values shown are averaged from a total of 60 combustion cycles of two test runs. In order to characterize the combustion process, numerical values are derived from the combustion images. For the calculation of the flame projected area, the combustion image is binarized. By addition of all pixels having a value of one and using the appropriate image scale x_A (pixels/mm) the flame projected area A_f is calculated (Fig. 4). An initial flame kernel development time ($T_{FC,300}$) is defined for the comparison of the flame kernel formation between the corona and the conventional ignition system. This time $T_{FC,300}$ is defined as the time range between ignition timing (IT) and the attainment of a flame projected area $A_f = 300\ mm^2$ (corresponding to an idealized flame projected radius of $R_f \approx 10$ mm).

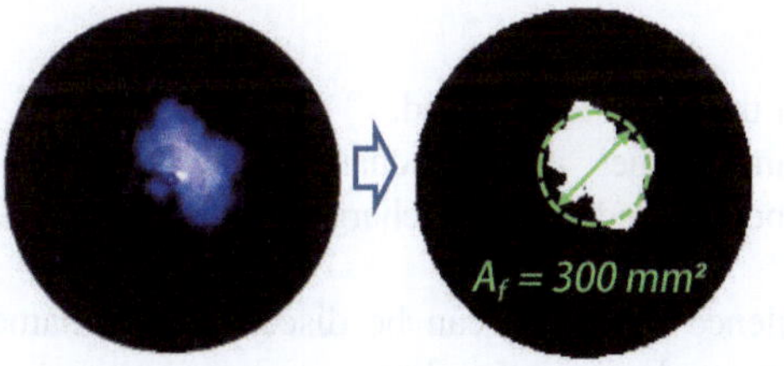

$$A_f = x_A \sum P_{=1}$$
$$x_A = image\ scale\ [Pixel/mm]$$
$$P_{=1} = Binary\ Pixel$$

Fig. 4. Calculation of optical parameters with numerical image processing

A first law pressure trace analysis is performed with AVL Indicom to determine the thermodynamic characteristics. The parameters used in this publication are the indicated mean effective pressure (IMEP), the position (MFB_{50}) and the standard deviation (σMFB_{50}) of the center of combustion (50 % mass fraction burnt). The coefficient of variance (COV) is determined for the characterization of the cyclical variation of the indicated mean effective pressure as a measure for the engine running smoothness. A threshold of COV = 2 % is accepted. This is defined as the ratio of the standard deviation of the indicated mean effective pressure σIMEP and the average value of IMEP:

$$\text{COV } [\%] = \sigma_{IMEP} / \overline{IMEP} * 100 \quad (2)$$

In case of the conventional spark-ignition system, the ignition timing is detected from the drop of the primary current in the coil (electric breakdown). This method reliably shows the spark flashover, as confirmed by the evaluation of the OH* signal, as well as the combustion images. With the corona ignition the first recognizable increase of the OH* -signal is a suitable way to determine the ignition timing. This signal is a good indicator of the first chemical reactions and also shows the initial formation of streamers reliably, as validated by the combustion images. With this method, a more precise indication of the ignition timing (IT) is possible compared to the ignition trigger signal. The time for the energy supply of the controller as well as the time for the initial ionization before the first dispersion of streamers is taken into consideration.

5 Experimental Results

5.1 Phenomenology of Ignition and Flame Propagation

The Fig. 5 shows a comparison of the initial flame propagation for the conventional and the corona ignition system of one representative combustion cycles for the A50 operating point without EGR. In the presented images, the area outside the piston bowl is covered by a mask. The first image corresponds to the point of ignition, it is followed by subsequent frames at a constant distance of 2.5 deg between frames; it should be noted that the frames in a row correspond to different crank angle positions due to different spark timing. The characteristic phases of ignition, initial flame kernel formation, as well as subsequent turbulent flame propagation are described below in keywords for both ignition systems:

1. *Conventional ignition system:*

- *IT:* spark flashover, bright light from the spark is reflected.
- IT + 2,5 deg: initial hemispheric flame kernel, glow discharge with lateral displacement of the plasma channel can be detected, spark discharge duration is up to 9 deg.
- IT + 5 deg: influence of large turbulence structures can be discerned, the flame kernel moves slightly off-center in the combustion chamber.
- IT + 7,5 deg: large scale wrinkling of the flame front surface increases the active flame front surface and thus enlarges and accelerates its propagation, increasing light intensity due to a magnified flame volume and higher pressure/temperature.

2. *Corona ignition system:*

- IT: initial streamer formation, length of streamers up to 15 mm, no swirl effect can be detected.
- IT + 2,5 deg: formation of five individual flame kernels around streamers, the streamer plasma channel is moved slightly in the direction of swirl, total flame projected areas similar at IT + 10 deg to the conventional spark ignition system.

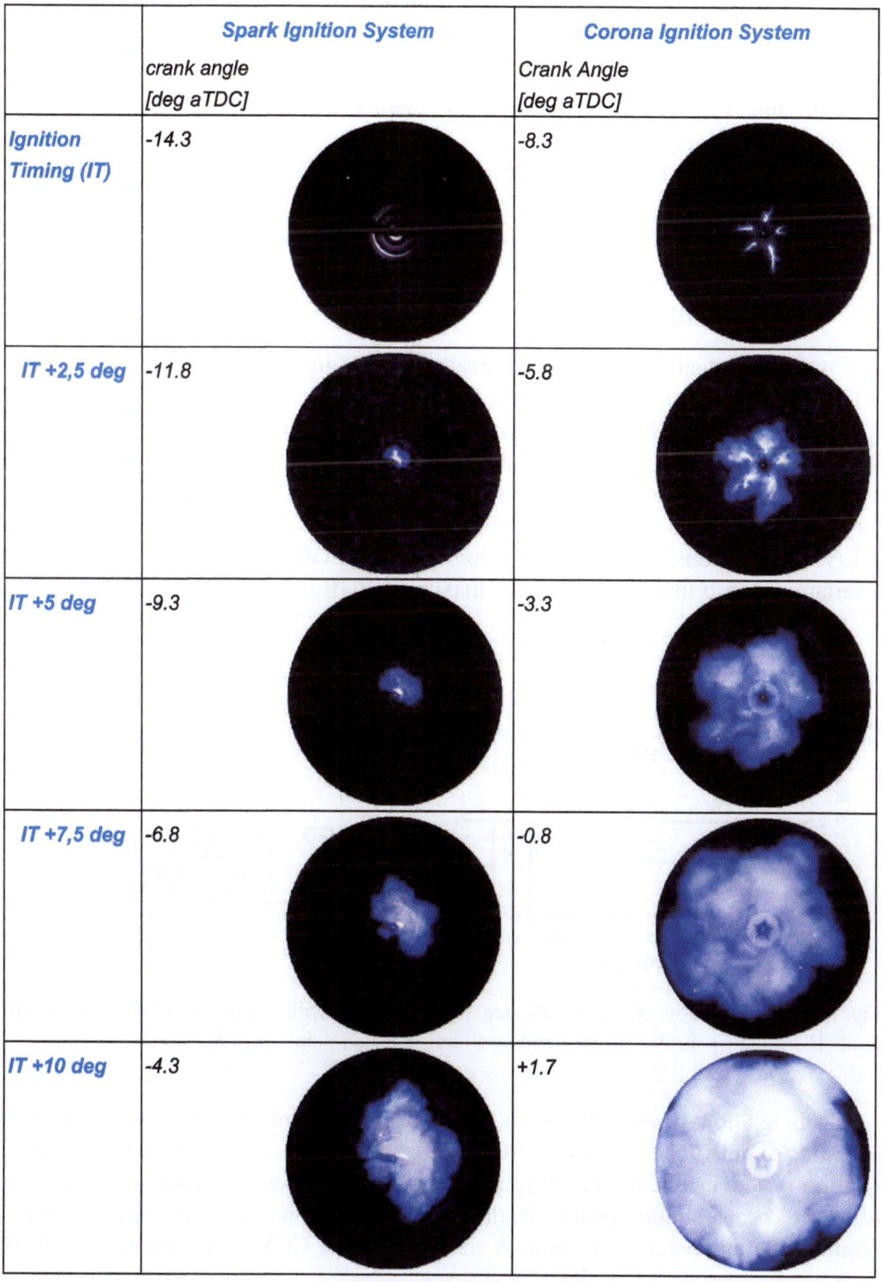

Fig. 5. Comparison of the ignition and the initial flame propagation with the conventional spark (left) and the corona (right) ignition system for the A50 operating point with 0 % EGR and MFB_{50} = 8 deg aTDC.

- IT + 5 deg: merging of individual flame kernels; flame fronts of the five individual flame kernels encounter each other to form one common large flame kernel, flame zone is very well centered in the combustion chamber – in contrast to the situation with the spark-ignition system, less influence of global in-cylinder flow in-homogeneities.
- IT + 7,5 deg: turbulent flame propagation: flame front strongly folded, former pentagonal shape (from five individual kernels) still visible.

5.2 Charge Dilution with EGR

The most important criterion for the evaluation of ignition systems in the case considered is their potential to increase the EGR compatibility limit. With the conventional ignition system the EGR limit is attained at X_{EGR} = 27 % for both load points A50 and A75, as shown in Fig. 6. A typical characteristic for the EGR limit is a slight increase up to COV = 2 % followed by a very abrupt rise in the cyclical variations. For instance for the spark ignition system misfire events already occur at X_{AGR} = 30 %. Due to this behavior in real transient engine operation either the EGR limit must be held exactly or a certain distance must be kept to the maximum EGR compatibility.

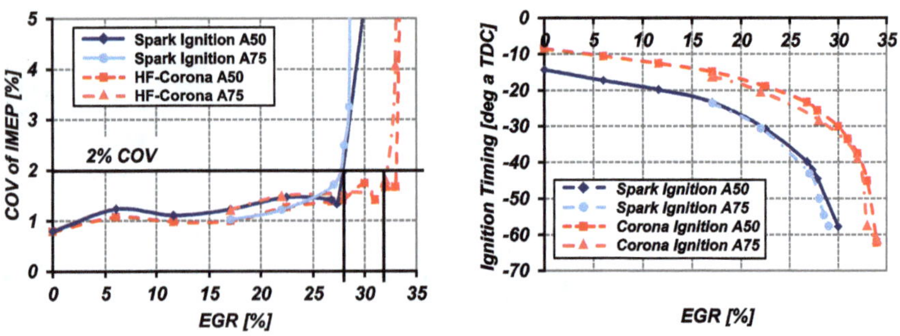

Fig. 6. Left: Variation of mean indicated pressure COV with maximum EGR compatibility limits indicated by arrows. Right: ignition timing for load points A50 & A75.

Each ignition system shows an individual EGR limit which is valid for both operating points. The maximum EGR compatibility limit of 27 % with the spark ignition system can be increased significantly by five percentage points up to 32 % by using the corona ignition system. Both ignition systems show the same behavior when reaching the individual EGR limit. A steep rise of the COV following the 2 % COV limit can be seen with both ignition systems.

With an increasing EGR rate the ignition timing has to be shifted considerably to earlier crank angles, as shown in Fig. 6. With the conventional spark ignition system a spark advance of 24.7 deg is necessary, when increasing the EGR rate from 0 % towards 25 %. When advancing the ignition timing, significantly worse conditions for ignition of the mixture arise due to decreasing temperatures and pressures at ignition

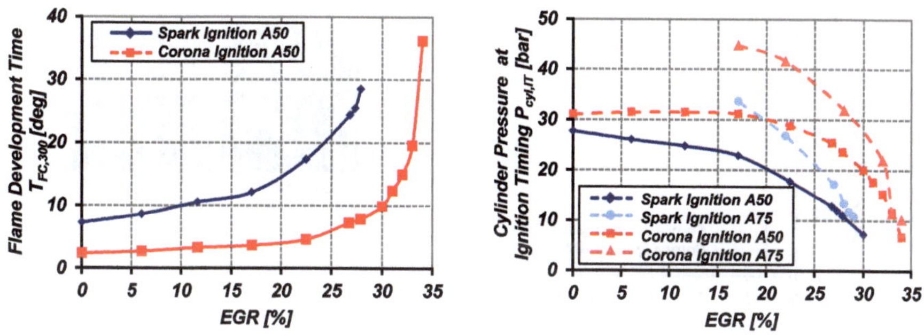

Fig. 7. Left: Optical flame kernel formation times $T_{FC,300}$ for the A50 operating point; Right: Cylinder pressure P_{cyl} at ignition timing for load points A50 and A75.

timing (IT), in addition to the dilution of the charge by EGR. For example, the pressure drops from P_{Cyl} = 28 bar to 15 bar at the A50 load point (right diagram in Fig. 7). The cause of the requirement of increasing spark advance is the combination of a reduction in flame propagation speed and an increase in flame kernel development time, due to the factors already stated above (charge dilution by EGR, reduction of pressure and temperature). For the conventional ignition system an increase in flame kernel formation time $T_{FC,300}$ by 12 deg (corresponding to 60 %) results when the EGR rate is increased from X_{EGR} = 0 % to 25 %. Thus, when increasing the EGR rate, one half of the required spark advance results only from the significantly slower initial flame propagation.

With the corona ignition system, the flame kernel formation time decreases significantly due to the initial inflammation of a larger mixture volume. This results in a reduced spark advance, by up to 6 deg at low EGR rates (Fig. 6). With high EGR rates of X_{EGR} > 20 %, an even significant advantage of the corona ignition system concerning flame kernel formation becomes evident. At an EGR rate of X_{EGR} = 25 % the difference in flame kernel development time between the two ignition systems is a time range of $\Delta T_{FC,300}$ = 16 deg. This corresponds to a reduction in the flame kernel formation time by 72 %. At very high EGR rates, the accelerated flame kernel formation with the corona ignition system allows significant later ignition timings and thus better ignition conditions. It is worth noting that at the EGR limit X_{EGR} = 32 % with the corona ignition system, the cylinder pressure at ignition timing attains identical values as at the ignition timing with the conventional ignition system at its EGR limit of X_{AGR} = 27 %. After exceeding the EGR limit X_{EGR} > 32 % the flame kernel formation time with the corona ignition system increases exponentially.

Effect of ignition voltage "corona buck-voltage". The maximum possible ignition voltage can be used in the two operating points up to an EGR rate of 27 % without getting a flashover towards the cylinder head (Fig. 8). The length of the corona streamers increases with decreasing cylinder pressure at early ignition timing, this increases the risk of flashovers to the electrical ground. However, when motoring the engine it was found that flashover already starts occurring at higher cylinder pressures compared to the pressure when flashover occurred in fired operation. The cause of the

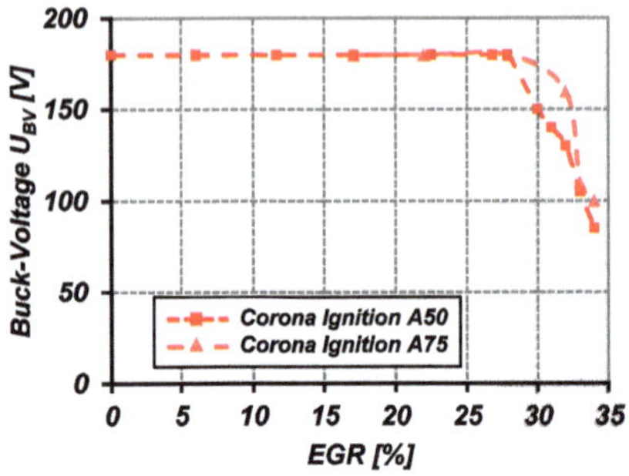

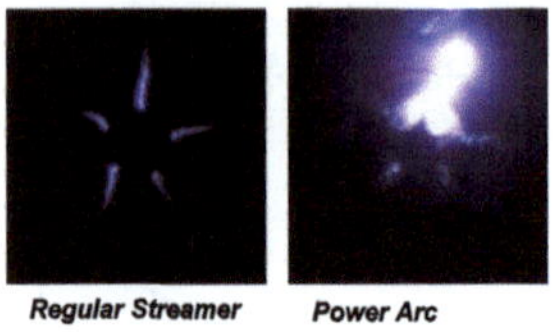

Fig. 8. Left: Maximum possible corona buck-voltage U_{BV} in EGR-variation at A50 and A75 load point; Right: Images of initial streamer formation followed by a flashover to the cylinder head.

higher flashover resistance in fired operation is an observed stabilization of the streamers inside the initial flame kernels. Here a streamer is being retained in its plasma channel which contains ionized combustion chemical reaction products. The streamer does not leave this ionized channel, as a consequence the likelihood of a flashover to the cylinder head is reduced. However, at EGR rates > 27 % this stabilizing effect does not apply due to a slow start of chemical reaction and hence a delayed flame kernel development. An increased length of the streamers, which would be advantageous especially at such high EGR operating points, cannot be used in this case.

Cyclical variations. In addition to the consideration of a maximum EGR compatibility, the reduction of cyclical variation of the center of combustion is also an important criterion in the development of combustion systems. There is a correlation between the variation of the center of combustion and a reduction in the knocking tendency due to less early cycles, as described as an example in [17]. The relationship between the cyclical variation of the center of combustion and the fluctuation of flame kernel formation time shall be considered in detail below. Figure 9 shows the correlations between the deviation in flame kernel formation time from the particular mean value $\Delta T_{FC,300}$ and the deviation between the center of combustion from the mean value ΔMFB_{50} for three EGR rates of 0 %, 17 % and 27 %.

The data from 60 single cycles are shown for each operating point and each ignition system. The cyclical variation of the center of combustion MFB_{50} increases with a rising EGR rate for both ignition systems. However in the case of the corona ignition system, the cyclical variations are significantly reduced. There is a linear relationship between MFB_{50} and flame kernel formation time $T_{FC,300}$ for both ignition systems. The slope of the trend lines in the diagram denotes the absolute influence of flame kernel formation time on the center of combustion. With an increasing EGR rate the proportional influence of flame kernel formation time on the center of combustion also increases for both ignition systems, the gradients of best-fit lines are reduced. At an EGR rate of 27 % a nearly one-to-one relationship results for the conventional ignition system, i.e. increasing the flame kernel formation time by 2.0 deg leads to a

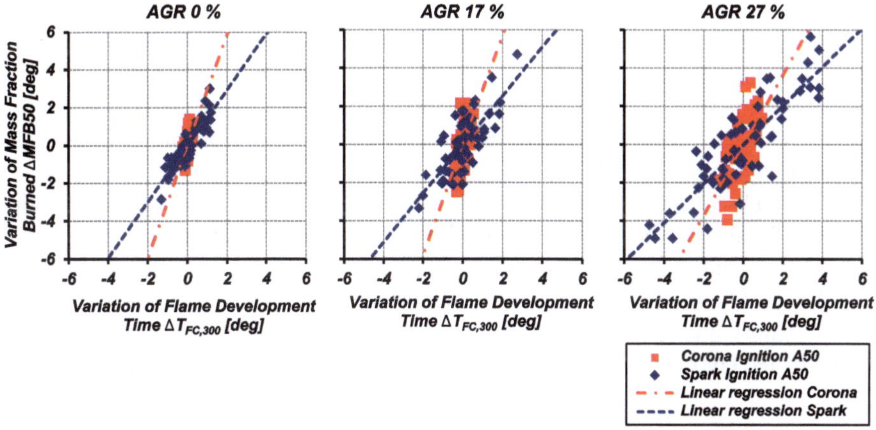

Fig. 9. Correlation between flame kernel development time $\Delta T_{FC,300}$ and center of combustion ΔMFB_{50}, each as a deviation from particular mean values; operating point A50 with EGR rates of 0 %; 17 % and 27 %.

approximately 2.0 deg later MFB_{50}. In case of the corona ignition system, on the one hand the variation in flame kernel formation time is reduced; on the other hand the absolute influence of the variation in flame kernel formation time on the center of combustion MFB_{50} is also reduced.

In Fig. 10 the particular standard deviations of the center of combustion and the flame kernel development time are shown for an EGR variation at the load point A50 for both ignition systems. Using the corona ignition system, the standard deviation of the combustion center σMFB_{50} is reduced by 50 % in a range of the EGR rate X_{EGR} = 0–25 % (Fig. 10, left chart). The steep increase towards very high standard deviations is shifted to higher charge dilution rates, when the corona ignition system is used.

The decreased fluctuation of the combustion center results from the fact that the ignition with the corona system is significantly more robust. As shown in the right graph of Fig. 10, the standard deviation of flame kernel formation time $\sigma T_{FC,300}$

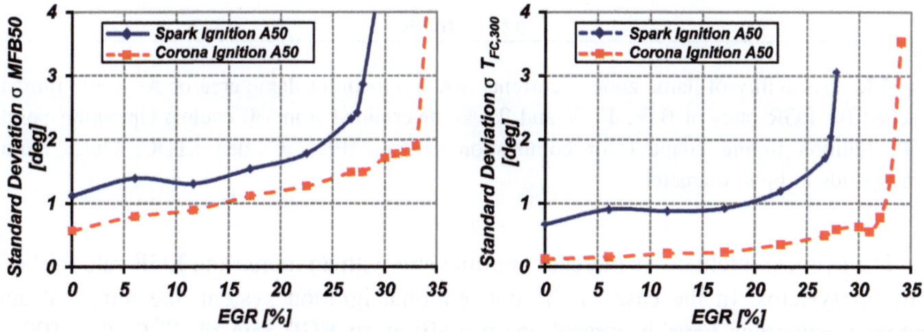

Fig. 10. Left: Cyclical standard deviation of the combustion center σMFB50; Right: optically determined flame kernel formation time $T_{FC,300}$; operating points A50 each.

decreases significantly when the corona ignition system is used. The cyclical fluctuations in the flame kernel formation in case of the corona ignition system are very low, with a standard deviation of $\sigma T_{FC,300} < 0.5$ deg up to an EGR rate of $X_{AGR} = 30$ %. This shows the robustness of the flame kernel development with the corona ignition system. Only at very high EGR rates > 30 % does the standard deviation rise exponentially.

In the subsequent Fig. 11 the cyclical fluctuations of the flame kernel formation time are visualized. Shown is the probability of flame zone occurrence at an point in time when a medium flame area of $A_f = 300$ mm^2 is reached. For every position, the probability of flame occurrence in each combustion cycle is summed up for that point in time. Those areas in which the flame occurs in all combustion cycles are characterized by red coloring; areas where no flame occurs are characterized by blue coloring.

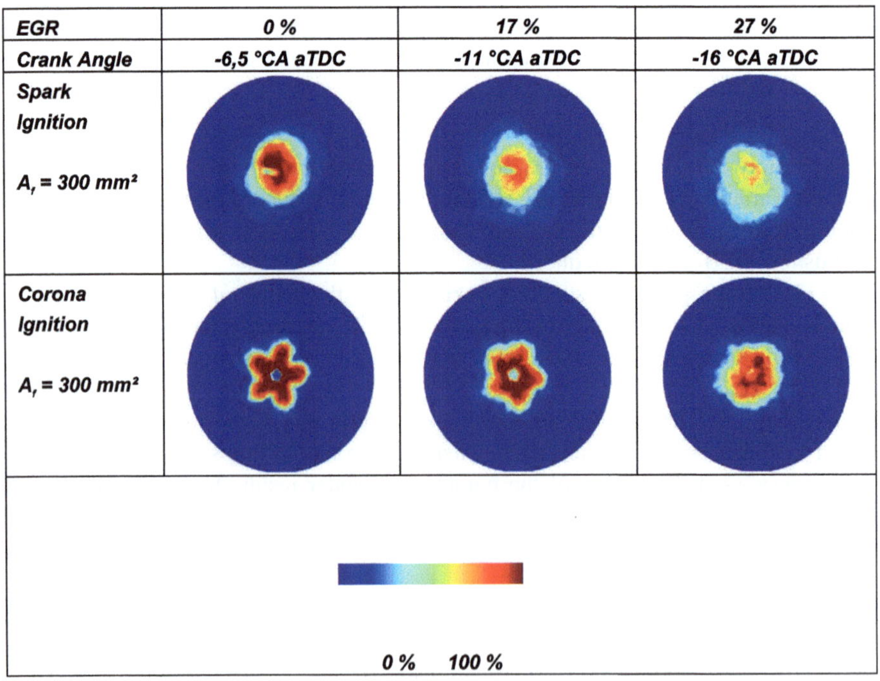

Fig. 11. Probability of flame zone occurrence when a medium flame area of Af = 300 mm^2 is reached for EGR rates of 0 %, 17 % and 27 %; determined from 30 cycles; Operating points A50; Ignition timing adapted for combustion center $MFB_{50} = 8$ deg aTDC; Outer radius corresponds to bowl diameter.

The cyclical fluctuations increase significantly with an increasing EGR rate for both ignition systems. In the case of the conventional ignition system, the temporal and spatial fluctuations have increased so strongly at an EGR rate of 27 % that 100 % probability of flame occurrence in each cycle only exists in the areas directly at the spark plug. Compared to the conventional spark-ignition system, the significant reduction in the cyclical fluctuations of the flame kernel formation, as already discussed on the basis of the standard deviation $\sigma T_{FC,300}$, is visually recognizable in Fig. 10. The five individual flame kernels are seen to be clearly separated at low EGR rates. At a high EGR rate of X_{EGR} = 27 % a significantly slower flame kernel formation time leads to a higher influence of the swirl flow. As a result the individual flame kernels unite at an earlier point in time.

5.3 Ignition Energy Variation

The variation of the buck-voltage U_{BV} alter the length of the streamers, as shown in Fig. 12. At maximum ignition voltage, a streamer length of up to 15 mm is attained at the initial streamer formation, whereas streamer length decreases significantly when the buck-voltage is reduced.

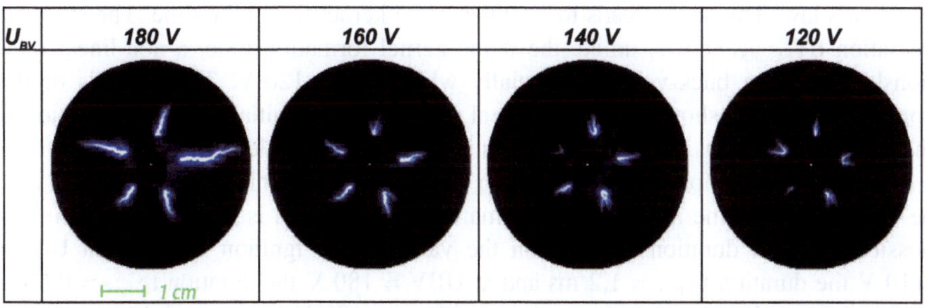

Fig. 12. Initial streamer formation for a variation of the ignition voltage; A50 operating point, 0 % AGR; constant IT - 9.5 deg aTDC with cylinder pressure P_{Cyl} = 30 bar at ignition timing.

Figure 12 shows how more streamers per prong seem to appear at lower buck-voltages. However, high speed recordings of the streamers with 120.000 fps,

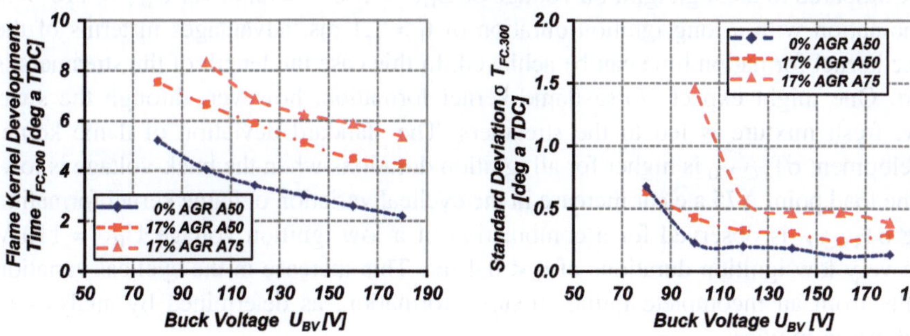

Fig. 13. Left: Influence of pressure, EGR and buck-voltage on flame kernel formation time $T_{FC,300}$; Right: Standard deviation of flame kernel formation time $\sigma_{TFC,300}$; operating points according to Table 2.

shows that this apparent branching results from a faster reformation of individual streamer channels, while the figure above shows an averaging effect of a long integration time.

Figure 13 shows the determined flame kernel formation times $T_{FC,300}$ as well as the standard deviation of flame kernel formation time $\sigma T_{FC,300}$ for three operating points. The flame kernel formation time $T_{FC,300}$ can be influenced by the buck-voltage, which directly influences the streamer length and thus the size of the initial flame kernel. There is an almost linear relationship between buck-voltage and flame kernel formation time.

An increase of the EGR rate from $X_{AGR} = 0$ % to 17 % at identical values of buck-voltage and pressure at ignition timing ($P_{Cyl} = 30$ bar) leads to a doubling of the flame kernel formation time. The effect of pressure on the flame kernel formation is shown in a comparison of load points, A50 and A75. At ignition timing and with an identical EGR rate of $X_{EGR} = 17$ %, the cylinder pressures for these two load points are $P_{Cyl} = 30$ bar and $P_{Cyl} = 45$ bar, respectively. At the load point A75 the length of the streamers are shorter due to the lower mean free length of path, according to Paschen's law. This factor leads to a longer flame kernel formation time. The standard deviation $\sigma T_{FC,300}$ shows, unlike the flame kernel formation time, a non-linear relationship with the buck-voltage especially when $U_{BV} < 120$ V. The analysis of the combustion movies shows a strong cyclical variation of the initial streamer formation at these operating points. Therefore, low ignition voltages should be avoided.

In addition to the parameter of ignition voltage U_{BV} the ignition duration t_I can also be adjusted. Since the maximum performance of the system is limited, the maximum possible ignition durations depend on the value of the ignition voltages: at U_{BV} = 110 V the duration $t_{I,max} = 1{,}2$ ms and at UBV = 180 V the duration $t_{I,max} = 0{,}6$ ms could be used. With an ignition voltage of $U_{BV} = 180$ V in none of the considered operating points, an effect of the ignition duration on the flame kernel formation is detected. Due to the initial long streamers, a large volume of the mixture is captured at the beginning of the ignition and the streamer remains within its incipient plasma channel. The ignition energy flows into this plasma channel, but no significant effect on the subsequent flame kernel growth can be observed. A short ignition duration of $t_I = 0.1$ ms already provides a good flame kernel formation.

Compared to the high ignition voltage of $U_{BV} = 180$ V with lower $U_{BV} = 110$ V in combination with a long ignition duration of $t_I > 1.1$ ms, advantages in terms of the flame kernel formation time can be achieved. In this case the length of the streamers is short. One might expect worse flame kernel formation, however, through the swirl flow, fresh mixture is fed to the streamers. The standard deviation of flame kernel development $\sigma T_{FC,300}$ is higher for all ignition durations when the buck-voltage is low. At the load point A75 a clear increase in the cyclical variation of flame kernel formation time $\sigma T_{FC,300}$ is observed for a combination of a low ignition voltage $U_{BV} = 110$ V with very low ignition durations of $t_I < 0.4$ ms. This increase in the cyclical variation results from an incomplete initial streamer formation, as determined by analysis of combustion movies.

From an energetic point of view a high ignition voltage U_{BV} in combination with a short ignition duration $t_I < 0.4$ ms is advantageous at low EGR rates with regard to energy consumption and igniter durability. This short ignition time reduces also the flow of energy into the existing plasma channel inside the flame kernel which would otherwise heat up the plasma (that is literally relatively cold). At high EGR rates, however, the highest possible energy should be chosen with maximum buck-voltage U_{BV} and maximum ignition duration t_I. Hence the flame kernel formation time is short and cyclical variations are low so a later ignition timing results in better conditions for ignition is possible.

6 Summary

Investigations are performed on a heavy-duty single-cylinder gas engine with optical access to the combustion chamber; comparisons are made between the corona ignition system EcoFlash® and a conventional spark ignition system. The combustion concept of the gas engine used is characterized by a stoichiometric fuel air ratio ($\lambda = 1$) with pure methane externally mixed in the intake mainfold, charge dilution by external cooled exhaust gas recirculation and turbulence generation by swirl. High loads and thus high mean indicated pressures of IMEP = 11 and 17 bar are investigated. The maximum available pressure at the ignition timing (IT) is P_{Cyl} = 45 bar. The installation conditions are optimal for the use of a corona ignition system. The spacious dimensioned spark plug shaft allows the use of a M14 igniter with high electric efficiency. A deep piston bowl in combination with a combustion chamber with a flat roof and a centrally located spark plug leads to a large distance of the corona ignitor tips to the piston as well as to the cylinder head. Thus, maximum energy efficiency and the use of high streamer length is possible in large operating ranges of the engine.

A variation of the EGR rate is examined up to the misfire limit at two load points with IMEP = 11 and 17 bar. The corona ignition system shows the following advantages compared to the conventional spark ignition system:

- Significant increase of EGR compatibility with corona ignition system around 5 % from X_{EGR} = 27 to 32 %.
- Clear reduction in flame kernel formation time due to a larger initially captured mixture volume, especially at high EGR rates. Thus, later and more favorable ignition timing possible.
- Halving of the standard deviation of the center of combustion as a result of very robust flame kernel formation is observed. Hence, a reduction of knock sensitivity is expected.
- Better centering of the flame in the combustion chamber as the cycle to cycle variations of large-scale turbulence have lower influence on the flame kernel.
- High EGR rates and maximum ignition voltage can be used due to a deep piston bowl and the observed retention of the streamers within the ionized initial combustion zone.

Parameter studies concerning the impact of the corona ignition system specific energy parameters: ignition duration (t_I) and ignition voltage (Buck-voltage U_{BV}) are performed with the following results:

- Linear relationship between flame kernel formation time and buck-voltage. Streamer length correlates with the buck-voltage. This forms a simple application parameter.
- Increase in the cyclical fluctuations at low buck-voltage resulting from fluctuations of the initial streamer formation.
- Doubling of the flame kernel formation time when EGR rate is increase from 0 % to 17 %.
- Increase of P_{Cyl} from 30 to 45 bar at ignition timing leads to shorter streamer length and, as a result to a 22 % increase in flame kernel formation time. Due to that, slightly earlier ignition timing is required for the A75 compared to the A50 load point.
- Combination of low U_{BV} = 110 V with a long ignition duration leads to a reduction in flame kernel formation time, however, compared to U_{BV} = 180 V the standard deviation is increased.
- Short ignition time t_I < 0.2 ms in combination with high ignition voltages is optimal from an energetic point of view.

Explanation of Abbreviations

A_f	Flame Projected Area
aTDC	After Top Dead Center
Cp	Heat Capacity[J/K]
CV	Commercial Vehicle
deg	Degrees Crank Angle
dQ	Heat Release Rate [J/kgK]
EGR	Exhaust Gas Recirculation
fps	Frames per Second
IMEP	Indicated Mean Effective Pressure [bar]
IT	Ignition Timing [deg CA]
LAM (λ)	Air-to-Fuel Ratio
MFB_{50}	Mass Fraction Burned, Center of Combustion [%]
OE	Optical Engine
P	Pixel
P_{Cyl}	In-Cylinder Pressure
R_f	Flame Projected Radii
rmp	Revolutions per Minute
SCE	Single Cylinder Engine
TDC	Top Dead Center
t_I	Ignition Timing [ms]
U_{BV}	Corona Buck-Voltage [V]
w_i	Mass Fraction [%]]
x_A	Image Scale [Pixel/mm]
X_{AGR}	EGR-Rate [%]
z_F	Spark Position [mm]
σ	Standard Deviation

Acknowledgement. The authors would like to thank all who have contributed to this publication, especially Mr. Daniel Pfeffer for supporting the processing of combustion images and Dr. Günter Karl for motivating us to write this paper.

References

1. Hoffmann, K., Benz, M., Weirich, M., Herrmann, H.-O.: Der neue Erdgasmotor für mittelschwere NFZ von Mercedes-Benz. MTZ **75**(11), 12–19 (2014)
2. Berndt, F., Rottenkolber, G., Herweg, R., Haase, D., Dieler, T.: Lean burn combustion for gasoline engines: potential of high frequency ignition and high pressure injection. In: Bargende, M., Reuss, H.-C., Wiedemann, J. (eds.) 13 Internationales Stuttgarter Symposium, Stuttgart, pp. 223–241 (2013)
3. Schenk, M., Schröter, F., Zellinger, F., Klaus, B., Pfeiffer, D., Fischer, H.: Corona-ignition vs. spark ignition: a fundamental comparison for varying thermodynamic conditions of modern turbocharged gasoline engines. In: International Conference: SIA Powertrain, Versailles (2015)
4. Hampe, C., Bertsch, M., Spicher, U., Bohne, S., Hadler, J.: Thermodynamic experiments of high-frequency ignition with different charge motions and valve timings on a supercharged gasoline DI engine. In: Günther, M., Tröger, R. (eds.) Ignition Systems for Gasoline Engines, Berlin (2014)
5. Schenk, A., Bohne, S., Rixecker, G., Budde, M.: Das Korona-Zündsystem EcoFlash: Grundlagen und Erprobungsergebnisse im Erdgasbetrieb. In: Proceedings - 9. Dessau Gas Engine Conference, Dessau, pp. 121–136 (2015)
6. Maly, R.R., Herweg, R.: Spark ignition and combustion in four-stroke gasoline engines. In: Arcoumanis, C., Kamimoto, T. (eds.) Flow and Combustion in Reciprocating Engines. Experimental Fluid Mechanics, pp. 1–66. Springer, Berlin, London (2009). ISBN 978-3-540-68901-0
7. Herweg, R.; Maly, R.R.: A fundamental model for flame kernel formation in S.I. engines. SAE Technical Paper 922243 (1992). doi:10.4271/922243
8. Paschen, F.: Ueber die zum Funkenübergang in Luft, Wasserstoff und Kohlensäure bei verschiedenen Drucken erforderliche Potentialdifferenz. Ann. Phys. **273**(5), 69–96 (1889)
9. Bohne, S., Rixecker, G., Brichzin, V., Becker, M.: Hochfrequenz-Zündsystem mittels Korona-Entladung. MTZ **75**(1), 50–55 (2014)
10. Heise, V., Farah, P., Husted, H., Wolf, E.: High frequency ignition system for gasoline direct injection engines. SAE Technical Paper 2011-01-1223 (2011). doi:10.4271/2011-01-1223
11. Bradley, D., Lung, F.-K.: Spark ignition and the early stages of turbulent flame propagation. Combust. Flame **69**(1), 71–93 (1987). doi:10.1016/0010-2180(87)90022-8
12. Gaydon, A.G.: The Spectroscopy of Flames. Chapman & Hall Ltd., London (1957)
13. Schöffler, T., Hoffmann, K., Koch, T.: Stoichiometric natural gas combustion in a single cylinder SI engine and impact of charge dilution by means of EGR. SAE Technical Paper 2013-24-0113 (2013)
14. Schöffler, T., König, G., Dinkelacker, F.: Optische Analyse der vorgemischten Methanverbrennung an einem Einzylinder-Transparentaggregat mit Ladungsverdünnung durch Abgasrückführung. In: Leipertz, A. (ed.) Motorische Verbrennung - aktuelle Probleme und moderne Lösungsansätze (XII. Tagung), Ludwigsburg, Berichte zur Energie-und Verfahrenstechnik, ESYTEC, Erlangen (2015)

15. Schwarz, V., König, G., Dittrich, P., Binder, K.: Analysis of mixture formation, combustion and pollutant formation in HD diesel engines using modern optical diagnostics and numerical simulation. SAE Technical Paper 1999-01-3647 (1999). doi:10.4271/1999-01-3647
16. Günther, M., Tröger, R., Nicklitzsch, S., Adolf, M.: Design criteria for the use of corona ignition systems in combustion engines. In: Günther, M., Tröger, R. (ed.) Ignition Systems for Gasoline Engines, Berlin (2012)
17. Heuser, G.: Der Einfluß der Entflammungs- und Verbrennungsphase auf das Klopfen in Ottomotoren, Dissertation, Aachen, RWTH Aachen (1993)

Potential of Advanced Corona Ignition System (ACIS) for Future Engine Applications

Cherian A. Idicheria(✉) and Paul M. Najt

Propulsion Systems Research Laboratory,
General Motors Global Research and Development, Warren, USA
cherian.idicheria@gm.com

Abstract. In this study, the advanced corona ignition system (ACIS) was evaluated. First, the system was evaluated in a direct-injected, single cylinder optical engine and high-speed imaging of enhanced combustion luminosity was used to visualize the ignition, flame kernel formation and flame propagation. The imaging results show that the ACIS promotes simultaneous ignition of the mixture at multiple locations in the combustion chamber as opposed to ignition being limited to the spark gap channel. Therefore, ignition delay is always shorter with the ACIS than with a traditional spark plug, inductive discharge system. The ACIS is able to support stable combustion (COV of IMEP < 3 %) in a leaner homogeneous mixture than the spark plug based system. Similarly, the ACIS has shorter ignition delay and faster burn rate for stoichiometric, rich and simulated EGR diluted mixtures. Second, the ACIS system was evaluated in a 2.0L, direct-injected, turbocharged, engine. Comparison of combustion results showed that significant reduction in 0–10 burn durations can be achieved over the production ignition system, enabling higher internal dilution limits. The ACIS was able to match the performance of the production ignition system for the idle spark sweep test. The test results showed that the close proximity of the piston to the ACIS electrode can transition the streamer discharge to arcing mode, which hampers optimal performance. Overall, ACIS has the potential to perform as well or better than the production ignition system and for best combustion performance the piston design must be optimized.

1 Introduction

Global fuel economy requirements, emission compliance, climate change, energy security, fuel prices, customer requirement and innovative technologies are among the main factors that drive the development of advanced internal combustion engines. With the need to improve engine efficiency and reduce fuel consumption while meeting emission requirements, next-generation gasoline engines are being downsized, boosted, diluted, and moving away from stoichiometric to lean combustion operation. Advancements in fuel injection systems, closed loop controls, and fast response boost systems are critical enabling technologies. All these enabling technologies create significant challenges to the conventional ignition system. For example, specific to the ignition system, the cylinder pressure at time of ignition increases with downsizing and boosting, and the required breakdown voltage increases as a result. As the breakdown

M. Günther and M. Sens (eds.), *Ignition Systems for Gasoline Engines*,
DOI 10.1007/978-3-319-45504-4_19

voltage increases, spark plugs become more susceptible to failure owing to increased electrode wear and the finite dielectric strength of the insulating material. A way to counter this issue is to decrease the spark gap (nominally 0.9 mm–1.1 mm) which in turn reduces the breakdown voltage requirement at elevated pressure. However, smaller spark plug gap size (0.5 mm–0.8 mm) has an adverse effect on combustion stability, especially in dilute (e.g. air dilution, cooled external EGR) environment. Therefore, alternate ignition technologies that can enable stable and robust ignition in dilute or lean environment and at high boost pressures is a key enabler for next generation engines and is an active area of research and development [1, 2].

Among advanced ignition systems, low-temperature plasma based systems are gaining in importance. Several alternatives to conventional spark plug systems are being explored and the most prominent among them are the radio frequency (RF) based systems like the corona ignition system and the Barrier Discharge Ignition (BDI) system [3, 4]. Other technologies being evaluated include multi-charge ignition, Dual-coil offset/ignition system and microwave high-frequency ignition system [1, 2] and laser ignition [5]. While the majority of these systems were evaluated for homogeneous lean and dilute combustion applications, low-temperature plasma ignition systems were explored for achieving advanced combustion strategies like lean stratified charge combustion and homogeneous charge compression ignition (HCCI).

GM Global Research and Development has evaluated the advanced corona ignition system (ACIS). The system was tested in the optical engines under different combustion strategies (homogeneous [stoichiometric, lean, rich, simulated EGR]) to understand its capabilities and limitations. The system was also evaluated in a turbocharged, direct-injected production multi-cylinder engine.

2 Optical Engine Specifications

The experiments were performed in a single-cylinder, four-stroke, direct-injected engine with optical access to the combustion chamber. It is a four-valve, pent-roof configuration with bore × stroke of 86.0 × 86.0 mm (0.5 L displacement) and a compression ratio of 10:1. Fuel is injected directly into the cylinder using a high-pressure (20 MPa) multi-hole solenoid injector. As sketched in Fig. 1, an injector and a spark plug are mounted between the intake valves and exhaust valves. An eight-hole valve-covered-orifice (VCO) stepped-hole injector supplies an eight-plume spray, and two spray plumes straddle the spark plug electrode. A spark plug with spark energy of 135 mJ was the ignition source. While using the ACIS, the electrodes were positioned to match the protrusion of the spark plug. The piston has a cylindrical bowl (44 mm clear aperture). The throttle plate in one of the intake runner is closed to generate swirl motion in this engine. A Bowditch-type extended-piston arrangement was used together with carbon-composite piston rings in the optical section. A quartz window forms the bottom of the piston bowl and a 45° mirror provides optical access, as shown in Fig. 1. The major engine specifications are presented in Table 1.

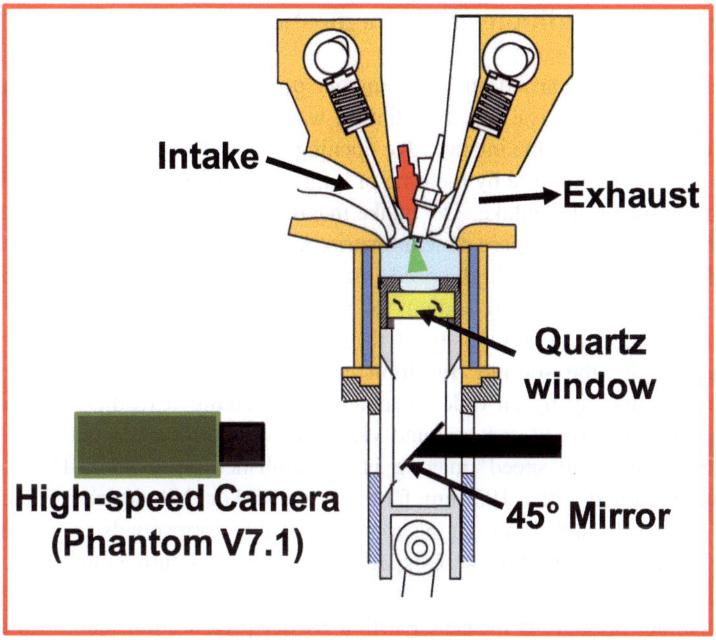

Fig. 1. Schematic of the optical engine and high-speed imaging setup

Table 1. Optical engine specifications and operating conditions.

Bore	86.0 mm
Stroke	86.0 mm
CR	10:1
Engine speed	800 rpm
Fuel mass/cycle	8 mg
Fuel injection pressure	20 MPa
Fuel	Indolene with sodium additive
Spark energy	135 mJ (secondary side)
ACIS energy	Variable (estimated 250 mJ–300 mJ)
Intake temperature	90°C
Coolant temperature	90°C
Intake manifold pressure	Variable depending on condition
Exhaust manifold pressure	100 kPa (not controlled)
End of injection (EOI) timing	−290 aTDC
Spark timing (Spark Plug)	−42° aTDC
Spark timing (ACIS)	−36° aTDC

3 Operating Conditions and Diagnostics

The experiments were carried out at an engine speed of 800 rpm with total fueling of 8 mg per engine cycle. The injection timing was kept constant at −290 aTDC. The injection event is early in the intake valve opening and is therefore expected to generate nearly homogeneous mixture by the time of ignition. The effect of EGR was simulated by adding nitrogen to the intake. The intake manifold pressure was adjusted to achieve the required air flow to prepare stoichiometric, lean, or rich mixtures.

The diagnostic used was high speed imaging of combustion luminosity. An additive (Lubrizol 5318) containing sodium (~50 ppm) was added to the Indolene fuel to enhance the combustion luminosity at these operating conditions. The combustion temperature excite the sodium transition centered at 589.9 nm, and increases the combustion luminosity by an order of magnitude, making it readily detectable with non-intensified high speed cameras and visible wavelength camera lenses [6]. In the current study, the high speed imaging is accomplished by using a Phantom V7.1 CMOS camera, fitted with a 105 mm, f/2.8, Micro-Nikkor (Nikon) lens. The camera is operated at 24 kHz and at an engine rpm of 800 this corresponds to acquiring five images every crank angle. For these experiments, 250 consecutive engine cycles were imaged.

The cylinder pressure is measured with a Kistler pressure transducer (6125B) and digitized through the high-speed data acquisition system. The pressure data is acquired at every crank angle for 300 consecutive engine cycles. Standard emission measurements are acquired by using a Horiba MEXA-7500EGR emission bench. The cylinder pressure data and the emission data is used for heat release rate analysis and other relevant thermodynamic calculations by using an in house tool.

4 Optical Engine Results and Discussion

4.1 Stoichiometric Mixture, a/F = 15.0, Same Spark Advance

The objective of this study is to evaluate the difference in the performance with ACIS and spark plug system, for homogenous combustion. Figures 2 and 3 shows a comparison of the ignition, early flame kernel formation and flame propagation processes for ACIS (~250 mJ) and spark plug system (~135 mJ), for an individual combustion cycle event. The energy per discharge for the ACIS is from an empirical correlation characterizing the energy required to generate the plasma and is not necessarily the secondary energy delivered.

The condition presented in Figs. 2 and 3 is at 800 rpm and a stoichiometric mixture. Even though a single cycle is shown, the cycles are representative of the trends as the cycle to cycle variability is very low for both ignition systems. The images are false colored. The spark timing is kept the same at −42 aTDC. Comparing the two sets of images, it is evident that for the same spark timing, the ACIS has earlier ignition, flame kernel formation and superior flame propagation when compared to the spark plug system. For example, while there is no indication of an arc discharge (−43 to −39 aTDC) in the spark plug case, the four streamer discharges (~5 mm) of the ACIS from the four prong tips is clearly visible during the same time. The arc discharge in the spark plug is confined to the spark gap (~1 mm) for this combustion system and thus not visible in this particular image view. Beyond −38 aTDC, the four streamer discharges develop into individual ignition spots and by −34 aTDC, they merge to form a robust flame kernel. The flame kernel then propagates to mostly encompass the visible portion of the combustion chamber by −24 aTDC. In contrast, as evident from Fig. 3, the ignition and early flame kernel formation events are significantly slower for the spark plug. For example, from −38 aTDC to −34 aTDC, the ignition and flame kernel formation has barely begun and the combustion phasing is retarded relative to the ACIS system. The line-of-sight flame size with the spark plug at −24 aTDC is similar to the flame size at −34 aTDC with the ACIS. Also, the ACIS is potentially capable of delivering higher total energy in the discharge (increase duration) than the spark plug and may offer additional advantage to igniting mixtures, especially with higher dilution.

As shown from Figs. 2 and 3, the combustion phasing is significantly advanced with the ACIS for same spark advance timing. For the rest of the study the spark advance for the ACIS was retarded (−36 aTDC) to match the fifty percent burn point (CA50) with the spark plug system (−42 aTDC) for a stoichiometric mixture. Also, mean combustion luminosity images computed from 250 consecutive cycles are presented to evaluate the effect of ignition system on the flame kernel formation and flame propagation.

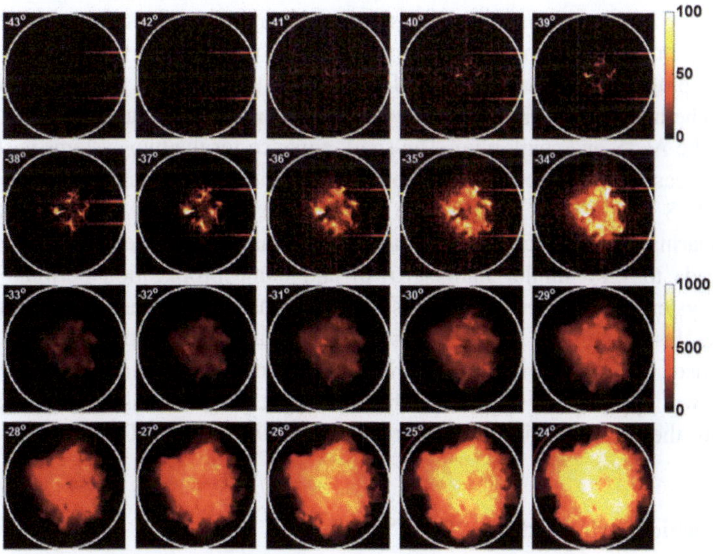

Fig. 2. Progress of ignition, flame kernel formation and flame propagation with ACIS. 800 rpm, Stoichiometric mixture, SA = −42 aTDC.

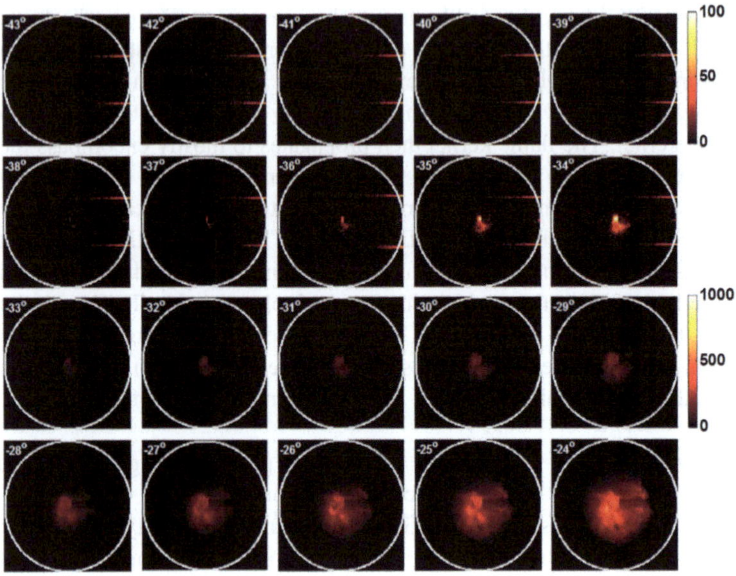

Fig. 3. Progress of ignition, flame kernel formation and flame propagation with spark plug. 800 rpm, Stoichiometric mixture, SA = −42 aTDC.

4.2 Stoichiometric Mixture, Matched CA50

In Fig. 4, the mean combustion luminosity images and heat release rates are shown for a stoichiometric mixture and different spark advance to match the CA50 between ACIS and spark plug. The estimated energy from the ignition systems are shown above the images. The heat release rate plot shows that the combustion phasing is well matched between the ACIS and spark plug. The spark advance for the two systems is shown on the heat release plot as two lightning bolts. As described previously, the spark advance for the ACIS is −36 aTDC and for the spark plug is −42 aTDC for matching CA50.

Comparing the mean combustion luminosity images, it is clear that matching the CA50 leads to very similar flame growth and flame propagation between the two ignition systems. However, the process of ignition to early flame kernel formation progresses faster with the ACIS because of the simultaneous, closely located ignition points. Also, careful examination of the outer boundary will indicate that the ACIS flame growth appears more spherical than the spark plug. This is more evident when comparing the corresponding images from −27 aTDC to −20 aTDC.

4.3 Stoichiometric Mixture, 20 % N_2 Dilution

In Fig. 5, the mean combustion luminosity images and heat release rates are shown for a stoichiometric mixture with 20 % N_2 dilution to simulate exhaust gas recirculation (EGR) conditions. This condition was included to investigate the capability of the ACIS when charge dilution is present. Figure 5 shows that the voltage setting of the

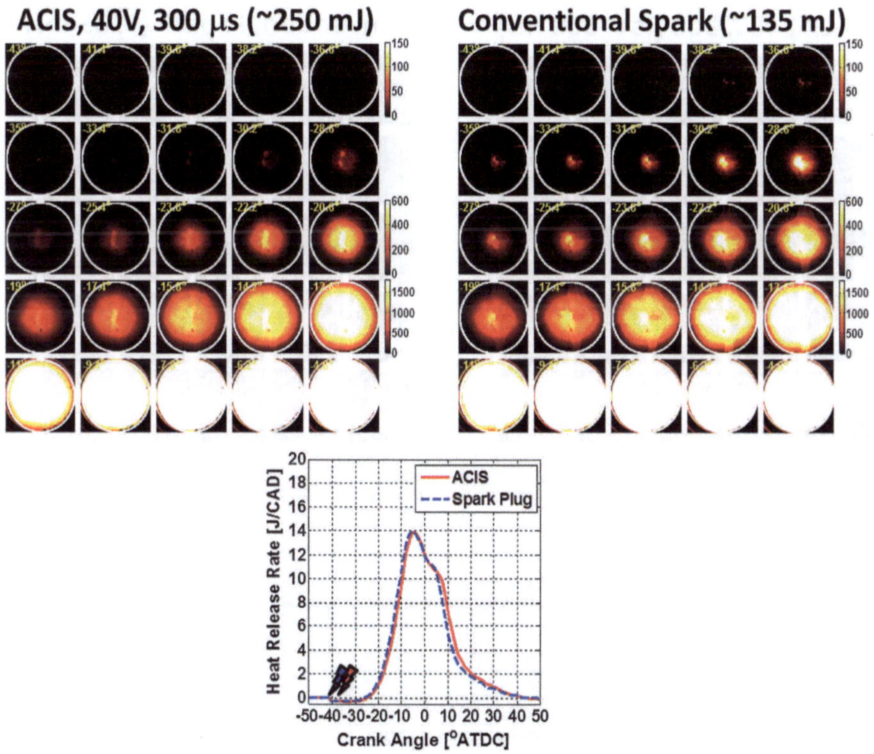

Fig. 4. Mean images of flame kernel formation and flame propagation with ACIS and spark plug. The heat release rates are shown below. 800 rpm, Stoichiometric mixture, SA = −42 aTDC for spark plug and SA = −36 aTDC for ACIS.

ACIS had to be increased for reliable ignition. If the same setting was used as previous condition without EGR, the combustion was unstable with partial burn or misfire cycles. Also, the ignition timing was kept the same for both the systems as the previous case. From the heat release it is evident that the combustion phasing is retarded for both ignition systems when compared to the stoichiometric mixture without EGR. However, the ACIS system resulted in faster overall burning with EGR than the spark plug. It is notable that while the ACIS was incapable of igniting the mixture reliably at a lower voltage (energy) setting, the spark plug system was able to perform without any problem. This is probably due to the increasing cylinder pressure with excess EGR and the increased voltage demand for a consistent streamer discharge with ACIS.

Comparing the mean combustion luminosity images, it is clear that the luminosity is significantly reduced because of the dilution. The onset of combustion is clearly visible with the spark plug at −36 aTDC and the flame kernel growth is slow and reaching a critical size by −20 aTDC. In contrast, even though there is no strong indication of ignition with ACIS, the flame becomes visible and is clearly larger than the spark plug flame by −20 aTDC. With increasing time the flame size continues to increase for both ignition systems with the ACIS flame volume being larger than the

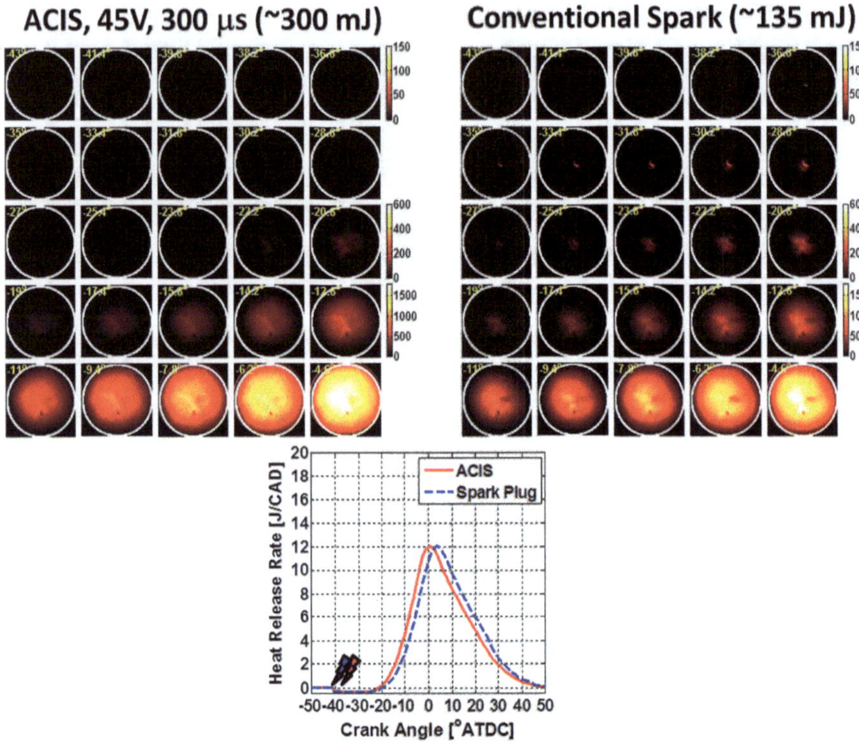

Fig. 5. Mean images of flame kernel formation and flame propagation with ACIS and spark plug. The heat release rates are shown below. 800 rpm, Stoichiometric mixture with 20 % N_2 dilution to simulate EGR, SA = −42 aTDC for spark plug and SA = −36 aTDC for ACIS.

spark plug flame. Thus the imaging result is consistent with and confirms the faster burn rate trend observed from heat release rates.

4.4 Lean Limit, a/F = 25.8

In this section the leanest mixture that is ignitable by either ignition system while maintaining stable combustion is presented. The criterion for stable combustion is defined as COV of IMEP of 3 %. The intake manifold pressure is increased to introduce excess air while keeping the fueling rate, spark timing, and other relevant parameters the same as before. The voltage setting for the ACIS was kept the same as the condition with the nitrogen dilution case. It was determined that the ACIS can sustain combustion for an A/F of 25.8 with COV of IMEP of 3.3 %. At the same lean mixture condition, the COV of IMEP for spark plug system was 6.1 %. Figure 6 shows the mean combustion luminosity images and the cycle averaged heat release rates for the ACIS and spark plug system.

The heat release rate plots show that even with retarded spark timing the ACIS has significantly advanced combustion phasing when compared to the spark plug. The peak heat release rate is also greater than the spark plug and as expected owing to the larger COV of IMEP with the spark plug. The images support the trends depicted in the heat release and also clearly show the difference in the ignition, flame kernel formation and flame propagation between the ACIS and spark plug. The ACIS owing to the multiple ignition sites is able to develop a larger flame kernel in a shorter time than the spark plug. For example, while luminosity from early ignition is clearly visible with the spark plug by −38 aTDC, the flame kernel has not grown to a significant size even at −20 aTDC. In contrast, the multiple streamer discharge from the ACIS is clearly visible at -35 aTDC and careful examination also reveals multiple simultaneous ignition kernels by −28 aTDC. The flame kernel grows to encompass a larger volume of the piston bowl by −20 aTDC when compared to the spark plug system. The ACIS flame continues to grow at a faster rate than the spark plug system, consistent with the heat release rate.

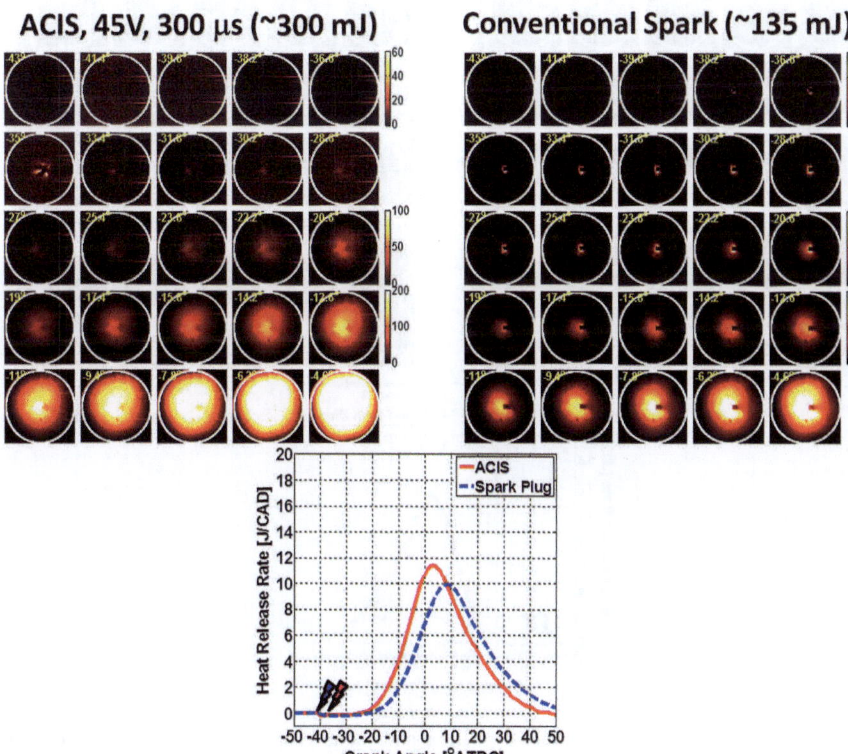

Fig. 6. Mean images of flame kernel formation and flame propagation with ACIS and spark plug. The heat release rates are shown below. 800 rpm, Lean mixture with excess air dilution for A/F = 25.8, SA = −42 aTDC for spark plug and SA = −36 aTDC for ACIS.

4.5 Lean Limit, a/F = 25.2, Similar COV_{IMEP}

This section describes the leanest A/F mixture where the two ignition systems have similar COV of IMEP. The A/F ratio of 25.2 corresponds to a COV of IMEP of 2.5 %, for both ACIS and spark plug system. Figure 7 shows the cycle averaged combustion luminosity images and the heat release rate plots. From the heat release rates, it is evident that the burn rates are very similar between the two ignition systems. However, there is still a marginal advance in combustion phasing with the ACIS. Consistent with the previous lean cases, the ignition delay is significantly shorter with the ACIS when taking into account the retarded spark advance. The images allow a direct comparison of the ignition, flame growth and flame propagation between the two ignition systems. It is again clear that while the ignition and flame growth takes significantly longer with the spark plug, the multiple streamer discharge and multiple ignition points with the ACIS promotes a faster ignition and flame growth in lean conditions.

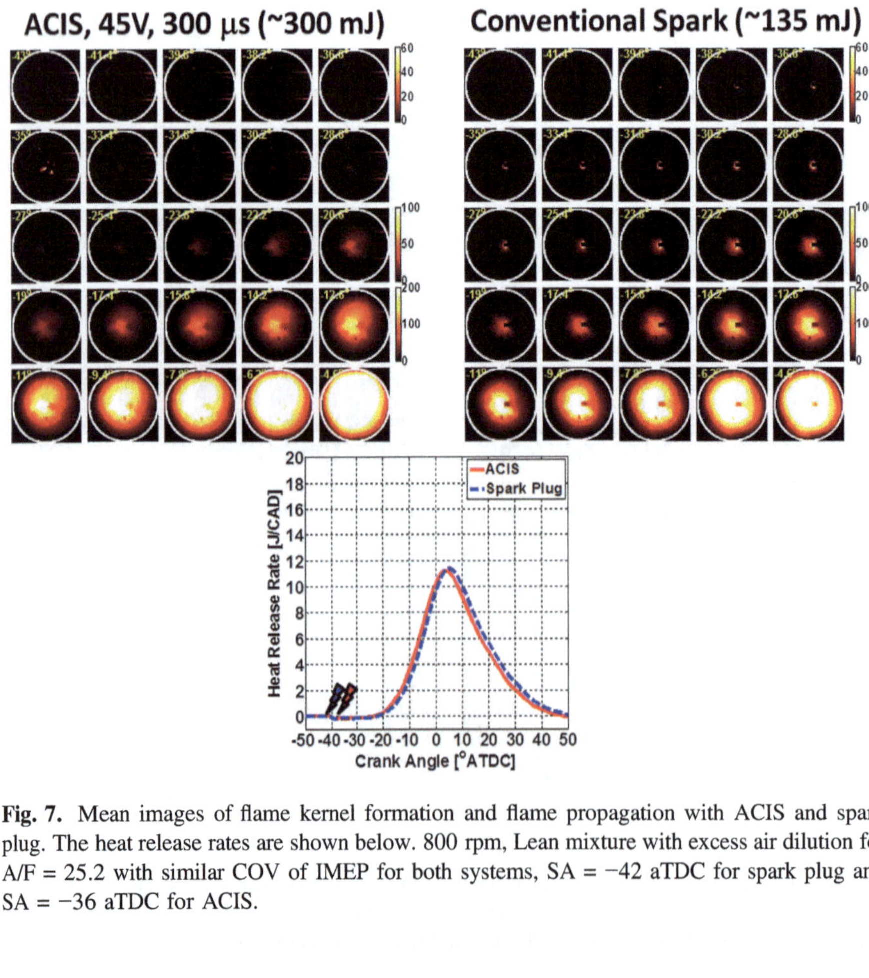

Fig. 7. Mean images of flame kernel formation and flame propagation with ACIS and spark plug. The heat release rates are shown below. 800 rpm, Lean mixture with excess air dilution for A/F = 25.2 with similar COV of IMEP for both systems, SA = −42 aTDC for spark plug and SA = −36 aTDC for ACIS.

4.6 Rich Condition, a/F = 13.1

The effect of the ignition system for homogenous rich condition was also explored. The intake was further throttled from stoichiometric to make the mixture rich. The ignition timing for the two systems were kept the same as the previous cases. From Fig. 8, the heat release rate shows that the ACIS is significantly advanced when compared to the spark plug system. The combustion phasing of the rich condition with the spark plug is similar to that of the stoichiometric mixture. However, the combustion phasing is advanced with the ACIS for the rich mixture when compared to the stoichiometric mixture, indicating an increased ease of igniting the mixture. It should also be noted that the voltage setting on the ACIS was decreased to 40 V for the rich condition. The images also clearly show the significant advance in flame propagation with the ACIS than the spark plug. The streamer discharge from the four prongs, the ignition and flame kernel formation from each streamer and the merging and flame propagation is clearly visible from these cycle averaged images. This process is in contrast to the slower ignition, flame kernel growth and flame propagation of the spark plug system.

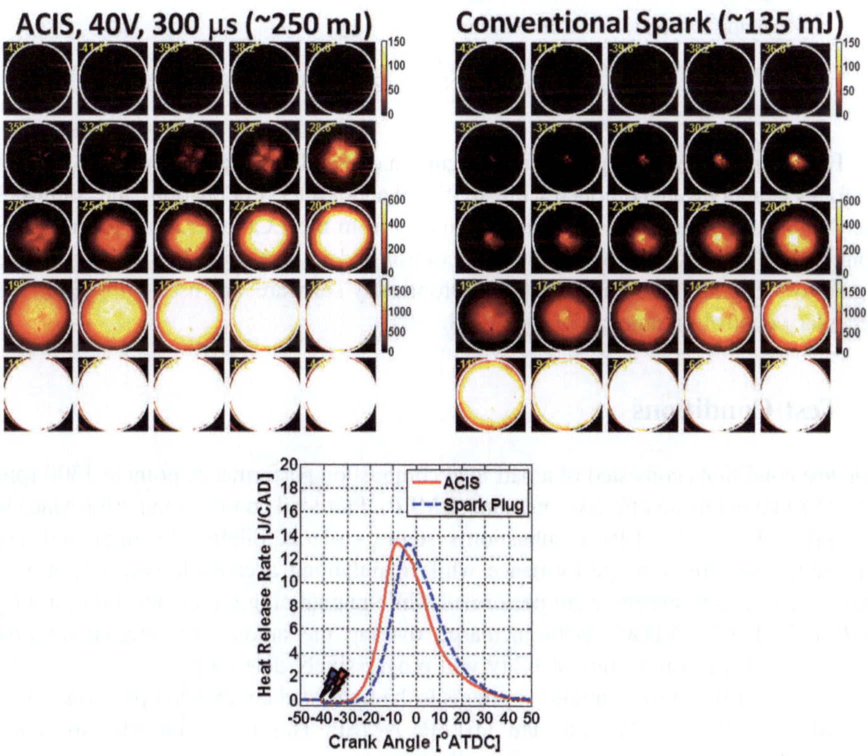

Fig. 8. Mean images of flame kernel formation and flame propagation with ACIS and spark plug. The heat release rates are shown below. 800 rpm, rich mixture with A/F = 13.1, SA = −42 aTDC for spark plug and SA = −36 aTDC for ACIS.

5 Multi-cylinder Engine Specifications

The experiments were performed in a 2.0L, direct-injected, turbocharged, in-line four-cylinder engine. The main specifications of the multi-cylinder engine is shown in Table 2.

Table 2. Specifications of the multicylinder engine

Type:	2.0L I4 Turbo
Displacement:	1998 cc (122 ci)
Compression Ratio:	9.2:1
Valve Configuration:	Dual overhead camshafts
Valves per Cylinder:	4
Cam Phasing:	Dual continuously variable cam phasing
Valve Lifters:	Hydraulic roller finger follower
Firing Order:	1–3–4–2
Bore x Stroke:	86.00 mm × 86.00 mm
Fuel System:	Direct Injection (side)
Ignition System:	Coil on Plug Ignition Coil: Denso 40 mJ/35 kV @2.72 ms Dwell Spark Plug: NGK HR5, 0.9 +0/−0.15 mm Gap

For testing the ACIS, the production ignition coil and spark plug were removed and all four cylinders were equipped with ACIS coil and igniter. For enabling the control of the ACIS igniter, the four spark timing signals from the ECU was input to the ACIS control box. The output from the ACIS control box in turn provided the appropriate signal to the ignition coil and igniter. A proprietary software was used to program the necessary control parameters of the ACIS.

6 Test Conditions

The test conditions consisted of a part load combustion performance point at 1300 rpm and 330 kPa net mean effective pressure (NMEP). Part load combustion performance is defined as the ability of the combustion system to tolerate dilution for improved fuel efficiency and emissions performance while maintaining acceptable combustion stability. Cam phaser sweeps were performed while maintaining the combustion phasing to CA50 of 8.5° ATDC. With increased overlap the amount of internal dilution increases and the combustion stability will progressively deteriorate.

Another critical test of ignition systems is the light load combustion performance at near-idle condition of 750 rpm and 110 kPa NMEP. This test is intended to define capacity for idle torque reserve. The idle torque reserve requirement defines spark retard requirement and, therefore, spark plug gap size and energy requirements. For these tests, the cam phasers are maintained in a parked position and spark timing sweeps are performed.

7 Multi-cylinder Engine Results

7.1 Part Load Combustion Performance

For the ACIS, a setting of 30 V and 500 μs was chosen as the operating parameter, delivering a similar energy when compared to the production ignition system. Figures 9, 10 and 11 shows the difference between the production ignition system and ACIS. The contour plots are generated from the cam sweep tests and shows the difference in COV of IMEP, 0–10 burn duration (ignition delay) and 10–75 burn duration (bulk burn), between the spark ignition system and ACIS. Also in Figs. 9 and 10, the 3 % COV of IMEP line for the production system and ACIS are shown in dashed and solid line, respectively. The amount of internal dilution increases along the diagonal of the plot and for these operating conditions, the ACIS is able to tolerate higher internal dilution. From these two contour plots, for the same amount of internal dilution, ACIS has improved COV of IMEP and a shorter 0–10 burn duration. The improvement in COV of IMEP and 0–10 burn duration is more significant at increasing dilution levels. However, as shown by Fig. 11, there is no significant improvement in 10–75 burn duration from the ACIS system. The maximum improvement of 4 to 5 degrees in 10–75 burn duration is seen at the higher dilution levels. Since the voltage and duration setting for ACIS is kept constant for this operating condition, the ACIS may suffer from deteriorated performance if it were to transition to arcing depending on the required spark timing to maintain the CA50. The reason for transition to arcing can be due to the in-cylinder pressure, temperature, mixture composition or due to the close proximity to an electrical ground (e.g. piston).

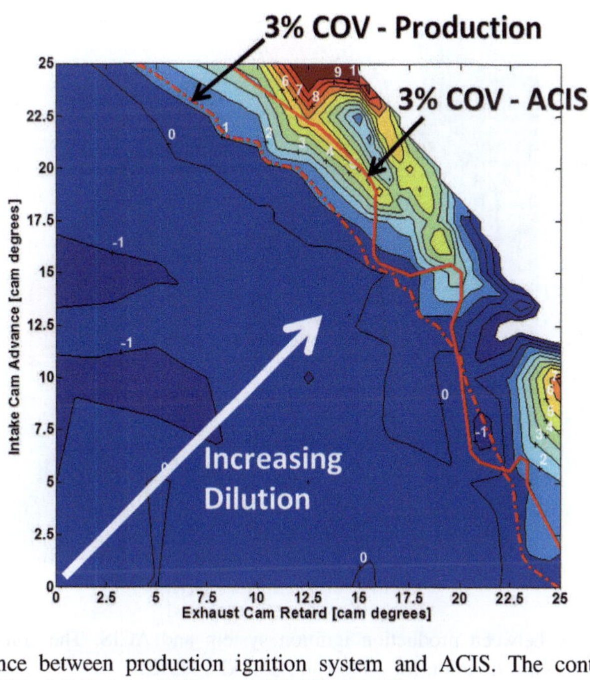

Fig. 9. Difference between production ignition system and ACIS. The contour plot shows difference in COV of IMEP for 1300 rpm, 330 kPa NMEP.

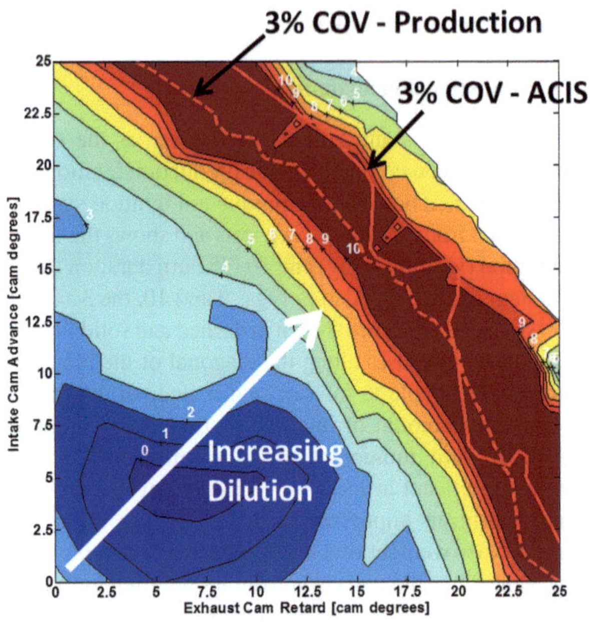

Fig. 10. Difference between production ignition system and ACIS. The contour plot shows difference in 0–10 burn duration for 1300 rpm, 330 kPa NMEP.

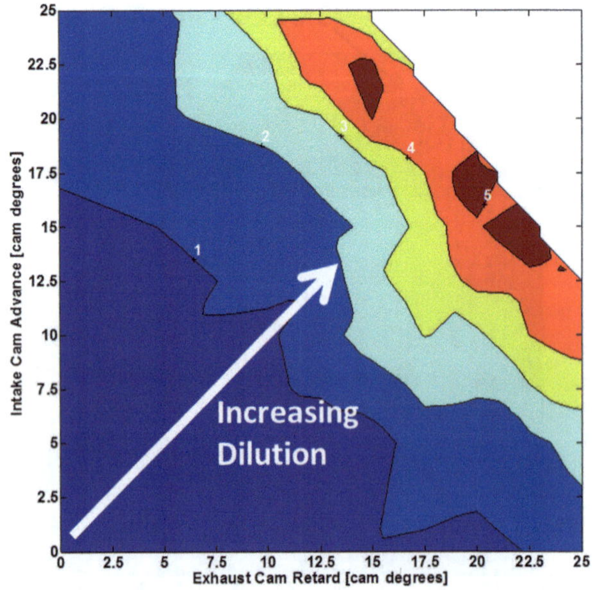

Fig. 11. Difference between production ignition system and ACIS. The contour plot shows difference in 10–75 burn duration for 1300 rpm, 330 kPa NMEP.

7.2 Light Load Combustion Performance

Figure 12 shows the light load combustion performance results at 750 rpm and 110 kPa NMEP for ACIS and the production ignition system. For the ACIS, a setting of 25 V and 500 µs was chosen. The top plot in Fig. 12 shows a comparison of engine average COV of IMEP as a function of 50 % burn location, which directly correlates to

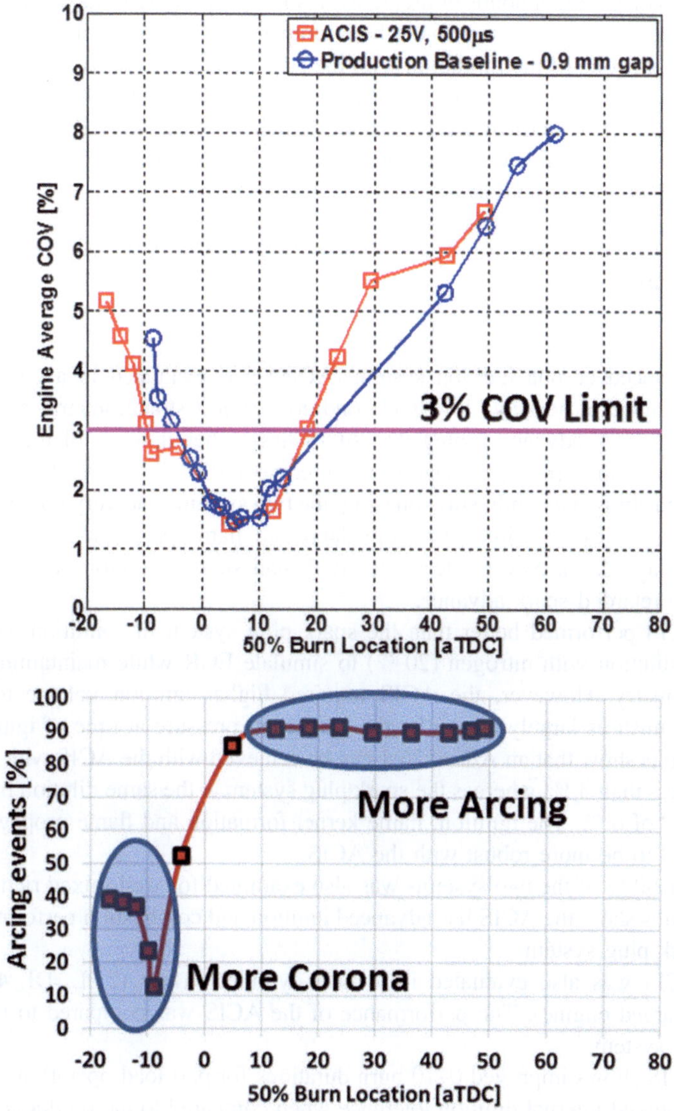

Fig. 12. Idle spark sweep test at 750 rpm and 110 kPa NMEP. The plot above shows engine average COV or IMEP as a function of 50 % burn location for ACIS and production ignition system. The plot below shows the arcing events for ACIS for the same tests.

the spark advance. For these tests, a wide range of spark advance timing is desired while maintaining combustion stability (COV of IMEP < 3 %). Comparing between the ACIS and production system, the performance is quite similar. The bottom plot in Fig. 12 shows the amount of arcing events the ACIS has as a function of the 50 % burn location. As the combustion phasing is more retarded with spark advance closer to TDC, the data shows that the ACIS has much more arcing events and the performance is worse than the spark ignition system. However, for advanced combustion phasing and spark advance, the amount of arcing events is substantially reduced and improved spark advance timing is achievable when compared to the production system. This result highlights one of the limitation with the ACIS and the need to avoid arcing. For these light load conditions, as the spark advance gets closer to TDC, the close proximity of the piston (electrical ground) to the igniter prongs, the ACIS discharge is in arcing mode than in streamer or corona mode and clearly degrades the combustion performance.

8 Summary

- The advanced corona ignition system (ACIS) was evaluated in a direct-injected optical engine at near-idle low-load conditions. For a stoichiometric mixture with the same spark advance timing, the ACIS has earlier ignition, flame kernel formation and flame front propagation when compared to the spark plug system.
- When matching the combustion phasing, the two systems had very comparable heat release rates (from cylinder pressure data) and flame propagation characteristics (from image data). Nevertheless, the ACIS had shorter ignition delay time, even with the retarded spark advance.
- The ACIS performed better than the spark plug system in condition with excess charge dilution with nitrogen (20 %) to simulate EGR while maintaining mixture stoichiometry. However, the ACIS required higher ignition voltage to perform without misfires largely driven by the increasing pressure at time of ignition.
- The results show that an A/F of 26:1 can be achieved with the ACIS with a COV of IMEP less than 3 %, whereas the spark plug system at the same dilution had a COV of IMEP of 6 %. The ignition, flame kernel formation and flame propagation was observed to be more robust with the ACIS.
- The capability of the two systems was also examined for well-mixed rich mixtures. The results show the ACIS has advanced ignition and combustion performance than the spark plug system.
- The ACIS was also evaluated in a multi-cylinder engine (2.0L, DI, 4-cylinder, turbocharged engine). The performance of the ACIS was compared to production ignition system.
- The ACIS shows improved 0–10 burn durations for part load operation. The ACIS had improved internal dilution tolerance when compared to the production ignition system. The 10–75 burn durations did not show any significant difference when compared to the production ignition system. Measurable difference in 10–75 burn duration was largely limited to the higher internal dilution operating regions.

- Spark sweeps performed at engine idle conditions were shown to match baseline production data. However, potential improvement in this test condition was limited by the propensity of the ACIS to transition to arcing mode because of proximity of the power electrode to the piston surface, depending on the spark advance timing. Potential for further improvement with ACIS is possible with optimizing the piston design.
- The most significant advantage for the ACIS was determined to be the closely located, multiple ignition sites as opposed to the single ignition location which is constrained in the spark channel. It is also important to have the ACIS operate in corona mode as opposed to the arcing mode to take full advantage of the system.
- The system must be evaluated at wide open throttle (WOT) conditions and specific tests to assess the reliability and durability of ACIS is also warranted. Continued investigations to understand the relevance of these systems to enable advanced combustion strategies (e.g. lean stratified, controlled autoignition) is also essential.

Acknowledgements. The authors would like to thank Mr. Gerald Silvas for his contributions in performing the optical engine experiments.

References

1. Kratzsch, M., Gunther, M. (eds.). 1st International Conference on Ignition Systems for Gasoline Engines, Berlin (2012)
2. Gunther, M., Troger, R. (eds.) 2nd International Conference on Ignition Systems for Gasoline Engines, Berlin (2014)
3. Suess, M., Guenthner, M., Schenk, M., Rottengruber, H.-S.: Investigation of the potential of corona ignition to control gasoline homogeneous charge compression ignition combustion. Proc. IMechE Part D: J. Automobile Eng. **226**, 275–286 (2011)
4. Shiraishi, T.: Possibility of the new ignition system using the low temperature plasma having dual functions of strengthening ignition for SI combustion and promoting and controlling autoignition of HCCI combustion. In: Proceedings of First International Conference: Advanced Ignition Systems for Gasoline Engines, Berlin, Germany. 12–13 November 2012
5. Schenk, M., Fessler, M., Wolf, T., Klaus, B., Fischer, H.: Comparison of the thermodynamic Potential of alternative Ignition systems for SI-engines In: Proceedings of First International Conference: Advanced Ignition Systems for Gasoline Engines, Berlin, Germany, 12–13 November 2012
6. Drake, M.C., Fansler, T.D., Peterson, K.H.: Stratified Ignition processes in spray-guided SIDI engines. In: 9th International Symposium Internal Combustion Diagnostics, Baden-Baden, 8–9 June 2010

MIX
Papier aus verantwortungsvollen Quellen
Paper from responsible sources
FSC® C105338

If you have any concerns about our products,
you can contact us on
ProductSafety@springernature.com

In case Publisher is established outside the EU,
the EU authorized representative is:
Springer Nature Customer Service Center GmbH
Europaplatz 3, 69115 Heidelberg, Germany

Printed by Libri Plureos GmbH
in Hamburg, Germany